T0250009

Wetlands and Habitats

The Handbook of Natural Resources, Second Edition

Series Editor:
Yeqiao Wang

Volume 1
Terrestrial Ecosystems and Biodiversity

Volume 2
Landscape and Land Capacity

Volume 3
Wetlands and Habitats

Volume 4
Fresh Water and Watersheds

Volume 5
Coastal and Marine Environments

Volume 6
Atmosphere and Climate

Wetlands and Habitats

Edited by
Yeqiao Wang

CRC Press
Taylor & Francis Group
Boca Raton London New York

CRC Press is an imprint of the
Taylor & Francis Group, an **informa** business

CRC Press
Taylor & Francis Group
6000 Broken Sound Parkway NW, Suite 300
Boca Raton, FL 33487-2742

First issued in paperback 2022

© 2020 by Taylor & Francis Group, LLC
CRC Press is an imprint of Taylor & Francis Group, an Informa business

No claim to original U.S. Government works

ISBN 13: 978-1-03-247438-0 (pbk)
ISBN 13: 978-1-138-33419-9 (hbk)

DOI: 10.1201/9780429445507

This book contains information obtained from authentic and highly regarded sources. Reasonable efforts have been made to publish reliable data and information, but the author and publisher cannot assume responsibility for the validity of all materials or the consequences of their use. The authors and publishers have attempted to trace the copyright holders of all material reproduced in this publication and apologize to copyright holders if permission to publish in this form has not been obtained. If any copyright material has not been acknowledged, please write and let us know so we may rectify in any future reprint.

Except as permitted under U.S. Copyright Law, no part of this book may be reprinted, reproduced, transmitted, or utilized in any form by any electronic, mechanical, or other means, now known or hereafter invented, including photocopying, microfilming, and recording, or in any information storage or retrieval system, without written permission from the publishers.

For permission to photocopy or use material electronically from this work, please access www.copyright.com (http://www.copyright.com/) or contact the Copyright Clearance Center, Inc. (CCC), 222 Rosewood Drive, Danvers, MA 01923, 978-750-8400. CCC is a not-for-profit organization that provides licenses and registration for a variety of users. For organizations that have been granted a photocopy license by the CCC, a separate system of payment has been arranged.

Trademark Notice: Product or corporate names may be trademarks or registered trademarks, and are used only for identification and explanation without intent to infringe.

Publisher's Note
The publisher has gone to great lengths to ensure the quality of this reprint but points out that some imperfections in the original copies may be apparent.

Library of Congress Cataloging-in-Publication Data

Names: Wang, Yeqiao, editor.
Title: Handbook of natural resources / edited by Yeqiao Wang.
Other titles: Encyclopedia of natural resources.
Description: Second edition. | Boca Raton: CRC Press, [2020] | Revised edition of: Encyclopedia of natural resources. [2014]. | Includes bibliographical references and index. | Contents: volume 1. Ecosystems and biodiversity — volume 2. Landscape and land capacity — volume 3. Wetland and habitats — volume 4. Fresh water and watersheds — volume 5. Coastal and marine environments — volume 6. Atmosphere and climate. | Summary: "This volume covers topical areas of terrestrial ecosystems, their biodiversity, services, and ecosystem management. Organized for ease of reference, the handbook provides fundamental information on terrestrial systems and a complete overview on the impacts of climate change on natural vegetation and forests. New to this edition are discussions on decision support systems, biodiversity conservation, gross and net primary production, soil microbiology, and land surface phenology. The book demonstrates the key processes, methods, and models used through several practical case studies from around the world" — Provided by publisher.
Identifiers: LCCN 2019051202 | ISBN 9781138333918 (volume 1 ; hardback) | ISBN 9780429445651 (volume 1 ; ebook)
Subjects: LCSH: Natural resources. | Land use. | Climatic changes.
Classification: LCC HC85 .E493 2020 | DDC 333.95—dc23
LC record available at https://lccn.loc.gov/2019051202

Visit the Taylor & Francis Web site at
http://www.taylorandfrancis.com

and the CRC Press Web site at
http://www.crcpress.com

Contents

SECTION I Riparian Zone and Management

SECTION II   Wetland Ecosystem

SECTION III   Wetland Assessment and Monitoring

Preface

Wetlands and Habitats is the third volume of *The Handbook of Natural Resources, Second Edition (HNR)*. This volume consists of 30 chapters authored by 52 contributors from 7 countries. The contents are organized in three sections: *Riparian Zone and Management* (13 chapters); *Wetland Ecosystem* (8 chapters); and *Wetland Assessment and Monitoring* (9 chapters).

Wetland is a universal term used to describe a variety of flooded or saturated environments. Wetlands include a diverse assemblage of highly complex ecosystems subjected to hydrologic regimes, climatic conditions, soil formation processes, geomorphologic settings, and support high rates of annual carbon sequestration. The variety of seasonal and perennial wetlands provides environmental conditions for highly distinctive fauna and flora.

Freshwater wetlands are habitats with vegetation especially adapted to prolonged water saturation or shallow flooding of the soils. Wetlands are usually highly productive, support high biodiversity, and provide numerous direct and indirect ecosystem services and societal benefits. Tidal wetlands are some of the most dynamic areas of the Earth and are found at the interface between the land and sea. Salinity, regular tidal flooding, and infrequent catastrophic flooding due to storm events result in complex interactions among biotic and abiotic factors for tidal marshes, mangroves, and freshwater forested wetlands. Riparian wetlands are critical ecosystems that perform functions and provide services disproportionate to their extent in the landscape. Vegetated riparian zones are a common management practice used to reduce nitrate delivery to coastal waters. Because of their position between terrestrial and aquatic ecosystems, vegetated riparian zones provide opportunities for nitrate transformation before emerging to surface water. River deltas form on shorelines where sediment can accumulate faster than it is taken away and are recognized by the presence of a distributary channel network that protrudes from the shoreline. Peatlands, although occupying only 3% of the global terrestrial surface, hold 10% of the world's drinking water, and contain a thick layer of peat that stores about one-third of the Earth's soil carbon and 10% of the world's soil nitrogen. Peatlands are sensitive to both climatic and anthropogenic disturbances, which could change the long-term peatland carbon sink into a source of carbon to the atmosphere. Wetlands are subject to alteration and loss. Humans have dramatically changed marshes for agriculture, development, and resource extraction. The future of wetland resources depends on conservation, restoration, and climate. Protection and conservation of wetland resources are among priorities in land management decision and practice in all scales.

With the challenges and concerns, the 30 chapters in this volume cover topics in *Wetland Ecosystem*, including wetland classification, ecosystems, biodiversity, fresh water and tidal wetlands, effects of wetlands in urban environment, and declining ecosystem service in responding of wetland loss; in *Riparian Zone and Management*, including riverine floods, riparian wetlands, riparian zone groundwater nitrate cycling, riparian wetland mapping, floodplain management, salt and brackish marshes, tidal wetlands, peatlands, perennial and seasonal streams, vernal pool, nonpoint source pollution, groundwater salinity, and assessment of environmental integrity of riparian zone; and in *Wetland Assessment and Monitoring*, including wetland economic value, indices of biotic integrity (IBI), assessment of wetland health status by vegetation-IBI, remote sensing in wetland mapping, monitoring of wetland water dynamics, and wetland conservation and policy impacts.

The chapters provide updated knowledge and information in general environmental and natural science education and serve as a value-added collection of references for scientific research and management practices.

Yeqiao Wang
University of Rhode Island

About The Handbook of Natural Resources

With unprecedented attentions to the changing environment on the planet Earth, one of the central focuses is about the availability and sustainability of natural resources and the native biodiversity. It is critical to gain a full understanding about the consequences of the changing natural resources to the degradation of ecological integrity and the sustainability of life. Natural resources represent such a broad scope of complex and challenging topics.

The Handbook of Natural Resources, Second Edition (HNR), is a restructured and retitled book series based on the 2014 publication of the *Encyclopedia of Natural Resources (ENR)*. The ENR was reviewed favorably in February 2015 by CHOICE and commented as *highly recommended for lower-division undergraduates through professionals and practitioners*. This HNR is a continuation of the theme reference with restructured sectional design and extended topical coverage. The chapters included in the HNR provide authoritative references under the systematic relevance to the subject of the volumes. The case studies presented in the chapters cover diversified examples from local to global scales, and from addressing fundamental science questions to the needs in management practices.

The Handbook of Natural Resources consists of six volumes with 241 chapters organized by topical sections as summarized below.

Volume 1. Terrestrial Ecosystems and Biodiversity
Section I. Biodiversity and Conservation (15 Chapters)
Section II. Ecosystem Type, Function and Service (13 Chapters)
Section III. Ecological Processes (12 Chapters)
Section IV. Ecosystem Monitoring (6 Chapters)

Volume 2. Landscape and Land Capacity
Section I. Landscape Composition, Configuration and Change (10 Chapters)
Section II. Genetic Resource and Land Capability (13 Chapters)
Section III. Soil (15 Chapters)
Section IV. Landscape Change and Ecological Security (11 Chapters)

Volume 3. Wetlands and Habitats
Section I. Riparian Zone and Management (13 Chapters)
Section II. Wetland Ecosystem (8 Chapters)
Section III. Wetland Assessment and Monitoring (9 Chapters)

Volume 4. Fresh Water and Watersheds
Section I. Fresh Water and Hydrology (16 Chapters)
Section II. Water Management (16 Chapters)
Section III. Water and Watershed Monitoring (8 Chapters)

Volume 5. Coastal and Marine Environments
Section I. Terrestrial Coastal Environment (14 Chapters)
Section II. Marine Environment (13 Chapters)
Section III. Coastal Change and Monitoring (9 Chapters)

Volume 6. Atmosphere and Climate
Section I. Atmosphere (16 Chapters)
Section II. Weather and Climate (16 Chapters)
Section III. Climate Change (8 Chapters)

With the challenges and uncertainties ahead, I hope that the collective wisdom, the improved science, technology, and awareness and willingness of the people could lead us toward the right direction and decision in governance of natural resources and make responsible collaborative efforts in balancing the equilibrium between societal demands and the capacity of natural resources base. I hope that this *Handbook of Natural Resources* series can help facilitate the understanding about the consequences of changing resource base to the ecological integrity and the sustainability of life on the planet Earth.

Yeqiao Wang
University of Rhode Island

Acknowledgments

I am honored to have this opportunity and privilege to work on *The Handbook of Natural Resources, Second Edition (HNR)*. It would be impossible to complete such a task without the tremendous amount of support from so many individuals and groups during the process. First and foremost, I thank the 342 contributors from 28 countries around the world, including Australia, Austria, Brazil, China, Cameroon, Canada, Czech Republic, Finland, France, Germany, Hungary, India, Israel, Japan, Nepal, New Zealand, Norway, Puerto Rico, Spain, Sweden, Switzerland, Syria, Turkey, Uganda, the United Kingdom, the United States, Uzbekistan, and Venezuela. Their expertise, insights, dedication, hard work, and professionalism ensure the quality of this important publication. I wish to express my gratitude in particular to those contributors who authored chapters for this HNR and those who provided revisions from their original articles published in the *Encyclopedia of Natural Resources*.

The preparation for the development of this HNR started in 2017. I appreciate the visionary initiation of the restructure idea and the guidance throughout the preparation of this HNR from Irma Shagla Britton, Senior Editor for Environmental Sciences and Engineering of the Taylor & Francis Group/ CRC Press. I appreciate the professional assistance and support from Claudia Kisielewicz and Rebecca Pringle of the Taylor & Francis Group/CRC Press, which are vital toward the completion of this important reference series.

The inspiration for working on this reference series came from my over 30 years of research and teaching experiences in different stages of my professional career. I am grateful for the opportunities to work with many top-notch scholars, colleagues, staff members, administrators, and enthusiastic students, domestic and international, throughout the time. Many of my former graduate students are among and/or becoming world-class scholars, scientists, educators, resource managers, and administrators, and they are playing leadership roles in scientific exploration and in management practice. I appreciate their dedication toward the advancement of science and technology for governing the precious natural resources. I am thankful for their contributions in HNR chapters.

As always, the most special appreciation is due to my wife and daughters for their love, patience, understanding, and encouragement during the preparation of this publication. I wish my late parents, who were past professors of soil ecology and of climatology from the School of Geographical Sciences, Northeast Normal University, could see this set of publications.

Yeqiao Wang
University of Rhode Island

Aims and Scope

Land, water, and air are the most precious natural resources that sustain life and civilization. Maintenance of clean air and water and preservation of land resources and native biological diversity are among the challenges that we are facing for the sustainability and well-being of all on the planet Earth. Natural and anthropogenic forces have affected constantly land, water, and air resources through interactive processes such as shifting climate patterns, disturbing hydrological regimes, and alternating landscape configurations and compositions. Improvements in understanding of the complexity of land, water, and air systems and their interactions with human activities and disturbances represent priorities in scientific research, technology development, education programs, and administrative actions for conservation and management of natural resources.

The chapters of *The Handbook of Natural Resources, Second Edition (HNR)*, are authored by world-class scientists and scholars. The theme topics of the chapters reflect the state-of-the-art science and technology, and management practices and understanding. The chapters are written at the level that allows a broad scope of audience to understand. The graphical and photographic support and list of references provide the helpful information for extended understanding.

Public and private libraries, educational and research institutions, scientists, scholars, resource managers, and graduate and undergraduate students will be the primary audience of this set of reference series. The full set of the HNR and individual volumes and chapters can be used as the references in general environmental science and natural science courses at different levels and disciplines, such as biology, geography, Earth system science, environmental and life sciences, ecology, and natural resources science. The chapters can be a value-added collection of references for scientific research and management practices.

Editor

Yeqiao Wang, PhD, is a professor at the Department of Natural Resources Science, College of the Environment and Life Sciences, University of Rhode Island. He earned his BS from the Northeast Normal University in 1982 and his MS degree from the Chinese Academy of Sciences in 1987. He earned the MS and PhD degrees in natural resources management and engineering from the University of Connecticut in 1992 and 1995, respectively. From 1995 to 1999, he held the position of assistant professor in the Department of Geography and the Department of Anthropology, University of Illinois at Chicago. He has been on the faculty of the University of Rhode Island since 1999. Among his awards and recognitions, Dr. Wang was a recipient of the prestigious Presidential Early Career Award for Scientists and Engineers (PECASE) in 2000 by former U.S. President William J. Clinton, for his outstanding research in the area of land cover and land use in the Greater Chicago area in connection with the Chicago Wilderness Program.

Dr. Wang's specialties and research interests are in terrestrial remote sensing and the applications in natural resources analysis and mapping. One of his primary interests is the land change science, which includes performing repeated inventories of landscape dynamics and land-use and land-cover change from space, developing scientific understanding and models necessary to simulate the processes taking place, evaluating consequences of observed and predicted changes, and understanding the consequences of change on environmental goods and services and management of natural resources. His research and scholarships are aimed to provide scientific foundations in understanding of the sustainability, vulnerability and resilience of land and water systems, and the management and governance of their uses. His study areas include various regions in the United States, East and West Africa, and China.

Dr. Wang published over 170 refereed articles, edited *Remote Sensing of Coastal Environments* and *Remote Sensing of Protected Lands*, published by CRC Press in 2009 and 2011, respectively. He served as the editor-in-chief for the *Encyclopedia of Natural Resources* published by CRC Press in 2014, which was the first edition of *The Handbook of Natural Resources*.

Contributors

Gayatri Acharya
World Bank Institute (WBI)
Washington, District of Columbia

Vicenç Acuña
Catalan Institute for Water Research
Girona, Spain

Andrew H. Baldwin
Department of Environmental Science and
    Technology
University of Maryland
College Park, Maryland

Virginie Bouchard
School of Natural Resources
The Ohio State University
Columbus, Ohio

Rebecca L. Caldwell
Department of Geological Sciences
Indiana University
Bloomington, Indiana

Edward Capone
Northeast River Forecast Center
National Oceanic and Atmospheric
    Administration (NOAA)
Taunton, Massachusetts

Caifeng Cheng
Department of Geography
Shandong University of Science and Technology
Qingdao, China

William H. Conner
Baruch Institute of Coastal Ecology and Forest
    Science
Clemson University
Georgetown, South Carolina

Ray Correll
Commonwealth Scientific and Industrial
    Research Organisation (CSIRO)
Adelaide, South Australia, Australia

Sherri DeFauw
New England Plant, Soil and Water Laboratory,
    Agricultural Research Service
U.S. Department of Agriculture (USDA-ARS)
University of Maine
Orono, Maine

Peter Dillon
Commonwealth Scientific and Industrial
    Research Organisation (CSIRO)
Adelaide, South Australia, Australia

Jamie A. Duberstein
Baruch Institute of Coastal Ecology and Forest
    Science
Clemson University
Georgetown, South Carolina

Jacob M. Dybiec
Institute for Great Lakes Research, CMU
    Biological Station
Department of Biology
Central Michigan University
Mt. Pleasant, Michigan

Douglas A. Edmonds
Department of Geological Sciences
Indiana University
Bloomington, Indiana

Bolin Fu
College of Geomatics and Geoinformation
Guilin University of Technology
Guilin, China

Dan Gao
School of Geography and Environmental Science
Ministry of Education's Key Laboratory of
    Poyang Lake Wetland and Watershed
    Research
Jiangxi Normal University
Nanchang, China

Brij Gopal
Centre for Inland Waters in South Asia
Jaipur, India

Sally D. Hacker
Department of Integrative Biology
Oregon State University
Corvallis, Oregon

Anna M. Harrison
Institute for Great Lakes Research, CMU
    Biological Station
Department of Biology
Central Michigan University
Mt. Pleasant, Michigan

Kristen C. Hychka
Atlantic Ecology Division
Office of Research and Development
U.S. Environmental Protection Agency (EPA)
Narragansett, Rhode Island

Mingming Jia
Key Laboratory of Wetland Ecology and
    Environment
Northeast Institute of Geography and
    Agroecology
Chinese Academy of Science
Changchun, China

D. Q. Kellogg
Department of Natural Resources Science
University of Rhode Island
Kingston, Rhode Island

Rai Kookana
Commonwealth Scientific and Industrial
    Research Organisation (CSIRO)
Adelaide, South Australia, Australia

Jean-Claude Lefeuvre
Laboratory of the Evolution of Natural and
    Modified Systems
University of Rennes
Rennes, France

Ying Li
College of Geomatics and Geoinformation
Guilin University of Technology
Guilin, China

Alison MacNeil
Northeast River Forecast Center
National Oceanic and Atmospheric
    Administration (NOAA)
Taunton, Massachusetts

Dehua Mao
Key Laboratory of Wetland Ecology and
    Environment
Northeast Institute of Geography and
    Agroecology
Chinese Academy of Science
Changchun, China

Alexandra C. Mattingly
Institute for Great Lakes Research, CMU
    Biological Station
Department of Biology
Central Michigan University
Mt. Pleasant, Michigan

Mallavarapu Megharaj
Commonwealth Scientific and Industrial
    Research Organisation (CSIRO)
Adelaide, South Australia, Australia

Niti B. Mishra
University of Texas at Austin
Austin, Texas

Ravendra Naidu
Commonwealth Scientific and Industrial
    Research Organisation (CSIRO)
Adelaide, South Australia, Australia

Peter W. C. Paton
Department of Natural Resources Science
University of Rhode Island
Kingston, Rhode Island

Shuhua Qi
Key Laboratory of Poyang Lake Wetland and
    Watershed Research
Jiangxi Normal University
Ministry of Education
Nanchang, China

Sergi Sabater
Catalan Institute for Water Research
Girona, Spain

William Saunders
Northeast River Forecast Center
National Oceanic and Atmospheric
    Administration (NOAA)
Taunton, Massachusetts

Daniel von Schiller
Catalan Institute for Water Research
Girona, Spain

Neil T. Schock
Institute for Great Lakes Research, CMU
    Biological Station
Department of Biology
Central Michigan University
Mt. Pleasant, Michigan

Paul Short
Canadian Sphagnum Peat Moss Association
St. Albert, Alberta, Canada

Ralph W. Tiner
National Wetlands Inventory (NWI)
U.S. Fish and Wildlife Service
Hadley, Massachusetts

Donald G. Uzarski
Institute for Great Lakes Research, CMU
    Biological Station
Department of Biology
Central Michigan University
Mt. Pleasant, Michigan

Arnold G. van der Valk
Ecology, Evolution and Organismal Biology
Iowa State University
Ames, Iowa

Dale H. Vitt
Department of Plant Biology and Center for
    Ecology
Southern Illinois University
Carbondale, Illinois

Yeqiao Wang
Department of Natural Resources Science
University of Rhode Island
Kingston, Rhode Island

Zongming Wang
Key Laboratory of Wetland Ecology and
    Environment
Northeast Institute of Geography and
    Agroecology
Chinese Academy of Science
Changchun, China

Yuming Wen
Water & Environmental Research Institute
University of Guam
Mangilao, Guam

W. W. Wenzel
Institute of Soil Research
University of Natural Resources and Life
    Sciences
Vienna, Austria

French Wetmore
French & Associates, Ltd.
Park Forest, Illinois

Ryan L. Wheeler
Institute for Great Lakes Research, CMU
    Biological Station
Department of Biology
Central Michigan University
Mt. Pleasant, Michigan

Jian Xu
School of Geography and Environmental Science
Ministry of Education's Key Laboratory of
    Poyang Lake Wetland and Watershed
    Research
Jiangxi Normal University
Nanchang, China

Zhenshan Xue
Northeast Institute of Geography and
    Agroecology
Chinese Academy of Science
Changchun, China

Fengqin Yan
State Key Laboratory of Resources and
    Environmental Information System
Institute of Geographic Sciences and Natural
    Resources Research
Chinese Academy of Sciences
Beijing, China

Jason Yang
Department of Geography
Ball State University
Muncie, Indiana

Wenjing Yang
Key Laboratory of Poyang Lake Wetland and
    Watershed Research
Jiangxi Normal University
Ministry of Education
Nanchang, China

Shuwen Zhang
Research Group of Land System by Remote
    Sensing
Northeast Institute of Geography and
    Agroecology
Chinese Academy of Sciences
Changchun, China

Tingting Zhang
Department of Geography
Jilin Normal University
Siping, China

Zhongsheng Zhang
Northeast Institute of Geography and
    Agroecology
Chinese Academy of Science
Changchun, China

I

Riparian Zone
and Management

1

Floodplain Management

French Wetmore
French & Associates, Ltd.

Historical Approaches

During the 1920s, the insurance industry concluded that flood insurance could not be a profitable venture because the only people who would want flood coverage would be those who lived in floodplains. As they were sure to be flooded, the rates would be too high to attract customers. Unlike other hazards, such as wind and hail, where the risk can be spread, private industry opted out of playing a role in flood protection.

With the great Mississippi River flood of 1927, the federal government became a major player in flooding. As defined by several Flood Control Acts, the role of government agencies was to build massive flood control structures to control the great rivers, protect coastal areas, and prevent flash flooding.

Until the 1960s, such structural flood control projects were seen as the primary way to reduce flood losses. In some areas, they still are. However, starting in the 1960s, people questioned the effectiveness of this single solution. Disaster relief expenses were going up, making all taxpayers pay more to provide relief to those with property in floodplains. Studies during the 1960s concluded that flood losses were increasing, in spite of the number of flood control structures that had been built.

One of the main reasons structural flood control projects failed to reduce flood losses was that people continued to build in floodplains. In response, federal, state, and local agencies began to develop policies and programs with a "non-structural" emphasis, ones that did not prescribe projects to control or redirect the path of floods.

A milestone in this effort was the creation of the National Flood Insurance Program (NFIP) in 1968. The NFIP is based on a mutual agreement between the Federal government [represented by the Federal Emergency Management Agency (FEMA)] and local governments. Federally guaranteed flood insurance is made available in those communities that agree to regulate development in their mapped floodplains.

If the communities do their part in making sure future floodplain development meets certain criteria, FEMA will provide flood insurance for properties in the community. The Federal government is willing to support insurance because, over time, local practices will reduce the exposure to flood damage.

Also during the 1960s and 1970s, interest increased in protecting and restoring the environment, including the natural resources and functions of floodplains. Coordinating flood loss reduction programs with environmental protection and watershed management programs has since become a major goal of federal, state, and local programs. This evolution is shown graphically in Figure 1.1. Now, we no longer depend solely on structural projects to control floodwater. Instead of "flood control," we now speak of "floodplain management."

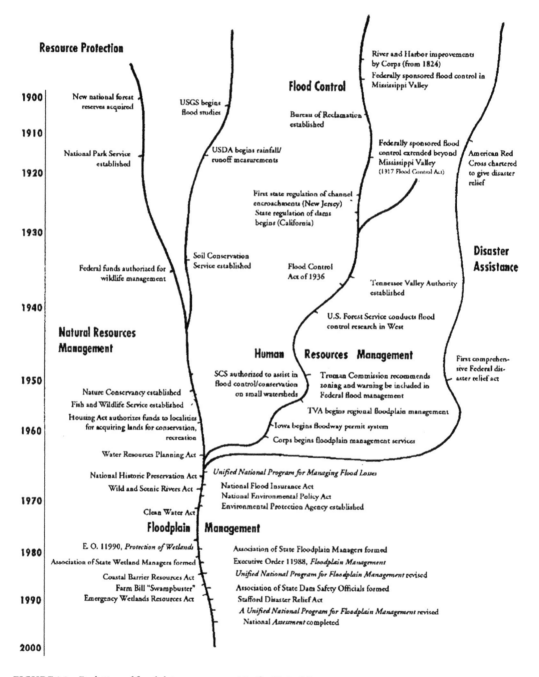

FIGURE 1.1 Evolution of floodplain management in the United States.

Floodplain Management

Floodplain management is officially defined by the Federal Government's *Unified National Program for Floodplain Management* as "a decision-making process that aims to achieve the wise use of the nation's floodplains." (see FEMA,[1] p. 8) "Wise use" means both reduced flood losses and protection of the natural resources and functions of floodplains. This is accomplished through different tools, including, but not limited to:

- Floodplain mapping.
- Land use regulations.
- Preservation of floodprone open space.
- Flood control (levees, reservoirs, channel modifications, etc.)
- Acquiring and clearing damaged or damage-prone areas.
- Floodproofing buildings to reduce their susceptibility to damage by floodwaters.
- Flood insurance.
- Water quality best management practices.
- Flood warning and response.
- Wetland protection programs.
- Public information.

There are a variety of Federal, state, and local programs that administer these tools. Private organizations and property owners also have roles.

The National Flood Insurance Program

The nation's focal floodplain management program is the NFIP. It has prepared floodplain maps for 22,000 communities. FEMA sets the minimum land use development standards that participating communities must administer within the floodplains designated on their Flood Insurance Rate Maps. These standards are summarized in Figure 1.2.

While participation is voluntary, communities that decide not to join or not to enforce those regulations do not receive Federal financial assistance for insurable buildings in their floodplains. Rather than face the loss of Federal aid (including VA home loans, HUD housing help, and disaster assistance), just about every community with a significant flood problem has joined. By 2002, 19,700 cities and counties were participating.

Within participating communities, Federal law requires the purchase of a flood insurance policy as a condition of receiving Federal aid, including mortgages and home improvement loans from Federally regulated or insured lenders. This requirement, coupled with personal experiences with flooding, has convinced over four million property owners to buy flood insurance. Unfortunately, it is estimated that only half of the properties in the FEMA mapped floodplains are insured.

Other Federal Programs

FEMA administers other floodplain management programs, including:

- Disaster assistance programs that help flooded communities and property owners recover after a flood.
- Mitigation assistance programs that fund local projects to acquire and clear floodprone properties.
- Research and technical assistance activities in the fields of mapping, planning, mitigation, and floodproofing.
- The National Dam Safety Program which assists state programs that regulate dams (dam failures were a factor in three of the four largest killer floods since 1970).

The National Flood Insurance Program (NFIP) is administered by the Federal Emergency Management Agency (FEMA). As a condition of making flood insurance available for their residents, communities that participate in the NFIP agree to regulate new construction in the area subject to inundation by the 100-year (base) flood.

There are four major floodplain regulatory requirements. Additional floodplain regulatory requirements may be set by state and local law.

1. All development in the 100-year floodplain must have a permit from the community. The NFIP regulations define "development" as any manmade change to improved or unimproved real estate, including but not limited to buildings or other structures, mining, dredging, filling, grading, paving, excavation or drilling operations or storage of equipment or materials.

2. Development should not be allowed in the floodway. The NFIP regulations define the floodway as the channel of a river or other watercourse and the adjacent land areas that must be reserved in order to discharge the base flood without cumulatively increasing the water surface elevation more than one foot. The floodway is usually the most hazardous area of a riverine floodplain and the most sensitive to development. At a minimum, no development in the floodway may cause an obstruction to flood flows. Generally an engineering study must be performed to determine whether an obstruction will be created.

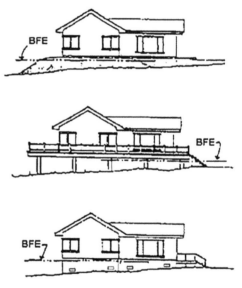

3. New buildings may be built in the floodplain, but they must be protected from damage by the base flood. In riverine floodplains, the lowest floor of residential buildings must be elevated to or above the base flood elevation (BFE). Nonresidential buildings must be either elevated or floodproofed.

4. Under the NFIP, a "substantially improved" building is treated as a new building. The NFIP regulations define "substantial improvement" as any reconstruction, rehabilitation, addition, or other improvement of a structure, the cost of which equals or exceeds 50 percent of the market value of the structure before the start of construction of the improvement. This requirement also applies to buildings that are substantially damaged.

Communities are encouraged to adopt local ordinances that are more comprehensive or provide more protection than the Federal criteria. This is especially important in areas with older Flood Insurance Rate Maps that may not reflect the current hazard. Such ordinances could include prohibiting certain types of highly damage-prone uses from the floodway or requiring that structures be elevated 1 or more feet above the BFE. The NFIP's Community Rating System provides insurance premium credits to recognize the additional flood protection benefit of higher regulatory standards.

FIGURE 1.2   Minimum National Flood Insurance Program regulatory requirements.

The U.S. Army Corps of Engineers is the second largest participant in Federal floodplain management programs. While it is best known as the builder of structural flood control projects, it has its own authority to regulate new development in navigable waterways and wetlands. It is also the leader in the technical aspects of floodproofing and river basin planning.

The U.S. Department of Agriculture's Natural Resources Conservation Service has a role in planning and building flood control projects, similar to the Corps,' but limited to smaller watersheds. Through local soil and water conservation districts, NRCS staff can be valuable advisors to local officials reviewing floodplain or watershed development proposals.

Just as rivers traverse many lands, floodplain management pervades many government programs. Other agencies with floodplain management responsibilities include:

- Tennessee Valley Authority (where floodplain management got its start)
- Bureau of Reclamation (water control projects in the west)
- U.S. Geological Survey (river data and mapping)
- Environmental Protection Agency (water quality programs)
- Small Business Administration (disaster assistance for private property owners)
- National Oceanic and Atmospheric Administration (coastal zone policies)
- National Weather Service (the lead in flood warning programs)

Other Programs

State and local agencies are also into a variety of floodplain management activities. Their regulatory programs often exceed the NFIP requirements. Many states set additional minimum standards for mapping, floodplain and wetland regulations and water quality. Some state agencies require their own permits, in addition to local permits, for new construction on waterways, lakes, shorelines, and floodplains.

In addition to being the lead regulators, most flood control projects are built and operated by local governments: cities, towns, counties, and special districts. The trend at the local level is toward special purpose authorities at the county or multicounty level to tackle problems holistically at the watershed level.

Private organizations have become more directly involved, too. Groups like the Nature Conservancy and land trusts work to preserve floodprone areas that have natural benefits. Others, like the National Wildlife Federation and American Rivers, are active on the political scene, reminding government agencies of their responsibilities and working to strengthen or expand their programs.

Over time, the distinction between what is done by what level of government has blurred. There are more and more cooperative and coordinated approaches, especially with increased non-federal cost sharing requirements and regional and river basin organizations. A recent example of this is FEMA's Cooperating Technical Partners program where a state or local government can contribute to the cost of floodplain mapping and have a say on the techniques and standards used to prepare their Flood Insurance Rate Maps.

Another reason for the blurring of the distinction is the increased professionalization of the field. Most people active in floodplain management are members of the Association of State Floodplain Managers. Private practitioners and staff from all levels of government work together on solving common problems, rather than debating authorities or funding. There is also a new program that certifies floodplain managers. In less than 3 years, over 1000 professionals have earned the right to put "CFM" after their names.

Progress

The impact of these efforts can be measured in three ways: threat to life, property damage, and the environment. Statistics have shown that the loss of life due to floods decreased during the last century, primarily due to better warning and public information programs.

Progress in the other two fields has not been as encouraging. Property damage is still increasing, although at a slower rate than if there were no NFIP and other floodplain management efforts (Figure 1.3). It is harder to see improvements in water quality and habitat protection, but it is generally concluded that while things are better than if there were no programs, we have a long way to go.

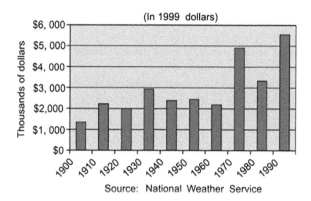

Source: National Weather Service

FIGURE 1.3   Dollar damage caused by flooding.

Agricultural Concerns

Farmers, ranchers, and other agricultural interests are likely to be involved in floodplain management in several different ways. First, as landowners, their freedom to develop the floodplain portions of their properties may be limited by floodplain management or wetland regulations.

Federal, state, and/or local regulations require permits for the following:

- Regrading in the floodway.
- Construction of a levee.
- Modifications to a channel.
- Filling in a wetland.
- Construction of a new building in the floodplain.

This is the controversial part of floodplain management: activities on one's own property are subject to government restrictions in order to prevent diverting flood flows to other properties or adversely affecting wetlands or habitat or to reduce government disaster response and assistance expenses. While many state laws exempt some agricultural activities from local zoning or building codes, FEMA has ensured that in every state, agricultural buildings will be regulated as a condition for a city or county to participate in the NFIP.

A loan or Federal financial assistance to purchase, improve or repair a building in the floodplain will likely be accompanied by a requirement to purchase a flood insurance policy on that building. However, by taking certain protection measures, such as elevating the building above flood levels, insurance premiums can be reduced.

Federal and state programs are not all about restrictive regulations. Federal disaster assistance, flood insurance and crop insurance can come to one's aid after a flood. After the Great Flood of 1993 in the Mississippi River basin, many farmers accepted Federal funds to set aside wetlands and marginal farmland as a start to allowing Mother Nature to reclaim the natural floodplains.

Hopefully, farmers, ranchers, and other agricultural interests will become involved in floodplain management activities voluntarily and in a broader extent. They can reduce their own exposure to flood losses, help their communities and neighbors protect themselves, and improve their environment. Good places to learn more are the following websites:

- FEMA— http://www.FEMA.gov
- Association of State Floodplain Managers— http://www.floods.org

Both have links to other agencies and organizations. The latter has links to state floodplain management associations.

Reference

1. *A Unified National Program for Floodplain Management*; FEMA 248, Federal Interagency Floodplain Management Task Force: Washington, DC, 1994.

Bibliography

Addressing Your Community's Flood Problems: A Guide for Elected Officials; *Association of State Floodplain Managers, Inc.: Madison, WI, 1996.*

Answers to Questions about the National Flood Insurance Program; FEMA-387, http://www.fema.gov/nfip/qanda.htm (accessed February 2002) Federal Emergency Management Agency: Washington, DC 2001.

Floodplain Management in the United States: An Assessment Report; FIA-18, Federal Interagency Floodplain Management Task Force: Washington, DC, 1992.

National Flood Programs in Review, http://www.floods.org/PDF%20files/2000-fpm.pdf (accessed February 2002) Association of State Floodplain Managers: Madison, WI, 2000.

Using Multi-Objective Management to Reduce Flood Losses in Your Watershed; *Association of State Floodplain Managers, Inc.: Madison, WI, 1996.*

2

Floods: Riverine

William Saunders,
Alison MacNeil,
and Edward Capone
*National Oceanic
and Atmospheric
Administration (NOAA)*

Introduction

Riverine flooding occurs whenever a river overtops its banks and water begins to spread laterally away from the river channel (Figure 2.1). This happens when the flow within the river channel exceeds the channel's capacity to carry water downstream. Primary factors that result in riverine flooding include (a) an overabundance of rainfall and surface runoff in the drainage area (or "watershed") upstream of the flood location, (b) a quickly melting snowpack in the watershed, (c) failures of riverine structures, such as dams, resulting in a sudden release of stored water, and (d) any combination of the above.

Categories based on the severity of flooding in a particular area have been defined by the National Weather Service (NWS), in coordination with local emergency management officials nationwide. The flood categories are defined as follows:

- **Minor Flooding:** Minimal or no property damage, but possibly some public threat (e.g., minor inundation of roads).
- **Moderate Flooding:** Some inundation of structures and roads near stream. Some evacuations of people or transfer of property or both to higher elevations.
- **Major Flooding:** Life-threatening. Extensive inundation of structures and roads. Significant evacuations of people or transfer of property or both to higher elevations.

River flooding occurs at different rates for different locations. Slow and steady river rises to flood thresholds commonly occur on larger rivers that have more significant drainage areas and wider channels with greater assimilative capacity. Conversely, more rapid flooding events, or "flash floods," generally occur on smaller rivers that drain smaller watersheds. Other characteristics of locations that are prone to frequent flash flooding include (a) steeply sloping upstream river channels and (b) a high percentage of low-permeable soils in the contributing watershed. Flash floods are common in mountainous regions, where exposed rock formations inhibit the infiltration of rainfall into the soil and tributary streambeds are steep. Flash floods are also common in urban areas, where much of the land cover is developed with impermeable surfaces (e.g., roadways, parking lots, and roof structures).

FIGURE 2.1   Overtopping river with lateral flooding.
Source: Adapted from NWS Central Region, http://www.crh.noaa.gov/cys/floodweekWY.php?wfo=riw.

According to the National Oceanic and Atmospheric Administration (NOAA), annual flood damages in the United States average over $8 billion (historical costs adjusted for 2012 inflation) and flood-related fatalities average over 90 persons per year.[1] Annual damages in any one year can be much higher, (e.g., flood damages over $52.5 billion in 2005). While these costs are impressive, it should be noted that floods from excessive rainfall and snowmelt are naturally occurring phenomena that can also bring tangible benefits, such as ground water recharge, soil fertilization, nutrient replenishment, and maintenance of biodiversity in riparian areas adjacent to river channels.

Different rivers and watersheds typically demonstrate unique responses to the atmospheric and hydrologic stresses placed on them (although some regional commonalities can be observed). As such, understanding how and when riverine floods occur is critical to the accurate prediction of future floods, the mitigation of flood-related damages, and the efficient realization of flood-related benefits.

Riverine Flooding due to Surface Runoff

Surface water runoff is one step in the hydrologic cycle (Figure 2.2) and is a leading cause of riverine flooding. During precipitation events, water will infiltrate the soil, evaporate, or flow directly overland to river channels. A number of factors determine the amount of surface runoff generated, including rainfall intensity, duration, antecedent soil moisture conditions, topography, soil type, vegetation, and land use. Other factors, such as the presence of lakes and reservoirs, can help delay surface runoff from reaching the river channels and decrease flooding intensity.

Surface runoff occurs when the rainfall intensity is greater than the infiltration rate of the soil or when the amount of rainfall exceeds the storage capacity of the soil or both. Excess water from these conditions then flows over the land surface toward river channels (this process is called overland flow). The infiltration rate of water into the soil is dependent on the soil's porosity and permeability, gravity, and capillary action, with porosity and permeability being the most important. Highly porous soils, such as sand, allow

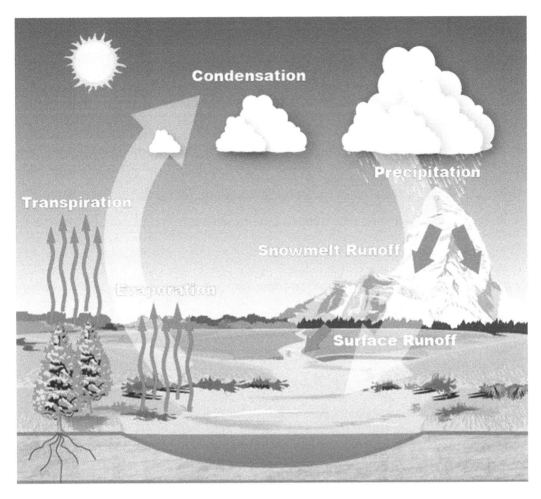

FIGURE 2.2 Hydrologic cycle.
Source: Adapted from NWS Southern Region, http://www.srh.noaa.gov/jetstream/atmos/hydro.htm.

water to infiltrate rapidly and surface runoff rarely occurs. Soils with low porosity and/or permeability, such as clay, have very slow infiltration rates and surface runoff can occur regularly.

Urbanization enhances flood potential by increasing the amount of surface runoff that occurs due to the impermeability of man-made surfaces, such as pavement, that inhibit water from infiltrating into the soil below. In colder regions, frozen soil can also increase the amount of surface runoff by preventing water from infiltrating into the soil column.

Antecedent soil moisture content is a measure of soil "wetness" prior to the onset of precipitation and is another factor that significantly impacts the amount and timing of surface runoff. How quickly the soil becomes saturated depends on the antecedent soil moisture content. The higher the moisture content at the start of a rainfall event, the faster the soil becomes saturated. Once the soil column is saturated, any additional rainfall results in surface runoff.

Not all surface runoff events result in riverine flooding. Flooding begins to occur once the channel capacity of a river is exceeded. The excess water then spills over the banks of the channel and spreads outward laterally into the floodplain. The amount of water in the river channel prior to a heavy rainfall event plays a significant role in whether flooding will occur. It takes more runoff at low flow or base flow conditions to cause flooding than for a river with elevated flow from recent rainfall.

Snowmelt Flooding

A riverine flood is deemed a snowmelt flood when the primary cause of the flood is from melting snow. Snowpacks have the potential to store large quantities of water for variable lengths of time, from days to months. Before a snowpack begins to warm up and "ripen," it can also act like a sponge and absorb large quantities of rainfall under certain conditions. Typically, snowmelt runoff occurs once the snowpack reaches 32°F and enough melt has occurred to saturate the snowpack. Once the snowpack becomes saturated, any additional melt is released and is then either absorbed by the soil column or becomes surface runoff, flowing downslope toward river and stream channels. Flooding then occurs as it does for surface runoff flooding, i.e., when the infiltration rate or storage capacity of the soil or both are exceeded, and river channels have exceeded channel capacity.

Six factors typically contribute to snowmelt flooding in winter and spring:[2]

- **High antecedent soil moisture conditions prior to snowmelt:** After the growing season has ended, heavy rains increase the potential for snowmelt flooding in the spring, because less evapotranspiration occurs, and there is less time for the soil to drain and dry before it freezes or a snowpack begins to form.
- **Ground frost or frozen soil:** A deep, hard frost layer will inhibit snowmelt from infiltrating into the soil, thus generating increased amounts of surface runoff during rainfall and snowmelt events. Frozen soil column conditions are aided by cold temperatures prior to a heavy snowfall and normal-to-above normal soil moisture. A heavy snowfall on top of a shallow ground frost will also insulate the ground and help prevent deep frosts from occurring.
- **Heavy winter snow cover:** Heavy snowfall naturally increases the amount of water typically stored and available for snowmelt. Widespread heavy snow cover also keeps air temperatures cooler and enhances nighttime radiational cooling, which extends cooler temperatures and delays spring warming. These conditions increase the potential for heavy rain events to coincide with snowmelt, leading to more rapid snowmelt in the later spring.
- **Widespread heavy rains during the snow season and melt period:** Heavy rain on a fresh, cold snowpack gets absorbed by the pack and increases the amount of water available during flooding. A warm, heavy rain also warms up cold snowpacks, causing them to begin melting earlier than normal. Heavy rain on a ripe snowpack enhances the immediate snowmelt rate. Some of the most devastating floods have occurred during "rain on ripe snowpack" events.
- **Rapid snowmelt:** Typically, the melting of large snowpacks occurs over an extended period of time, from days to weeks and even months. Snowmelt in the absence of rainfall follows a diurnal pattern along with air temperature, with snowmelt rates increasing during the day and then dropping off at night as temperatures cool. Rapidly melting snowpacks occur under certain conditions, such as during unusually warm periods with high dew point temperatures (humidity), increased wind speeds, and when nighttime temperatures remain above freezing.
- **Ice jams in rivers:** Ice jams can occur either during the early winter at the onset of ice formation (freeze-up jams) or during the melting period when ice begins to break up and move downstream. These ice jams act like temporary dams and restrict the flow through the river, thus causing the potential for flooding upstream as the water backs up behind the jam.

Flooding due to Dam/Structure Failure

The purpose of a dam is to retain or store water for any of several reasons, such as consumptive water supply, irrigation, flood control, energy generation, recreation, pollution and sediment control, and low flow augmentation. Many dams fulfill a combination of these functions.

In the United States, dams are classified by their hazard potential (i.e., high, significant, or low). The failure of a high hazard dam would probably result in loss of human life, economic and environmental losses, and disruption of basic lifeline facilities. A significant hazard dam failure would not be expected to cause loss of human life but would cause economic and environmental losses and disruption of lifeline facilities. Losses due to a low hazard dam failure would generally be limited to the owner of the dam/structure. The nation has approximately 84,000 dams with an overall average age of 52 years. Of these, nearly 14,000 dams are high hazard structures.[3]

Dams can fail with little warning (Figure 2.3). While intense storms may produce flash flooding in a few hours, dam break flooding can occur within a half hour for downstream locations closest to the breach. Other catastrophic failures and breaches can take days or weeks to occur, as a result of debris jams, the accumulation of melting snow, or increasing water pressure on a damaged or weakened dam after days of heavy rain. Flooding can also occur when a dam operator releases excess water downstream to relieve pressure on the dam.

Some common causes of dam failure inducing significant riverine flooding include:

- Substandard construction materials/techniques (e.g., Gleno Dam, Bergamo, Italy, 1923)
- Spillway design errors (South Fork Dam, Johnstown, Pennsylvania, USA, 1889)
- Geological instability (Malpasset, Cote d'Azur, France, 1959)
- Poor maintenance, especially of outlet pipes (Lawn Lake Dam, Estes Park, Colorado, USA, 1982)
- Extreme inflow (Shakidor Dam, Balochistan, Pakistan, 2005)
- Human, computer, or design error (Buffalo Creek Flood, Logan County, West Virginia, USA, 1972; Dale Dike Reservoir, South Yorkshire, England, 1864)
- Internal erosion, especially in earthen dams (Teton Dam, eastern Idaho, USA, 1976)
- Reservoir volumetric displacement (Vajont Dam, Monte Toc, Italy, 1963)
- Earthquakes

Failures of high hazard dam structures commonly result in floods of much greater magnitude than normal floods due to surface runoff or snowmelt. The area downstream of a dam that would flood in the event of a breach (or failure) is referred to as the "dam breach inundation zone" and is generally much larger than the area that would be flooded for a normal river or stream flood event. The dam breach

FIGURE 2.3   Dam breach at Hadlock Pond Dam, Fort Ann, NY.
Source: Adapted from Hom, V. IWRSS Dam Failure Collaboration. National HIC Meeting, July 8–10, 2009. NWS Central Region, http://www.crh.noaa.gov/Image/lmrfc/about/HIC_july_2009/Hom_IWRSS_Dambreak.ppt.

inundation zones for high hazard dams are frequently larger than the flood zones that are depicted on Flood Insurance Rate Maps of the Federal Emergency Management Agency (FEMA). Some people who live in these zones may be completely unaware of the potential hazard lurking upstream.

Common Riverine Flooding Statistics

As noted in the introduction, the NWS identifies minor, moderate, and major flood categories associated with discrete elevations at individual forecast sites. These flood categories are based on the specific impacts at the individual locations, (i.e., not United States Geological Survey [USGS] river gage data statistics or FEMA floodplain return period criteria). In the northeastern part of the nation, some forecast locations have seen an increase in the number of minor floods over the past 50 years, while the magnitude of major flooding in the region has also increased. Some studies[4] have indicated that a combination of basin urbanization/suburbanization and possibly climate change are partly responsible for the increase in the number of flood events.

The USGS, FEMA, and many other federal, state, local, and private agencies categorize the magnitude of a flood event according to its annual exceedance probability. The following is the present terminology used for flood magnitudes that replaces the common "nth" year return period terminology:

- 10% annual exceedance probability flood=10-year flood
- 2% annual exceedance probability flood=50-year flood
- 1% annual exceedance probability flood=100-year flood
- 0.2% annual exceedance probability flood=500-year flood

Accordingly, what was once commonly referred to as a "100-year flood" is now referred to as a 1% annual exceedance probability flood, because it has a 1% probability of occurring in any given year. This change in terminology helps to address a common misunderstanding that a 100-year flood can occur only once in a 100-year period. In fact, there is an approximate 63.4% chance of one or more 1% annual exceedance probability floods occurring in any 100-year period.[5]

The 1% annual exceedance probability flood can be expressed as a river-stage level or a flow rate. For a majority of their river gages, the USGS establishes a "rating curve," which provides a correlation of flow rate to river elevation at the gage location. Using this relationship, the 1% annual exceedance probability flood flow rate can also be expressed as an elevation, which can then be delineated on a detailed map (e.g., USGS quadrangle map) as an area of potential inundation. The resulting mapped floodplain is now referred to as the 1% annual exceedance probability floodplain, which figures prominently for building permits, environmental regulations, and flood insurance.

Monitoring and Forecasting of Riverine Floods in the United States

To monitor existing river conditions and deliver accurate predictions of future riverine flooding, the NWS employs watershed and river analysis computer models to help provide daily forecasts of water surface elevations and river flows at many locations across the United States. The models, which are used by hydrologists at the NWS River Forecast Centers (RFCs), incorporate observations and future predictions of precipitation and temperature, soil moisture and snowpack conditions, and routed flows from upstream locations and reservoir operations. The models are executed at least once per day and more frequently during high flow conditions.

The watershed and river analysis computer models are applied over a period of time that includes the recent past and the near future. Models are adjusted to ensure that simulated flows and elevations align with the recent hydrologic conditions and observations. By achieving this agreement between simulations and observations, modelers also establish better confidence that the near-future portion of the

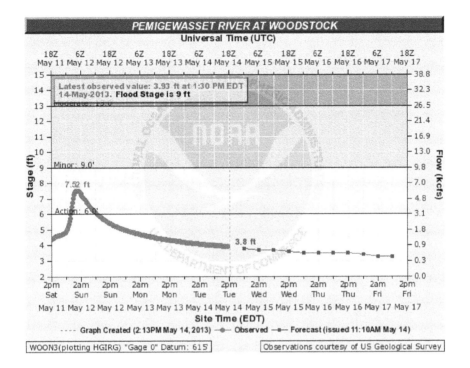

FIGURE 2.4 Forecast hydrograph for Woodstock, NH.
Source: Adapted from NWS AHPS, http://water.weather.gov/ahps2/index.php?wfo=gyx.

simulation provides an accurate forecast. The adjusted model results are then disseminated to Weather Forecast Offices (WFOs), which are, in turn, responsible for issuing flood watches and warnings. Other users of river forecast information include federal and state Emergency Management Agencies, municipal water providers, and the general public.

River forecasts are distributed in the form of hydrographs, which are graphical plots of elevation (or flow) vs. time. The NWS-generated hydrographs (e.g., Figure 2.4) can be found at the Advanced Hydrologic Prediction Service (AHPS) website.[6]

Future of Flood Forecasting in the United States

Current RFC operational procedures require at least daily issuance of forecasts for most locations. Using the current suite of watershed and river analysis models, these forecasts are generally limited to discrete locations that demonstrate monodirectional flow. In the future, as computer processing times decrease, more complicated algorithms (such as those in hydrodynamic models) may be added to RFC operational modeling suites, allowing for the expansion of forecasting services to more points, particularly at tidally influenced locations. These advanced models will also allow for operational simulation of in-line and lateral structures (such as dams and levees), so that their potential failure consequences may be considered as part of a real-time forecast.

An NWS effort is currently underway to couple discrete in-stream forecast locations with detailed near-shore terrain models. Automated inundation maps are generated through this coupling, based on the forecast elevations from the model simulations. As this coupling gets established for more forecast locations nationwide, the inundation maps may become part of the routine forecasts.

Finally, the future of riverine forecasting within the NWS also includes the use of multiple atmospheric and hydrologic models that can produce a family of hydrologic simulations, rather than a single deterministic forecast. These simulation families will comprise an ensemble forecast, which will include an envelope of maximum and minimum elevation values, as well as higher probability simulations in between. This Hydrologic Ensemble Forecast System (HEFS) will also expand the forecast product to include probabilistic river forecasts over a time horizon of days to months to one year. These ensemble forecasts will help local emergency managers to plan for multiple scenarios of an upcoming event and for longer-term (e.g., seasonal) contingencies.

References

1. National Oceanic and Atmospheric Administration (NOAA) National Weather Service (NWS). Hydrologic Information Center – Flood Loss Data. http://www.nws.noaa.gov/hic/ (accessed April 2013).
2. NOAA NWS. Snowmelt Flooding. http://www.floodsafety.noaa.gov/snowmelt.shtml (accessed April 2013).
3. American Society of Civil Engineers. 2013 Report Card for America's Infrastructure: Dams. http://www.infrastructurere-portcard.org/a/#p/dams/overview (accessed April 2013).
4. Vogel, R.M.; Yaindl, C.; Walter, M. Nonstationarity: Flood magnitude and recurrence reduction factors in the United States. J. Am. Water Resour. Assoc. **2011**, *47* (3), 464–474.
5. California Department of Water Resources. California Water Plan Update 2005, Bulletin 160–05, Volume 4; California Department of Water Resources, Sacramento, CA, 2005; 633–634.
6. NOAA NWS. Advanced Hydrologic Prediction Service. http://water.weather.gov/ahps/forecasts.php (accessed April 2013).

3

Marshes: Salt and Brackish

Sally D. Hacker
Oregon State University

Introduction

Salt and brackish marshes are vegetated coastal habitats that occur along the edges of estuaries, defined as places where rivers meet oceans. These highly productive habitats are dominated by terrestrial vegetation rooted in sediment and exposed to daily tidal inundation. An example of a salt marsh plant community is shown in Figure 3.1. Coastal marshes are created when vegetation captures terrestrial sediments carried by rivers, thereby elevating the habitat above sea level. At high tide, a variety of estuarine

FIGURE 3.1 A salt marsh along the Pacific Coast of North America. (Courtesy of Sally D. Hacker.)

animals, including crabs, snails, and fish, use the marsh for foraging and habitat protection. At low tide, terrestrial animals, including insects, birds, rodents, and even ungulates such as elk, are common visitors. Coastal marshes link marine and terrestrial ecosystems, making them unique and highly valuable.

Salt marshes differ from brackish marshes in their location within the estuary. Salt marshes are closer to the ocean and are thus flooded by nearly full-strength seawater compared to brackish marshes, which occur upstream and are exposed to a mixture of seawater and freshwater. In reality, salt and brackish marshes are exposed to contrasting salinities because they occupy different locations along an estuarine gradient. Given the substantial influence of seawater on the physiology of terrestrial plants, the location of a particular marsh within an estuary has important consequences for the types of organisms that can live in these habitats, the types of interactions they experience, and their overall productivity.

Salt and brackish marshes, because they occur at the interface between terrestrial and marine environments, are important arbiters of critical ecosystem services, including coastal protection, maintenance of fisheries, and water purification [1]. The interface nature of coastal marshes also places them at great risk from human impact and climate change. Here, basic information about the ecology of salt and brackish marshes, including patterns of plant zonation, the engineering attributes of marsh plants, and the types of ecological interactions that shape them, is provided. Then, the important ecosystem services of coastal marshes and the impacts of humans on these services are reviewed, and in conclusion, the conservation and restoration of these unique habitats is described.

Basic Ecology

Zonation Patterns of Salt and Brackish Marshes

Marsh plant distributions vary across estuarine gradients and also across intertidal zones, areas between the highest and the lowest extent of the tides. As one moves up the intertidal, from the water's edge to the terrestrial border, there are distinct intertidal zones characterized by different plant assemblages and physical conditions. In salt marshes, the lowest zone is covered with seawater on a daily basis, making it stressful for most terrestrial plants. As a result, this zone is dominated by salt- and flood-tolerant plants, including grasses such as *Spartina* spp. and succulents such as *Salicornia* spp. In some cases, macroalgae may live attached to, or among, the stems of plants in this zone. In the middle zones of most salt marshes, tides cover the marsh surface less frequently, which can lead to highly variable salinity and oxygen conditions within the sediment. Plant species richness is typically greater in the middle marsh with mixtures of grasses (e.g., *Distichlis* spp.), rushes (e.g., *Juncus* spp.), succulents (e.g., *Salicornia* spp.), and forbs (e.g., *Atriplex* spp., *Plantago* spp., *Triglochin* spp., and *Potentilla* spp.), as shown in Figure 3.2. In the high-elevation zones of salt marshes, where tides rarely reach and freshwater runoff is common, less salt- and flood-tolerant grasses (e.g., *Deschampsia* spp.) and shrubs (e.g., *Iva* spp.) typically dominate.

In brackish marshes, both the tides and the rivers influence the salinity of the water that floods these habitats. Although these marshes experience low-salinity water (i.e., >15 g/kg NaCl), they may experience the same or even more flooding due to the influence of the river. Brackish marshes have greater overall biomass than salt marshes [2]. In the low marsh, low-salinity but flood-tolerant sedges and rushes (e.g., *Carex* spp. and *Scirpus* spp.) dominate. In higher intertidal zones, tidal flooding might only occur a few days a year, with most water coming from rivers and terrestrial runoff. Here, freshwater species such as cattails (*Typha* spp.) can be common.

Coastal Marsh Plants as Ecosystem Engineers

Salt and brackish marshes are the result of the interaction between vegetation, sediments, and the tides. Marsh plants act as ecosystem engineers (i.e., organisms that create, modify, or maintain physical habitat such as trees, kelps, or beavers) by increasing the drag of water that moves across the surface of

FIGURE 3.2   A brightly colored forb, *Potentilla pacifica*, among a background of the rush, *Juncus balticus*, in a Pacific Coast salt marsh. (Courtesy of Sally D. Hacker.)

the marsh, causing sediments to settle out and be deposited on the marsh surface [3]. The roots of plants may contribute to this process by binding sediments and reducing erosion, but it is the upright stems of marsh plants that control sediment deposition [4]. Plants vary in their ability to accrete sediment depending on their morphology, density, and even stiffness [5,6]. With the deposition of sediment, plant growth is stimulated, creating a positive feedback between growth and deposition. The ability of a marsh to maintain a constant elevation with sea level is sensitive to plant productivity, sediment supply, and rates of sea-level rise [7]. As sea level rises with climate change, there are concerns about whether coastal marshes will be able to keep pace via sediment accretion.

Beyond accreting sediment, coastal marsh plants are masters at modifying the sediment chemistry via two main mechanisms [8]. First, they can passively shade the soil surface and reduce salt accumulation that occurs when seawater evaporates at the marsh surface. Figure 3.3 shows an example of salt accumulation on the surface of a large patch of salt marsh lacking vegetation. Second, some marsh plants, particularly grasses, rushes, and sedges, can use specialized tissue called aerenchyma to oxygenate their roots and rhizomes (underground stems) when exposed to flooding.

Species Interactions in Coastal Marshes

The distribution of marsh organisms across estuarine and elevation gradients is not only a product of salinity and tidal inundation but also of species interactions. Conceptual models [9,10] and numerous empirical studies in coastal marshes [2,11–13] show that species interactions can shift from mostly competitive (negative) under low stress and high productivity conditions (i.e., higher intertidal zones and more brackish marshes) to mostly facilitative (positive) under high stress and low productivity conditions (i.e., lower intertidal zones and more salty marshes). The main mechanisms behind the facilitative interactions were identified as those described earlier: shading and oxygenation of sediment by neighboring vegetation [8]. One well-studied example showed that positive interactions between the rush, *Juncus gerardii*, and a number of plants and their insect herbivores controlled the species richness of a New England marsh [10].

FIGURE 3.3   A high-salinity patch formed when vegetation was disturbed. The white on the surface is salt crystals. (Courtesy of Sally D. Hacker.)

Trophic interactions are also important to the distribution of organisms within coastal marshes. Plants are eaten by a variety of herbivores (e.g., snails, insects, rodents, and ungulates), which, in turn, are fed on by carnivores (e.g., crabs, predatory insects, fish, and even turtles). For a long time, it was assumed that trophic interactions played a minor role in structuring marsh plant communities [14]. However, Silliman and colleagues [15,16] showed that so-called top-down effects of predators could dramatically affect the major marsh builder, the grass *Spartina alterniflora*. *Spartina* is indirectly killed by snails (*Littoraria irrorata*), which scrape the grass surface with their mouthparts and cause wounds that become infected by a fungus. Normally, snail populations are controlled by the predatory effects of the blue crab, *Callinectes sapidus*; however, recent disease and overharvesting of the crabs has released the snails to their destructive ways. Large diebacks of *Spartina* have occurred in some parts of the southeast United States due in part to the loss of snail predators.

Human Importance and Impact

Coastal and estuarine ecosystems are some of the most heavily impacted systems on the Earth [17]. In the United States alone, nearly 40% of the population lives on the coast and this is expected to rise by nearly 10% each decade [18]. Humans have drained, diked, and diverted water away from coastal marshes both for agriculture and for development; it is estimated that 50% of the coastal marshes in the United States are gone [19] and the loss has accelerated recently [20]. In addition to direct human impact, marshes are highly susceptible to climate change via sea-level rise, terrestrial flooding, and extreme disasters in the form of hurricanes and tsunamis. Thus, coastal marshes occur at the nexus that includes important ecosystem services, chronic and extreme climate effects, and the sustainability of human health and welfare. In the following section, some of the important ecosystem services, some of the ongoing human threats, and some of the attempts at restoration of coastal marshes are described.

Ecosystem Services of Coastal Marshes

Salt and brackish marshes provide important and valuable ecosystem services to humans, because they are the result of coastal and terrestrial processes. Barbier et al. [1] reviewed these services in some detail and provided economic values, when available, from the literature. They identified seven main services, namely, raw materials and food, coastal protection, erosion control, water purification, maintenance of fisheries, carbon sequestration, and tourism and recreation. One of the most understudied and likely undervalued services provided by coastal marshes is protection from waves and storm surge caused by large storms and hurricanes. Marshes are able to reduce the velocity, height, and duration of incoming waves by increasing the drag of water across the vegetated surface [21]. The value of this service will surely increase given that more frequent and intense storms are projected to occur with climate change [22].

Salt marshes can also act as natural filters by purifying the water entering estuaries from watersheds [23]. This water often contains high nutrients and pollutants that can cause "dead zones" or extremely low oxygen conditions. When water passes over the vegetation in marshes, it traps sediment, stimulating plant growth and the uptake of nutrients. In some cases, coastal marshes have been used for wastewater treatment, saving significant amounts of money compared to conventional municipal treatment [24]. Finally, because coastal marshes are one of the most productive ecosystems on the Earth, they are also important for carbon sequestration [25], an ecosystem service that is likely to become more important as CO_2 continues to rise globally [26]. Because marshes have extremely low oxygen soils, dead plant matter does not easily decompose and can be stored in the form of peat for hundreds of years [25].

Human Impacts and Restoration

Despite the variety of ecosystem services provided by salt and brackish marshes, humans have had considerable negative impacts on these systems [19,27,28]. Loss of coastal marshes can be attributed to a number of factors that vary over local, regional, and global scales [28]. At local scales, marshes have been converted for agriculture and development purposes through draining, diking, or diverting water away from the marsh. Some major cities, including Boston, San Francisco, and London, are built on filled or drained wetlands. Another local-scale impact from humans is that of non-native invasions. For example, *Spartina* spp. have been introduced into estuaries worldwide, transforming mudflats into marshes [29,30]. This reduces habitat for seagrass, oysters, fish, and foraging birds. At regional scales, extraction of groundwater, oil, and gas has caused subsidence, leading to the submergence and erosion of hundreds of square kilometers of salt marsh habitat in the Chesapeake Bay, San Francisco Bay, and Gulf of Mexico. Finally, at global scales, sea-level rise caused by warming temperatures threatens coastal marshes if they are unable to maintain their elevation above sea level through sediment accretion and/ or if brackish marshes become more saline.

Generally, success at restoring coastal marshes once they are damaged is difficult and for the most part, has only been attempted at local scales [31,32]. Restoration of old marshes that were converted for agricultural use typically involves reestablishing the tidal inundation regimes that were diverted or blocked. For example, the removal of dikes allows saltwater to flow back into the fields; killing the vegetation; and reestablishing the connection with estuarine sediment, seeds, and animals. Slowly, coastal marsh vegetation can establish itself and start to build the marsh back to its former elevation. However, the restoration of marshes can be a very slow process, even with the intentional planting of vegetation, because the biological engineering of the habitat requires a certain density of plants that ameliorates the harsh physical conditions and promotes sediment accretion [32]. The establishment of marsh plants as ecosystem engineers in the restoration of marshes is particularly important to provide the foundation for the colonization of other species and thus the promotion of species diversity [33]. For this reason, if marshes are destroyed at very large spatial scales due to subsidence, erosion, or sea-level rise, their restoration will be virtually impossible without the reestablishment of the proper functional dynamics between vegetation and sediment [34].

Conclusion

Salt and brackish marshes are the product of both terrestrial and ocean processes. This unique position makes them important arbiters of highly valued services such as the reduction of coastal vulnerability and the buffering of marine environments from nutrient loading and pollution. Yet, coastal marshes also are at the forefront of threats from humans and climate. As human populations disproportionately increase on the coast, and climate change threatens to alter sea level and the frequency of extreme storms, the sustainability of marshes will be an important indicator of the overall health of our coasts.

References

1. Barbier, E.B.; Koch, E.W.; Silliman, B.R.; Hacker, S.D.; Wolanski, E.; Primavera, J.; Granek, E.F.; Polasky, S.; Aswani, S.; Cramer, L.A.; Stoms, D.M.; Kennedy, C.J.; Bael, D.; Kappel, C.V.; Perillo, G.M.; Reed, D.J. Coastal ecosystem-based management with nonlinear ecological functions and values. *Science* **2008**, *319* (5861), 321–323.
2. Crain, C.M.; Silliman, B.R.; Bertness, S.L.; Bertness, M.D. Physical and biotic drivers of plant distribution across estuarine salinity gradients. *Ecology* **2004**, *85* (9), 2539–2549.
3. Gutiérrez, J.L.; Jones, C.G.; Byers, J.E.; Arkema, K.; Berkenbusch, K.; Commito, J.A.; Duarte, C.M.; Hacker, S.D.; Hendriks, I.E.; Hogarth, P.J.; Lambrinos, J.G.; Palomo, G.; Wild, C. Physical ecosystem engineers and the functioning of estuaries and coasts. In Treatise on Estuarine and Coastal Science; Wolanski, E., McLusky, D.S., Eds. Academic Press: Waltham, 2011, Vol. 7; 53–81.
4. Feagin, R.A.; Lozada-Bernard, S.M.; Ravens, T.M.; Möllerb, I.; Yeager, K.M.; Baird, A.H. Does vegetation prevent wave erosion of salt marsh edges? *Proc. Nat. Acad. Sci. USA*, **2009**, *106* (25), 10109–10113.
5. Bouma, T.J.; De Vries, M.B.; Low, E.; Peralta, G.; Tanczos, I.C.; Van de Koppel, J.; Herman, P.M.J. Trade-offs related to ecosystem-engineering: A case study on stiffness of emerging macrophytes. *Ecology* **2005**, *86* (8), 2187–2199.
6. Neumeier, U.; Amos, C.L. The influence of vegetation on turbulence and flow velocities in European salt-marshes. *Sedimentology* **2006**, *53* (2), 259–277.
7. Morris, J.T.; Sundareshwar, P.V.; Nietch, C.T.; Kjerfve, B.; Cahoon, D.R. Responses of coastal wetlands to rising sea level. *Ecology* **2002**, *83* (10), 2869–2877.
8. Hacker, S.D.; Bertness, M.D. Morphological and physiological consequences of a positive plant interaction. *Ecology* **1995**, *76* (7), 2165–2175.
9. Bertness, M.D.; Callaway, R. Positive interactions in communities. *Trends Ecol. Evol.* **1994**, *9* (5), 191–193.

10. Hacker, S.D.; Gaines, S.D. Some implications of direct positive interactions for community species diversity. *Ecology* **1997**, *78* (7), 1990–2003.

11. Bertness, M.D.; Hacker, S.D. Physical stress and positive associations among marsh plants. *Am. Nat.* **1994**, *144* (3), 363–372.

12. Hacker, S.D.; Bertness, M.D. Experimental evidence for factors maintaining plant species diversity in a New England salt marsh. *Ecology* **1999**, *80* (6), 2064–2073.

13. Guo, H.; Pennings, S.C. Mechanisms mediating plant distributions across estuarine landscapes in a low-latitude tidal estuary. *Ecology* **2012**, *93* (1), 90–100.

14. Teal, J.M. Energy flow in the salt marsh ecosystem of Georgia. *Ecology* **1962**, *43* (4), 614–624.

15. Silliman, B.R.; Zieman, J.C. Top-down control of *Spartina alterniflora* production by periwinkle grazing in a Virginia salt marsh. *Ecology* **2001**, *82* (10), 2830–2845.

16. Silliman, B.R.; Van de Koppel, J.; Bertness, M.D.; Stanton, L.E.; Mendelssohn, I.A. Drought, snails, and large-scale die-off of southern U.S. salt marshes. *Science* **2005**, *310* (5755), 1803–1806.

17. Halpern, B.S.; Walbridge, S.; Selkoe, K.A.; Kappel, C.V.; Micheli, F.; D'Agrosa, C.; Bruno, J.F.; Casey, K.S.; Ebert, C.; Fox, H.E.; Fujita, R.; Heinemann, D.; Lenihan, H.S.; Madin, E.M.P.; Perry, M.T.; Selig, E.R.; Spalding, M.; Steneck, R.; Watson, R. A global map of human impact on marine ecosystems. *Science* **2008**, *319* (5865), 948–952.

18. Crossett, K.M.; Ache, B.; Pacheco, P.; Haber, K. National coastal population report: Population trends from 1970 to 2020, National Oceanic and Atmospheric Administration. National Ocean Service, Management and Budget Office, Special Projects: Washington, DC, 2013.

19. Kennish, M.J. Coastal salt marsh systems in the U.S.: A review of anthropogenic impacts. *J. Coastal Res.* **2001**, *17* (3), 731–748.

20. Dahl, T.E.; Stedman, S.M. *Status and Trends of Wetlands in the Coastal Watersheds of the Conterminous United States 2004 to 2009.* U.S. Department of Interior, Fish and Wildlife Service and National Oceanic and Atmospheric Administration, National Marine Fisheries Service: Washington, DC, 2013.

21. Gedan, K.B.; Kirwan, M.L.; Wolanski, E.; Barbier, E.B.; Silliman, B.R. The present and future role of coastal wetland vegetation in protecting shorelines: Answering recent challenges to the paradigm. *Clim. Change* **2001**, *106* (1), 7–29.

22. Patricola, C.M; Wehner, M.F. Anthropogenic influences on major tropical cyclone events. *Nature* **2018**, *563*, 339–346.

23. Mitsch, W.J.; Gosselink, J.G.; Zhang, L.; Anderson, C.J. *Wetland Ecosystems.* Wiley: Hoboken, NJ, 2009.

24. Breaux, A.; Farber, S.; Day, J. Using natural coastal wetlands systems for wastewater treatment: an economic benefit analysis. *J. Environ. Manag.* **1995**, *44* (3), 285–291.

25. Mcleod, E.; Chmura, G.L.; Bouillon, S.; Salm, R.; Bjork, M.; Duarte, C.M.; Lovelock, C.E.; Schlesinger, W.H.; Silliman, B.R. A blueprint for blue carbon: Toward an improved understanding of the role of vegetated coastal habitats in sequestering CO_2. *Ecol. Environ.* **2011**, *9* (10), 552–560.

26. Kirwin, M.L.; Mudd, S.M. Response of salt-marsh carbon accumulation to climate change. *Nature* **2012**, *489*, 550–553.

27. Bertness, M.; Silliman, B.R.; Jefferies, R. Salt marshes under siege: Agricultural practices, land development and overharvesting of the seas explain complex ecological cascades that threaten our shorelines. *Am. Sci.* **2004**, *92* (1), 54–61.

28. Silliman, B.R.; Grosholz, E.D.; Bertness, M.D.; Editors. *Human Impacts on Salt Marshes.* University of California Press: Berkeley, CA, 2009.

29. Daehler, C.C.; Strong, D.R. Status, prediction and prevention of introduced cordgrass *Spartina* spp. invasions in Pacific estuaries, USA. *Biol. Conserv.* **1996**, *78* (1), 51–58.

30. Hacker, S.D.; Heimer, D.; Hellquist, C.E.; Reeder, T.G.; Reeves, B.; Riordan, T.J.; Dethier, M.N. A marine plant (*Spartina anglica*) invades widely varying habitats: Potential mechanisms of invasion and control. *Biol. Invasions* **2001**, *3* (2), 211–217.

31. Broome, S.W.; Seneca, E.D.; Woodhouse, W.W. Tidal salt marsh restoration. *Aquat. Bot.* **1988**, *32* (1), 1–22.
32. Zedler, J.B. Progress in wetland restoration ecology. *Trends Ecol. Evol.* **2000**, *15* (10), 402–407.
33. Halpern, B.S.; Silliman, B.R.; Olden, J.D.; Bruno, J.P.; Bertness, M.D. Incorporating positive interactions in aquatic restoration and conservation. *Front. Ecol. Environ.* **2007**, *5* (3), 153–160.
34. Moreno-Mateos, D.; Power, M.E.; Comin, F.A.; Yockteng, R. Structural and functional loss in restored wetland ecosystems. *PLoS Biol.* **2012**, *10* (1), e1001247, doi:10.1371/ journal.pbio.1001247.

4

Peatlands

Dale H. Vitt
Southern Illinois University

Paul Short
*Canadian Sphagnum
Peat Moss Association*

Introduction

Peatlands contain abundant natural resources and are ecosystems that over the centuries have net primary production of organic matter exceeding losses from decomposition and export, and at millennial timescales have large stores of carbon [1]. As a result, peatlands develop a deep layer of organic soil, or peat, that is composed of at least 30%, but usually more than 90% organic matter [2]. Deposits of peat that accumulate in-place and contain a historical record of community change over time serve as proxies for past environmental and climatic changes [3], and in some places can hold a record of past human activities on the site. These records indicate that nearly all peatlands have initiated and developed over the past 8,000–12,000 years and since the last glaciation (ca. 15,000 years ago). Globally, peatlands represent a major carbon stockpile, storing approximately one-third of the world's soil carbon (436 Pg [2]; petagram $= 1 \times 10^{-15}$ g) and about 10% of the global drinkable water, yet cover only 3% of the world's terrestrial area [4]. As compared to ocean and lake sediments that also contain large reservoirs of stored carbon and are relatively unaffected by atmospheric processes, peatland ecosystems, and the carbon stored in the layers of peat, are extremely vulnerable to climate change and land-use changes. Additionally, peatlands are a sink for nitrogen, and it has been estimated that boreal peatlands contain about 10% of the global pool of N [2].

While all peatlands are wetlands, not all wetlands are peatlands, and the need to distinguish the difference is important to the recognition and management of peatland resources. In general, non-peat-forming wetlands are either treed or shrubby (swamps) or without woody vegetation (marshes). Both swamps and marshes have fluctuating water levels and relatively high amounts of nutrient inputs; hence, organic matter produced in these systems decomposes rapidly and peat formation is limited. Peatlands (sometimes called "mires" in Europe) are usually classified as either fens or bogs (see Figure 4.1 for examples) depending on the source of the water. Fens receive waters from precipitation as well as from the surrounding uplands (i.e., geogenous water sources), whereas bogs are somewhat raised above the surrounding area and receive water only from precipitation (i.e., ombrogenous). The ionic quality of these incoming waters greatly influences the structure and functioning of the ecosystem and dictates

FIGURE 4.1   (See color insert.) Right side: A poor fen, dominated by sedges and Sphagnum in the foreground and treed bog in the background. Left side: Continental bog of western Canada characterized by a dense tree layer of black spruce (*Picea mariana*). (Courtesy of Kimberli Scott.)

the amount and type of peat produced. Bogs and poor fens are dominated by the moss genus *Sphagnum*, as shown in Figure 4.2, and it has been estimated that there is more carbon stored in *Sphagnum* than in any other plant genus, whereas rich fens have ground layers dominated by true mosses [5].

Peatlands are estimated to have originally occupied 4,258,000 km², of which about 74% are in nontropical parts of the world. North America has 46%, Asia 35%, Europe 13%, South America 4%, Africa 1%, and Australia and Antarctica less than 1%. Tropical peatlands are most abundant in Indonesia (47%) and Malaysia (6%) [6]. Peatlands are concentrated in the boreal and arctic regions of the world where precipitation usually exceeds potential evapotranspiration. Countries with more than 50,000 km² of peatland are Russia (1,410,000 km²), Canada (1,235,000 km²), the United States (625,000 km²), Indonesia (207,000 km²), Finland (96,000 km²), Sweden (70,000 km²), and Peru (50,000 km²). Loss of natural peatlands to anthropogenic disturbance is variable and ranges from 52% of the natural peatlands being lost in Europe and 50% in nontropical Africa, 20% in nontropical South America, 8% in nontropical Asia, and 5% in North America. About 16% of nontropical peatlands have been lost to disturbance, with agriculture accounting for 50%; forestry 30%; peat extraction 10%; and the remainder from urbanization, inundation, and erosion. Restoration efforts, especially in peat-harvested bogs of North America, have been developed to return some of these areas to functioning peatlands [7]. In tropical regions, large areas of virgin tropical swamp forest have been and are currently being logged and converted to rice fields or palm oil plantations. For example, in 2000, it has been recorded that some 106,000 km² has been deforested, drained, and converted to some other land uses [8].

FIGURE 4.2 Ground cover of *Sphagnum lenense* with young leaves of *Rubus chamaemorus* (cloudberry) from the Yukon Territory, Canada. (Courtesy of Dale Vitt.)

Sensitivity to Disturbance

The large reservoirs of carbon and nitrogen found in peatlands have developed over the past 8,000–12,000 years. Over this period of time, it is estimated that peatlands accumulated carbon at a rate of about 22.9 g C/m²/yr [2] and that globally peat accumulation is about 40–70 Tg C/yr (teragram = 1×10^{-12}g). As a natural resource, peatlands are subject to both anthropogenic and natural disturbance impacts. In North America, particularly in the western Canadian provinces, wildfires account annually for the largest emissions of carbon, estimated at of carbon, estimated at 6,300 Gg C/yr (gigagram = 1×10^{-9}g), while anthropogenic impacts that include flooding from hydroelectric construction (−100 Gg C/yr), peat harvesting (−140 Gg C/yr), and mining (−50 Gg C/y) account for −290 Gg C/yr. Considering that these western Canadian peatlands sequester ±8,940 Gg C/yr and the results of permafrost melting contribute +160 Gg C/yr, the net estimate of annual carbon sequestration is +1,320 Gg C/yr [9]. In tropical Asia, the conversion of tropical virgin peat swamp forest into palm oil plantations has resulted in a total carbon loss of biomass and peat 17.1 Mg C/ha/yr (megagram = 1×10^{-6}g) over the past 25 years [8].

Globally, the conversion of peatlands to agriculture and forestry is estimated to contribute 100–200 Tg of C/yr to the atmosphere, and this combined with global peat extraction of approximately 15 Tg of C/yr would account for peatlands at the global scale to be a source rather than a sink for carbon [10]. However, these estimates are rapidly changing, due to changes in forestry practices along with regional differences in anthropogenic activities and the application of appropriate responsible peatland management practices.

Natural Resources of Peatlands

Although peatlands have often been considered as wastelands, even cancers on the landscape (illustrated in *The Lord of the Rings*, "Meres of Dead Faces" NW of Mordor), they actually contain a wealth of natural resources and, perhaps more importantly, carry out valuable ecosystem functions on the landscape. This view of peatlands as wastelands is exemplified by their use in prisons (e.g., the Gulags of the former Soviet Union and the German concentration camps such as Dachau), their use as landfills, and their use as military training grounds, with the latter allowing some peatlands to be preserved in areas of concentrated human occupation (such as Western Europe).

Given the poor regard for these natural peatland resources, responsible peatland management has until recently been a challenge. *The Wise Use of Mires and Peatlands*, published in 2002 [10], provided an excellent framework for decision-making when considering the use of peatlands. More recently, the *Strategy for Responsible Peatland Management*, published by the International Peat Society in 2011 [11], has provided a comprehensive management construct for peatlands worldwide. It provides those involved in or responsible for peatland management with strategic objectives and actions for implementation that for the first time defines objectives and actions for the conservation, management, and rehabilitation of peatlands globally, based on the principles of Wise Use. According to the Strategy, peatland management includes organizing, controlling, regulating, and utilizing peatland and peat for specific purposes, which should be appropriate to the peatland type and use while respecting cultural, environmental, and socioeconomic conditions. The Strategy is applicable to all types of peatlands under every use, including nonuse, and it is directed to everyone responsible for or involved in the management of peatlands, or in the peat supply chain.

Water Resources

In areas where large portions of watersheds are covered by peatlands, these ecosystems provide important source areas for drinking water. For example, in general 70% of Britain's drinking water comes from peat-dominated upland catchments and 45% of Yorkshire's public supply of water is from watersheds draining peatlands. Interestingly, Haworth Moor of *Wuthering Heights* fame was owned by a company providing drinking water. In the northern Andes, peatlands are source areas for drinking water for millions of people living in the large cities of Colombia and Peru. In western Canada's boreal forest, lake chemistry and function are influenced by watersheds having a large cover of peatlands. These peatland-dominated watersheds provide filtered waters to the downstream lakes, and in some cases acidifying the lakes [12]. Peatland-dominated watersheds with thawing permafrost in boreal western Canada have a substantial influence on increased water yields and may be the main cause of increased hydrologic flows in these pristine areas [13]. Nutrient chemistry of lakes downstream of peatland-dominated watersheds is significantly different from lakes with only upland inputs [14]. Thus, peatlands can provide a number of modifying influences on downstream ecosystems. The large dissolved organic carbon (DOC) component of peatlands, especially bogs, colors downstream lakes and streams. With warming and/or drying climatic conditions, coupled with increased atmospheric deposition and pasturing of animals, bogs can degrade, thus releasing large amounts of DOC to downstream drainages. This degradation leads to a loss of carbon from the long-term storage found in these peatlands and has been reported as significant in the British Isles.

Soil Resources

Peat has been extracted from peatlands for centuries for a number of uses, primarily as an organic substrate in agriculture and gardening. Other uses include biodegradable pots for cultivation of seedlings in tree nurseries, soil amendments for gardens, and growing media for both professional and amateur flower and vegetable growers, and as a substrate for mushroom growers. Among its unique properties for use as a soil amendment and as a growing medium are (i) its excellent water-holding capacity—holding up to 20 times its weight in water and retaining water for exceptionally long periods of time; (ii) a low pH that functions to reduce the microbial activity, thus lessening the number of pathogens; and (iii) its ability to lower the rate of decomposition, thus resulting in the amendment remaining in the soil medium longer. Additionally, peat is easy to harvest, as shown in Figure 4.3, handle, and ship to market, which is why peat harvesting is a small, but important, industry in Canada ($10.3 \times 10^{-6} m^3$ harvested in 1999) and Germany ($9.5 \times 10^{-6} m^3$ in 1999).

In Canada, the horticultural peat industry is a well-established, although relatively recent, commercial industrial use of peatlands. Peat for energy was the initial focus of peat use in Canada, but at present, there is no industrial use of peat for energy production. Horticultural peat has become, since the postwar era, a major resource industry in rural areas of Canada, particularly in Québec, and the Maritime Province of New Brunswick. Western Canada's Prairie Provinces (Alberta, Saskatchewan, and Manitoba) are areas of expansion for the industry and currently account for approximately 40% of the national production. The Canadian production principally services the horticultural industry of both Canada and the United States.

The Canadian industry, through policy and practice, is committed to the return of postharvest sites to functioning peatland ecosystems. Recognized throughout the world as leaders in restoration of peatlands are the Industrial Chair for Peatland Management and the Peatland Ecology Research Group at the Université Laval in Québec City, Québec. Beyond the applied level of peatland restoration, the industry continues to actively promote responsible management of peatland resources throughout the provincial and federal agencies accountable for natural resource management. Peatlands (bogs and fens) need to be acknowledged as natural biological resources and managed to ensure their environmental, social, and economic values are sustained. The key to responsible management is to set aside areas for protection and conservation and puts in place management practices that ensure the retention of ecosystem goods and services following development. The peat-harvesting industry has a commitment to support continued leading-edge peatland research, application of science-based restoration techniques, and improved responsible peatland management in order to provide sustainable peatland management that recognizes the environmental, social, and economic values of peatland resources.

Historically, peat has also been used in a number of fascinating ways. During the 1800s and early 1900s, peat was used as litter in horse stables, especially for cavalry, railways, and transportation companies that stabled large numbers of horses. In fact, this use led to the rapid development of peat

FIGURE 4.3 Vacuum harvesting of a drained peat field in Québec, Canada. (Courtesy of Dale Vitt.)

extraction in Western Europe. It has been estimated that one French army with a 13,500-horse cavalry needed 22,000 tons of peat in 1 year [10]. *Sphagnum* peat was used extensively during the Napoleonic and the Franco-Prussian Wars, by the Japanese in the 1904–1905 war with Russia, and in World War I by both sides as a substitute for cotton in surgical dressings. In Britain, "War Work for Women" included collecting and processing *Sphagnum* in the extensive British open bogs. Aboriginal people in North America used *Sphagnum* moss for diapers, and in the 1980s, *Sphagnum* was used in the commercial production of feminine hygiene products. Peat also forms a component of filtering systems for water purification. The cation-exchange properties of *Sphagnum* allow the exchange of hydrogen ions for all base cations, including cationic forms of heavy metals. The absorbent properties of peat allow not only the absorption of large quantities of water, but when dried to less than 8% moisture, peat becomes hydrophobic and absorbs large quantities of petroleum products; as such, it has been used in oil spill cleanup and as floor mats in industries with petroleum spillage. The use of peat for building and insulation was widespread in Europe in the past. In Ireland, canal banks were lined with peat; in Finland, it was used as a liner for roadways; and in Russia and Belarus, it was pressed into sheets and used as insulation in buildings and industry.

Peat is also utilized for energy production. The use of peat for electrical generation is an important national strategic use of peat in Ireland, Finland, and Sweden. Currently, within the Nordic countries of the European Union, energy generation of peat is one of the most important uses of peat and will almost certainly remain important for countries with large peat reserves and in producing energy for isolated locations. The use of peat briquettes for burning in household stoves, furnaces, and fireplaces is also common in rural Ireland, but currently, peat is mixed with *Miscanthus* biomass to form more energy-efficient renewable products. Bord da Mona reports that overall, about four megatons (=4 Tg) of peat is burned for energy generation each year, of which one megaton (=1 Tg) is used for local burning in households.

Biotic Resource

Since the early 1800s, and originating in Germany, peat has been used in balneology (baths). It is reported that owing to the biologically active substances in peat, such as humic acids, peat may influence the immune system and is effective against microorganisms. Peat is an important component in the distillation of Scotch whiskey. The special peaty flavor of Scotch whiskey is imparted by slowly drying the "green malt" over a smoldering peat fire.

Fens have long been cut for hay in both Canada and Eurasia. This "slough hay" is used for both fodder and straw for domestic animals. Also, drier peatlands have been long used as pasture. Wild plants are actively collected and eaten by local and indigenous people across the boreal zone. Cranberries (*Vaccinium oxycoccos* and *V. macrocarpon*) are commercially grown and processed for a variety of juice products in eastern United States and Canada. In Fennoscandia and Québec, Canada, the berries of cloudberry (*Rubus chamaemorus*, as shown in Figure 4.2, and e.g., Chicoutai, Société des Alcools du Québec) and lingonberry (*Vaccinium vitis-idaea*; illustrated in Figure 4.4) are distilled for liquors as well as used in preserves and jellies. A large number of medicinal preparations are produced from sundews (*Drosera* spp.). First Nations peoples of Canada extensively used plants from peatlands for both food and medicinal preparations; for example, leaves of *Ledum groenlandicum* (Labrador Tea—shown in Figure 4.5) were boiled and drank for nausea [5].

Drainage of peatlands to improve forest growth has long been practiced in Fennoscandia, Great Britain, and Russia. Current estimates are that 15×10^5 ha of northern peatlands has been drained for forestry use. In oceanic areas where peatlands are treeless, often nonindigenous tree species have been planted with subsequent fertilization, whereas in more continental areas that have stunted trees growing directly on the peat, the soils have been enhanced with drainage and fertilization. These practices, common in Europe and Asia, have not been implemented on a large scale in North America; however, black spruce (*Picea mariana*) is harvested from natural peatlands in eastern Canada.

FIGURE 4.4 **(See color insert.)** *Vaccinium vitis-idaea* (lingonberry) with red berries, growing on a mat of the reindeer lichen, *Cladonia arbuscula/mitis*. (Courtesy of Kimberli Scott.)

FIGURE 4.5 Flowering branch and characteristic leaves of *Ledum groenlandicum* (labrador tea). (Courtesy of Kimberli Scott.)

Biodiversity

Peatlands are among the last large undisturbed ecosystems in the world and serve as habitat for many animals and birds. Although there are no vertebrate species that occur exclusively in peatlands, peatlands do serve as primary habitat for caribou in North America, especially bogs with an abundance of reindeer lichens (mainly *Cladonia arbuscula/mitis*; as seen in Figure 4.3). Both moose and black bear utilize peatlands for food and shelter. In northern areas, peatland specialists include the arctic shrew, northern bog lemming, and southern bog lemming. Birds also occupy peatlands, but few are exclusive to these areas. Sandhill cranes nest in fens in western Canada; American bittern, Wilson's snipe, and upland sandpipers also use peatlands as major habitats in eastern Canada. Beavers, both in Europe and in North America, may disturb peatlands, but the lack of suitable substrate and food resources limits their activities to the edges. Surprisingly, little is known about the apparent rich invertebrate fauna of peatlands, where many undescribed species may occur. In the few studies completed, the invertebrate diversity is extremely high, yet little studied. In comparison, such environmental indicators as testate amoebae have been well studied and have been extensively utilized in studies that reconstruct past environmental conditions of the local area [15].

Carbon Resources

Over the long term, peatlands extract large amounts of carbon dioxide from the atmosphere and store it in deposits of peat. Presently, peatlands have stored about the same amount of carbon as there is in the atmosphere. Some believe that carbon extracted from the atmosphere by peatlands during the interglacial periods reduced atmospheric CO_2 concentrations and resulted in the development of the glacial periods. There is some evidence that global CO_2 concentrations are coupled to peatland initiation and global climate [16]. Although pristine peatlands have served as a sink for atmospheric CO_2, peatlands also are a source for methane (a potent greenhouse gas [GHG]) and also nitrous oxides. Wetlands, rice paddies, and animal livestock dominate global methane production, and emissions from peatlands are variable, but of less importance. While bogs produce little or no methane, fens may provide a rich source. Methane is a product of anaerobic decomposition and is often highest in wet habitats with relatively high rates of organic matter turnover (such as rich fens and eutrophic marshes). Nitrous oxide emissions from pristine peatlands are low.

Net GHGs from the horticultural peat harvesting process reveal that the entire life cycle of peat extraction emitted 0.54 Gg of CO_2 in 1999 rising to 0.89 Gg CO_2 in 2000. Seventy-one percent of the emissions were associated with peat decomposition, 15% with land-use change, 10% with transportation to market, and 4% with processing [17]. Canadian peat horticultural emissions from all sources (0.89 Gg) represent 0.03% of all degraded peatlands (3 Pg) worldwide. Emissions are 0.006% of all total global net anthropogenic emissions (15.7 Pg). Within the Canadian context (given the national total GHG in Canada at 771 Pg CO_2 in 2006), the peat industry represented 0.1% of total GHGs. Environmental life cycle analysis, such as this, has led both the Canadian Sphagnum Peat Moss Association and European Peat and Growing Medium Association to recognize that peat harvesting is a significant source of carbon emissions and peat harvesting companies are currently examining their emissions' footprint base and identifying opportunities to reduce their impacts.

Nitrogen

The biogeochemical cycle of nitrogen (N) in northern temperate, boreal, and subarctic peatland ecosystems has been altered substantially over much of North America and Europe due to anthropogenic activity [18]; however, in remote regions with historically low atmospheric N deposition (<2 kg N/ha/yr), both bogs and fens function with very limited inputs of inorganic nitrogen. This low atmospheric deposition of inorganic N is not sufficient to sustain the annual N requirements of the ecosystem (reported in the range of 34 kg N/ha/yr for bogs [19]), and other N sources are required for annual plant growth in the form of biological N_2 fixation. Increased atmospheric deposition, including that from anthropogenic sources, serves to decrease N fixation and shifts the balance of new N inputs from organic (biologically fixed) to inorganic, changing the plant community composition by favoring shrub production; shifting the balance of the assimilation of new N inputs from *Sphagnum* to vascular plants; and decreasing ecosystem-wide *Sphagnum* abundance, production, and the important role that these plants have in assimilation of new N inputs. Decreases in the abundance of *Sphagnum* mosses will serve to decrease peat accumulation and substantially increase CO_2 emissions.

Peatland Restoration

Following peat harvest, peat fields are bare and without vegetation. Decomposing peat produces GHGs, and these areas provide little in the form of suitable habitat or ecosystem services. Beginning in the late 1990s, however, techniques were developed (mainly in eastern Canada) to restore a viable living vegetative cover. Using the moss transfer technique developed through the research program at Université Laval's Peatland Ecology Group, a *Sphagnum*-dominated plant cover has been reestablished within 3–5 years following restoration procedures, with biodiversity and hydrology approaching preharvest conditions. It is predicted that carbon sequestration should become a net sink within 15–20 years [7].

Forestry practices result in a steady decrease in the carbon store due to increased aerobic conditions in the upper peat profile leading to increased decomposition. Even though greater initial tree growth of recently drained peatlands may increase carbon storage and exceed the losses from peat decomposition, longer-term cumulative losses from the peat result in an increase in carbon emissions to the atmosphere [20].

Conclusion

Peatlands occupy about 3% of the terrestrial surface of the world and are characterized by organic soils that contain about one-third of the world's soil carbon—about equal to the CO_2 in the atmosphere. This carbon store has accumulated gradually over the past 8,000–12,000 years and is extremely sensitive to land-use changes and climate change. Peatlands provide valuable ecosystem services in that they function as habitats for animals, store and filter water, and provide food and medicine for local as well as aboriginal peoples from the variety of peatland plant species. *Sphagnum* moss has been used in a variety of commercial applications. Peat has been harvested for horticultural purposes and is a valuable soil amendment, and peatlands have been drained for agricultural and forestry practices, releasing CO_2 to the atmosphere. In tropical regions, wholesale deforestation and degradation of virgin tropical peatlands contribute greatly to habitat loss, CO_2 emissions, and carbon loss from existing peat; however, in many parts of the world, restoration of harvested and degraded peatlands currently provides a successful avenue for future sustainable management of this valuable resource.

References

1. Yu, Z.; Loisel, J.; Brosseau, D.P.; Beilman, D.W.; Hunt, S.J. Global peatland dynamics since the last glacial maximum. *Geophys. Resear. Lett.* 2010, 37, L13402.
2. Loisel, J.; Yu, Z.; Beilman, D.W.; Camill, P.; Alm, J.; Amesbury, M.J.; Anderson, D.; Andersson, S.; Bochicchio, C.; Barber, K.; Belyea, L.R. A database and synthesis of northern peatland soil properties and Holocene carbon and nitrogen accumulation. *Holocene* 2014, 24, 1028–1042.
3. Kuhry, P.; Nicholson, B.J.; Gignac, L.D.; Vitt, D.H.; Bayley, S.E. Development of Sphagnum-dominated peatlands in boreal continental Canada. *Can. J. Bot.* 1993, 71, 10–22.
4. Holden, J. Peatland hydrology and carbon release: Why small-scale process matters. *Philos. T. Math. Phys. Eng. Sci.* 2005, 363, 2891–2913.
5. Vitt, D.H. Peatlands: Ecosystems dominated by bryophytes. In Bryophyte Biology; Shaw, A.J., Goffinet, B. Eds.; Cambridge University Press: Cambridge, 2000; 312–343.
6. Page, S.E.; Rieley, J.; Banks, C. Global and regional importance of the tropical peatland carbon pool. *Global Change Biol.* 2011, 17, 798–818.
7. Rochefort, L.; Lode, E. Restoration of degraded boreal peatlands. In Boreal Peatland Ecosystems; Wieder, R.K., Vitt, D.H. Eds.; Springer Verlag: Berlin, 2006; 381–423.
8. Hergoualc'h, K.; Verchot, L.V. Stocks and fluxes of carbon associated with land use change in Southeast Asian tropical peatlands: A review. *Global Biogeochem. Cycles* 2011, 25. doi:10.1029/2009GB003718.

9. Turetsky, M.R.; Wieder, R.K.; Halsey, L.; Vitt, D. Current disturbance and the diminishing peatland carbon sink. *Geophys. Res. Lett.* 2002, 29 (11), 21-1–21-4. doi:10.1029/2001GL014000.

10. Joosten, H.D.; Clarke, D. *Wise-Use of Mires and Peatlands*; International Mire Conservation Group and International Peat Society: Helsinki, Finland, 2002.

11. Clarke, D.; Rieley, J., Eds. *Strategy for Responsible Peatland Management*; International Peat Society: Jyväskylä, Finland, 2011.

12. Halsey, L.; Vitt, D.H.; Trew, D. Influence of peatlands on the acidity of lakes in northeastern Alberta, Canada. *Water, Air, Soil Pollut.* 1997, 96, 17–38.

13. Gibson, J.J.; Birks, S.J.; Yi, Y.; Vitt, D.H. Runoff to boreal lakes linked to land cover, watershed morphology and permafrost thaw: A 9-year isotope mass balance assessment. *Hydrol. Process.* 2015, 29, 3848–3861.

14. Prepas, E.E.; Planas, D.; Gibson, J.J.; Vitt, D.H.; Prowse, T.D.; Dinsmore, W.P.; Halsey, L.A.; McEachern, P.M.; Paquet, S.; Scrimgeour, G.J.; Tonn, W.M.; Paszkowski, C.A.; Wolfstein, K. Landscape variables influencing nutrients and phytoplankton communities in Boreal plain lakes of northern Alberta: A comparison of wetland- and upland-dominated catchments. *Can. J. Fish. Aquat. Sci.* 2001, 58, 1286–1299.

15. Charman, D. *Peatlands and Environmental Change*; John Wiley & Sons Ltd.: Chichester, UK, 2002.

16. Yu, Z.; Campbell, I.D.; Campbell, C.; Vitt, D.H.; Bond, G.C.; Apps, M.C. Carbon sequestration in western Canadian peat highly sensitive to Holocene wet-dry climate cycles at millennial timescales. *Holocene* 2003, 13, 801–808.

17. Cleary, J.; Roulet, N.T.; Moore, T.R. Greenhouse gas emissions from Canadian peat extraction, 1990–2000: A life-cycle analysis. *Ambio* 2005, 34, 456–461.

18. Limpens, J.; Heijmans, M.M.; Berendse, F. The peatland nitrogen cycle. In Boreal Peatland Ecosystems; Wieder, R.K., Vitt, D.H. Eds.; Springer Verlag: Berlin, 2006, 195–230.

19. Wieder, R.K.; Vitt, D.H.; Vile, M.A.; Graham, J.A.; Hartsock, J.A.; Fillingim, H.; House, M.; Quinn, J.C.; Scott K.D.; Petix, M.; McMillen K.J. Experimental nitrogen addition alters structure and function of a boreal bog: Critical loads and thresholds revealed. *Ecol. Monogr.* 2019, 89. doi:10.1002/ecm.1371.

20. Laine, J.; Laiho, R.; Minkkinen, K.; Vasander, H. Forestry and boreal peatlands. In Boreal Peatland Ecosystems; Wieder, R.K., Vitt, D.H. Eds.; Springer Verlag: Berlin, 2006, 331–358.

Bibliography

Bauerochse, A.; Haßmann, H., Eds. *Peatlands Archaeological Sites - Archives of Nature - Nature Conservation - Wise Use*; Verlag Marie Leidorf: Rahden, 2003.

Crum, H. *A Focus on Peatlands and Peat Mosses*; University of Michigan Press: Ann Arbor, MI, 1992.

Gore, A.J.P., Ed. *Mires—Swamp, Bog, Fen and Moor; General Studies; 4B – Regional Studies*; Elsevier Publishing Co.: Amsterdam, 1983.

Rydin, H.; Jeglum, J. *The Biology of Peatlands*; Oxford University Press: Oxford, 2006.

Wieder, R.K.; Vitt, D.H., Eds. *Boreal Peatland Ecosystems*; Springer Verlag: Berlin, 2006.

5

Pollution: Nonpoint Source

Ravendra Naidu,
Mallavarapu
Megharaj, Peter
Dillon, Rai Kookana,
and Ray Correll
*Commonwealth Scientific
and Industrial Research
Organisation (CSIRO)*

W. W. Wenzel
*University of Natural
Resources and Life Sciences*

Introduction

Non-point source pollution (NPSP) has no obvious single point source discharge and is of diffuse nature (Table 5.1). An example of NPSP includes aerial transport and deposition of contaminants such as SO_2 from industrial emissions leading to acidification of soil and water bodies. Rain water in urban areas could also be a source of NPSP as it may concentrate organic and inorganic contaminants.

TABLE 5.1   Industries, Land Uses, and Associated Chemicals Contributing to Non-Point Source Pollution

Industry	Type of Chemical	Associated Chemicals
Agricultural activities	Metals/metalloid	Cadmium, mercury, arsenic, selenium
Non-metals	Nitrate, phosphate, borate	
Salinity/sodicity	Sodium, chloride, sulfate, magnesium, alkalinity	
Pesticides	Range of organic and inorganic pesticides including arsenic, copper, zinc, lead, sulfonylureas, organochlorine, organophosphates, etc., salt, geogenic contaminants (e.g., arsenic, selenium, etc.)	
Irrigation	Sodium, chloride, arsenic, selenium	
Automobile and industrial emissions	Dust	Lead, arsenic, copper, cadmium, zinc, etc.
Gas	Sulfur oxides, carbon oxides	
Metals	Lead and lead organic compounds	
Rainwater	Organics	Polyaromatic hydrocarbons, polychlorobiphenyls, etc.
Inorganic	Sulfur oxides, carbon oxides acidity, metals and metalloids	

Source: Adapted from Barzi et al.[8]

Examples of such contaminants include polycyclic aromatic hydrocarbons, pesticides, polychlorinated biphenyls that could be present in urban air due to road traffic, domestic heating, industrial emissions, agricultural treatments, etc.[1–3] Other examples of NPSP include fertilizer (especially Cd, N, and P) and pesticide applications to improve crop yield. Use of industrial waste materials as soil amendments have been estimated to contaminate thousands of hectares of productive agricultural land in countries throughout the world.

Contaminant Interactions

Non-point pollution is generally associated with low- level contamination spread at broad acre level. Under these circumstances, the major reaction controlling contaminant interactions are sorption–desorption processes, plant uptake, surface runoff, and leaching. However, certain contaminants, in particular, organic compounds are also subjected to volatilization, chemical, and biological degradation. Sorption–desorption and degradation (both biotic and abiotic) are the two most important processes controlling organic contaminant behavior in soils. These processes are influenced by both soil and solution properties of the environment. Such interactions also determine the bioavailability and/ or transport of contaminants in soils. Where the contaminants are bioavailable, risk to surface and groundwater and soil, crop, and human health are enhanced.

Implications to Soil and Environmental Quality

Environmental contaminants can have a deleterious effect on non-target organisms and their beneficial activities. These effects could include a decline in primary production, decreased rate of organic matter break-down, and nutrient cycling as well as mineralization of harmful substances that in turn cause a loss of productivity of the ecosystems. Certain pollutants, even though present in very small concentrations in the soil and surrounding water, have potential to be taken up by various micro-organisms, plants, animals, and ultimately human beings. These pollutants may accumulate and concentrate in the food chain by several thousand times through a process referred to as biomagnification.

Urban sewage, because of its nutrient values and source of organic carbon in soils, is now increasingly being disposed to land. The contaminants present in sewage sludge (nutrients, heavy metals, organic compounds, and pathogens), if not managed properly, could potentially affect the environment adversely. Dumping of radioactive waste (e.g., radium, uranium, plutonium) onto soil is more complicated because these materials remain active for thousands of years in the soil and thus pose a continued threat to the future health of the ecosystem.

Industrial wastes, improper agricultural techniques, municipal wastes, and use of saline water for irrigation under high evaporative conditions result in the presence of excess soluble salts (predominantly Na and Cl ions) and metalloids such as Se and As in soils. Salinity and sodicity affect the vegetation by inhibiting seed germination, decreasing permeability of roots to water, and disrupting their functions such as photosynthesis, respiration, and synthesis of proteins and enzymes.

Some of the impacts of soil pollution migrate a long way from the source and can persist for some time. For example, suspended solids can increase water turbidity in streams, affecting benthic and pelagic aquatic ecosystems, filling reservoirs with unwanted silt, and requiring water treatment systems for potable water supplies. Phosphorus attached to soil particles, which are washed from a paddock into a stream, can dominate nutrient loads in streams and down-stream water bodies. Consequences include increases in algal biomass, reduced oxygen concentrations, impaired habitat for aquatic species, and even possible production of cyanobacterial toxins, with series impacts for humans and livestock consuming the water. Where waters discharge into estuaries, N can be the limiting factor for eutrophication; estuaries of some catchments where fertilizer use is extensive have suffered from excessive sea grass and algal growth.

More insidious is the leaching of nutrients, agricultural chemicals, and hydrocarbons to groundwater. Incremental increases in concentrations in groundwater may be observed over long periods of time resulting in initially potable water becoming undrinkable and then some of the highest valued uses of the resource may be lost for decades. This problem is most severe on tropical islands with shallow relief and some deltaic arsenopyrite deposits, where wells cannot be deepened to avoid polluted groundwater because underlying groundwater is either saline or contains too much As.

Sampling for Non-Point Source Pollution

The sampling requirements of NPSP are quite different from those of the point source contamination. Typically, the sampling is required to give a good estimate of the mean level of pollution rather than to delineate areas of pollution. In such a situation, sampling is typically carried out on a regular square or a triangular grid. Furthermore, gains may be possible by using composite sampling.[4] However, if the pollution is patchy, other strategies may be used. One such strategy is to divide the area into remediation units, and to sample each of these. The possibility of movement of the pollutant from the soil to some receptor (or asset) is assessed, and the potential harm is quantified. This process requires an analysis of the bioavailability of the pollutant, pathway analysis, and the toxicological risk. The risk analysis is then assessed and decisions are then made as to how the risk should be managed.

Management and/or Remediation of Non-Point Source Pollution

The treatment strategies used for managing NPSP are generally those that modify the soil properties to decrease the bioavailable contaminant fraction. This is particularly so in the rural agricultural environment where soil–plant transfer of contaminants is of greatest concern. Soil amendments commonly used include those that change the ion-exchange characteristics of the colloid particles and those that enhance the ability of soils to sorb contaminants. An example of NPSP management includes the application of lime to immobilize metals because the solubility of most heavy metals decreases with increasing soil pH. However, this approach is not applicable to all metals, especially those that form oxyanions—the bioavailability of such species increases with increasing pH. Therefore, one of the prerequisites for remediating contaminated sites is a detailed assessment of the nature of contaminants present in the soil. The application of a modified aluminosilicate to a highly contaminated soil around a zinc smelter in Belgium was shown to reduce the bioavailability of metals thereby reducing the Zn phytotoxicity.[5] The simple addition of rock phosphates to form Pb phosphate has also been demonstrated to reduce the bioavailability of Pb in aqueous solutions and contaminated soils due to immobilization in the metal.[6] Nevertheless, there is concern over the long-term stability of the processes. The immobilization process appears attractive currently given that there are very few cheap and effective in situ remediation techniques for metal-contaminated soils. A novel, innovative approach is using higher plants to stabilize, extract, degrade, or volatilize inorganic and organic contaminants for in situ treatment (cleanup or containment) of polluted topsoils.[7]

Preventing Water Pollution

The key to preventing water pollution from the soil zone is to manage the source of pollution. For example, nitrate pollution of groundwater will always occur if there is excess nitrate in the soil at a time when there is excess water leaching through the soil. This suggests that we should aim to reduce the nitrogen in the soil during wet seasons and the drainage through the soil. Local research may be needed to demonstrate the success of best management techniques in reducing nutrient, sediment, metal, and chemical exports via surface runoff and infiltration to groundwater. Production figures from the same experiments may also convince local farmers of the benefits of maintaining nutrients and chemicals where needed by a crop rather than losing them off site, and facilitate uptake of best management practices.

Global Challenges and Responsibility

The biosphere is a life-supporting system to the living organisms. Each species in this system has a role to play and thus every species is important and biological diversity is vital for ecosystem health and functioning. The detection of hazardous compounds in Antarctica, where these compounds were never used or no man has ever lived before, indicates how serious the problem of long-range atmospheric transport and deposition of these pollutants is. Clearly, pollution knows no boundaries. This ubiquitous pollution has had a global effect on our soils, which in turn has been affecting their biological health and productivity. Coupled with this, over 100,000 chemicals are being used in countries throughout the world. Recent focus has been on the endocrine disruptor chemicals that mimic natural hormones and do great harm to animal and human reproductive cycles.

These pollutants are only a few examples of contaminants that are found in the terrestrial environment.

References

1. Chan, C.H.; Bruce, G.; Harrison, B. Wet deposition of organochlorine pesticides and polychlorinated biphenyls to the great lakes. J. Great Lakes Res. **1994**, *20*, 546–560.
2. Lodovici, M.; Dolara, P.; Taiti, S.; Del Carmine, P.; Bernardi, L.; Agati, L.; Ciappellano, S. Polycyclic aromatic hydrocarbons in the leaves of the evergreen tree *Laurus Nobilis*. Sci. Total Environ. **1994**, *153*, 61–68.
3. Sweet, C.W.; Murphy, T.J.; Bannasch, J.H.; Kelsey, C.A.; Hong, J. Atmospheric deposition of PCBs into green bay. J. Great Lakes Res. **1993**, *18*, 109–128.
4. Patil, G.P.; Gore, S.D.; Johnson, G.D. *Manual on Statistical Design and Analysis with Composite Samples*; Technical Report No. 96-0501; EPA Observational Economy Series Center for Statistical Ecology and Environmental Statistics; Pennsylvania State University, 1996; Vol. 3.
5. Vangronsveld, J.; Van Assche, F.; Clijsters, H. Reclamation of a bare industrial area, contaminated by non-ferrous metals: in situ metal immobilisation and revegetation. Environ. Pollut. **1995**, *87*, 51–59.
6. Ma, Q.Y.; Logan, T.J.; Traina, S.J. Lead immobilisation from aqueous solutions and contaminated soils using phosphate rocks. Environ. Sci. Technol. **1995**, *29*, 1118–1126.
7. Wenzel, W.W.; Adriano, D.C.; Salt, D.; Smith, R. Phytoremediation: a plant-microbe based remediation system. In *Bioremediation of Contaminated Soils*; Soil Science Society of America Special Monograph No. 37, Adriano, D.C., Bollag, J.M., Frankenberger, W.T., Jr., Sims, W.R., Eds.; Soil Science Society of America: Madison, USA, 1999; 772 pp.
8. Barzi, F.; Naidu, R.; McLaughlin, M.J. *Contaminants and the Australian Soil Environment. In Contaminants and the Soil Environment in the Australasia-Pacific Region*; Naidu, R., Kookana, R.S., Oliver, D., Rogers, S., McLaughlin, M.J., Eds.; Kluwer Academic Publishers: Dordrecht, the Netherlands, 1996; 451–484.

6

Riparian Wetlands: Mapping

Kristen C. Hychka
U.S. Environmental
Protection Agency (EPA)

Introduction

Riparian wetlands are critical systems that perform functions and provide services disproportionate to their extent in the landscape. Mapping wetlands allows for better planning, management, and modeling, but riparian wetlands present several challenges to effective mapping. However, there are promising approaches to improve the accuracy of mapping riparian wetlands.

Characteristics and Significance of Riparian Wetlands

Riparian ecosystems are biophysical communities that border rivers, streams, or lakes and are some of the most diverse and dynamic systems on Earth.[1,2] They encompass the stream channel and adjacent land that are influenced by flooding or elevated water tables and have soils that hold water[3] and are comprised of aquatic, wetland, and upland components. This entry focuses specifically on riparian wetlands associated with fluvial or lotic systems. A wetland is an area where soil formation and the composition of the biota of the system are primarily determined by consistent or periodic saturation or inundation of the soils.[4] Riparian wetlands are predominantly associated with headwater streams and often drier than other wetland types.[5] Though they have a range of cover types, riparian wetlands are predominantly forested,[5] as the majority of vegetated freshwater wetlands in the continental United States are forested[6] and greater than 80% of stream networks are comprised of small streams,[7,8] which are more likely to be forested than streams lower in the network.[9] Riparian wetlands perform a suite of functions including sediment retention, provision of wildlife habitat, carbon sequestration, temperature regulation, biogeochemical processing, and are particularly critical for water quality and quantity regulation, because of their position in the landscape.[10–12] However, approximately 80% of riparian corridors in North America and Europe have been severely degraded in the past 200 years.[3] Though total wetland area has shown no significant change in the Continental United States recently, there have been significant losses of vegetated freshwater wetlands and particularly of forested wetlands.[6]

Importance of and Challenges in Mapping Riparian Wetlands

Mapping wetlands is important for spatial planning; water resources management;[13] and climatic, hydrologic, and coupled hydroclimatic models.[14] Also mapping extent, characteristics, and stressors associated with wetlands are important for monitoring, particularly in these rapidly changing systems.

The approaches to mapping wetlands outlined below apply to all wetland types, however riparian wetlands are especially "controversial wetlands," as they often function as wetlands but do not always meet regulatory definitions of wetlands[15] and are difficult wetlands to map for a number of reasons. First, their morphology is often thin and long, and therefore smaller than the detection limits of some mapping efforts.[16] They are also often small, because a large proportion of riparian wetlands are associated with headwater streams.[5] Additionally, they are generally forested, so they are particularly difficult to identify through optical remote sensing approaches often used to map wetlands.[17] Riparian wetlands are also often ecotones that do not have abrupt boundaries.[15] Some mapping approaches handle gradient boundaries or fuzzy classification in mapping, but traditionally wetlands are mapped with crisp boundaries between wetland types and between wetlands and uplands. Finally, riparian wetlands, particularly those associated with extreme headwaters, are often "drier end" wetlands,[5] so they are more difficult and less accurate to map when soil saturation and inundation are criteria in wetland mapping.[15]

There are extensive bodies of literature focusing on remote sensing of wetlands or of riparian systems, but much less attention has been paid to specifically riparian wetlands. This entry seeks to identify and summarize references that focus specifically on riparian wetlands and summarize some of the relevant information from more general papers on riparian systems or wetlands based on the characteristics of riparian wetlands.

Mapping Quality and Standards

The output and accuracy of wetland maps vary greatly based not only on the characteristics of the wetland described above but also on the approach, classification, validation, and base data used. First, the criteria used to define a wetland vary. For example, wetlands mapped through the detection of standing water versus the presence of mapped hydric soils will result in very different wetland extents for the same area.[18] Also, classification approaches may be visual, supervised (commonly maximum likelihood), or unsupervised (commonly clustering)[17] and delineate wetlands based on crisp or fuzzy classes.[19] Validation can be spatial or aspatial (aggregated validation by region), and the accuracy of a map can also be evaluated in terms of errors of omission (producer's accuracy) and errors of commission (user's accuracy).[16] Finally, maps are generated from different types of base data and this is discussed in more detail in the coming sections.

Mapping Approaches

For the reasons outlined earlier, some approaches to mapping riparian wetlands are more successful than others. There are three general approaches to mapping wetlands: on-site, remotely sensed, or ancillary data. Specific applications of these approaches have trade-offs in terms of time, cost, resolution, accuracy (omission and commission), and repeatability. It is important to know the strengths and weaknesses associated with an approach to mapping, particularly the type and amount of error associated with different mapping approaches.[18]

On-Site Mapping

On-site mapping typically follows ecological or jurisdictional protocol and involves identifying characteristics of vegetation, soils, and hydrology. Reviews of these approaches are covered in other texts.[20–22] On-site mapping provides the highest level of confidence in the data (low user's and producer's errors), but can be labor intensive and logistically difficult, so this approach is often used on small spatial extents for a single period.

Remote Sensing

Remote sensing approaches to mapping rely on the visual or automated classification of reflectance data from panchromatic, multi- or hyperspectral imagery, and other wavelengths including radar.

Pan-chromatic data are primarily in the visible light range and are either aerial photography or satellite imagery. Aerial photo interpretation was the first remote sensing approach to mapping wetlands and is still considered to be one of the most accurate approaches given cloud-free, high-resolution imagery collected at appropriate time of year based on phenology and hydroperiod.[16,23] Another advantage to using aerial photo interpretation is if older images are available, they can be used to create historical time series of wetland change. Optical images can have a wide range of horizontal spatial resolutions (sub-meter to hundreds of meters), which greatly influence the detection of small and, or narrow riparian wetlands.[18] Drawbacks to using this approach are that the interpretation is typically manual, so it is labor intensive and subjective, and forested wetlands are often underrepresented.[24] Automated methods for analyzing optical images are also important in detecting riparian wetlands; for example, leaf area index and other vegetation indices have also been used to map both the extent and the characteristics (such as biomass) of riparian wetlands.[25] Repeatability is an advantage of automated mapping over manual delineation. The National Wetlands Inventory of the United States primarily relies on the manual interpretation of color IR imagery to map wetlands.[6,23] This inventory, typical of optically derived wetlands maps, is fairly conservative and has high producer's accuracy, but does miss a lot of wetlands (low user's accuracy), particularly dry, narrow, or forested wetlands.[18,20]

Radar is a powerful tool in mapping riparian wetlands, because it is not constrained by weather conditions, does not require sunlight, and is useful in detecting both standing water and saturated soils.[26,27] Radar has been successfully used to detect soil saturation below herbaceous vegetation, but is less successful under woody vegetation, and consistently flooded wetlands are easier to detect than periodically flooded systems.[25,26] Radar data are available in a wide range of spatial resolutions, though most continental- or global-scale has horizontal resolution in hundreds of meters,[26] which is not accurate enough to detect small or narrow riparian wetlands.

LiDAR intensity returns have also been used to map inundation under canopy. In some cases this approach was better than topographic indices or reflectance in a panchromatic image[28] and particularly when used in combination with statistical approaches to account for spatial autocorrelation.[29] These studies were conducted primarily on isolated wetlands, but their approaches can be applied to mapping riparian wetlands. LiDAR data have particularly fine spatial resolution (meters to sub-meter),[30] though until recently the spatial extent of available LiDAR-derived data has been limited.

Additionally using a multitemporal approach, which relies on imagery from different times of the year, is also an effective way to accurately capture riparian wetland extent. Most commonly, this approach is used to determine vegetation from images during leaf off conditions to detect the ground below the canopy and to determine if vegetation is deciduous.[17,31] Also, it is useful in evaluating images from the wettest times of the year to determine soil saturation for drier systems and flooding in periodically flooded systems.[32]

Ancillary Data

Ancillary data are used to model processes or hydromorphic settings that support wetland formation. Some advantages to using an ancillary data approach to mapping riparian wetlands are that these can identify "wetland supportive settings" even in a disturbed landscape where wetlands have been drained and can be useful in scenario development. However, a key negative to using ancillary data is that though there may be a high user's accuracy there is often a low rate of producer's accuracy. Some of the approaches that use ancillary data include topographic indices, climato-topographic indices, and directly modeled hydrologic processes.

Many studies have used topographic indices to predict wetland occurrence primarily by using the terrain to model the occurrence of soil saturation.[28] There is a suite of topographic indices that are based on combinations of the contributing area and slope, and they differ mostly in their calculation of upslope contribution.[33] These approaches are helpful in identifying "cryptic" or forested riparian wetlands, as canopy cover has a smaller influence on the results compared to optical approaches.[34] Climato-topographic indices refine DEM-derived (digital elevation model) topographic indices by adding measured or modeled climate data, accounting for differences in the availability of water,[14,35] which is particularly important in large study areas that cross several climatic zones.[36] A drawback to both topographic and climato-topographic indices is that they do not work well in areas with complex geology where soil permeability is highly variable. An emerging approach to riparian wetland mapping is to use mechanistic modeling or to directly model processes that drive wetland occurrence, such as groundwater level and flooding.[37,38] This is a particularly useful approach when making projections of wetland change under differing scenarios; however, this approach is data intensive and combines data from multiple sources, which can have unaccounted error propagation.[39,40] As with other mapping approaches, the spatial resolution of the base data (DEMs in this case) influences the detection limits of mapping, and the horizontal resolution of DEMs ranges widely (sub-meter to kilometers).[41]

Mixed Method Approaches

Because different sensors can detect different characteristics of wetlands, the use of a multisensor approach can be the most fruitful for accurately mapping riparian wet-lands.[17,27,42] Using multiple sensors can draw on the strongest aspects of each approach, for example using radar to determine the extent and timing of flooding and optical data to determine vegetative characteristics[43] Another promising approach to riparian wetland mapping is the coupled use of remotely sensed data with data from in situ sensors.[44,45] For example, in periodically flooded systems, in situ sensors can identify potentially useful imagery for mapping wetland extents. Using participatory sensing or local knowledge can also inform riparian wetland mapping. One study coupled remotely sensed data with local ecological knowledge to map isolated wetlands,[46] and the same approach could be used for riparian systems.

Conclusion

Though riparian wetlands are difficult to map because of their morphology, size, typical forest cover, and often drier hydrology, several mapping approaches and combinations of approaches have improved or shown promise in improving the accuracy of mapping riparian wetlands.

References

1. Naiman, R.J.; DeCamps, H. The ecology of interfaces: riparian zones. Ann. Rev. Ecol. Syst. **1997**, *28* (1), 621–658.
2. Malanson, G.P. *Riparian Landscapes;* Great Britain: Cambridge University Press: Cambridge, 1993.
3. Naiman, R.J.; DeCamps, H.; Pollock, M. The role of riparian corridors in maintaining regional biodiversity. Ecol. Appl. **1993**, *3* (2), 209–212.
4. Cowardin, L.M.; Carter, V.; Golet, F.C. LaRoe, E.T. *Classification of Wetlands and Deepwater Habitats of the United States;* Washington, D.C., 1979.
5. Whigham, D.F. Ecological issues related to wetland preservation, restoration, creation and assessment. Sci. Total Environ. **1999**, *240* (1–3), 31–40.
6. Dahl, T.E. *Status and Trends of Wetlands in the Conterminous United States 2004 to 2009;* U.S. Department of the Interior, Fish and Wildlife Service: Washington, D.C. 2011; 108 pp.
7. Peterson, B.J.; Wollheim, W.M.; Mulholland, P. J.; Webster, J.R.; Meyer, J.L.; Tank, J.L.; Martí, E.; Bowden, W.B.; Valett, H.M.; Hershey, A.E.; McDowell, W.H.; Dodds, W.K.; Hamilton, S.K.; Gregory, S.; Morrall, D.D. Control of nitrogen export from watersheds by headwater streams. Science **2001**, *292* (5514), 86–90.

8. Naiman, R.J.; DeCamps, M.E. McClain, M.E. *Riparia: Ecology, Conservation, and Management of Streamside Communities;* Academic Press: Elsevier, 2005.

9. Vannote, R.L.; Minshall, G.W.; Cummins, K.W.; Sedell, J.R.; Cushing, C. E. The river continuum. Canadian J. Fisheries Aquat. Sci. **1980**, *37* (1), 130–137.

10. Mitsch, W.J. Gosselink, J.G. The value of wetlands: Importance of scale and landscape setting. Ecol. Econ. **2000**, *35* (1), 25–33.

11. Brinson, M.M. A *Hydrogeomorphic Classification for Wetlands*; WRP-DE-4, U.S. Army Corps of Engineers, Vicksburg, MS, USA 1993; 101 pp.

12. Mitsch, W.J. Landscape design and the role of created, restored, and natural riparian wetlands in controlling nonpoint source pollution. Ecol. Eng. **1992**, *1* (1–2), 27–47.

13. Baker, C.; Lawrence, R.; Montagne & D. Patten Mapping wetlands and riparian areas using Landsat ETM+ imagery and decision-tree-based models. Wetlands **2006**, *26* (2), 465–474.

14. Merot, P.; Squividant, H.; Aurousseau, P.; Hefting, M.; Burt, T.; Maitre, V.; Kruk, M.; Butturini, A.; Thenail, C.; Viaud, V. Testing a climato-topographic index for predicting wetlands distribution along an European climate gradient. Ecol. Model. **2003**, *163* (1–2), 51–71.

15. National Research Council. *Wetlands: Characteristics and Boundaries;* National Academy Press: Washington, D.C., 1995.

16. FGDC Wetlands Subcommittee. *Wetlands Mapping Standard.* Federal Geographic Data Committee: Reston, VA, 2009; 39 pp.

17. Ozesmi, S.L., Bauer, M. E. Satellite remote sensing of wetlands. Wetlands Ecol. Manag. **2002**, *10* (5), 381–402 http:// dx.doi.org/10.1023/A:1020908432489

18. Shapiro, C. *Coordination and Integration of Wetland data for Status and Trends and Inventory Estimates*; Federal Geographic Data Committee: Reston, VA, 1995.

19. Adam, E.; Mutanga, O.; Rugege, D. Multispectral and hyperspectral remote sensing for identification and mapping of wetland vegetation: a review. Wetlands Ecol. Manag. **2010**, *18* (3), 281–296.

20. Tiner, R.W. *Wetland Indicators: A Guide to Wetland Identification, Delineation, Classification, and Mapping*; CRC Press: Boca Raton, FL, 1999.

21. Richardson, J.L. Vepraskas, M.J. *Wetland soils: Genesis, Hydrology, Landscapes, and Classification;* Lewis Publishers: Boca Raton, FL, 2001.

22. Lyon, J.G.; Lyon, L. K. *Practical Handbook for Wetland Identification and Delineation,* 2nd Ed.; CRC Press, Taylor and Francis Group: Boca Raton, FL, 2011.

23. Dahl, T.E. *Status and Trends of Wetlands in the Conterminous United States 1986 to 1997*; U.S. Department of the Interior, Fish and Wildlife Service: Washington, D.C., 2000; 82 pp.

24. Yang, X. Integrated use of remote sensing and geographic information systems in riparian vegetation delineation and mapping. Int. J. Remote Sensing **2007**, *28* (2), 353–370.

25. Ramsey, E.W., III. Radar remote sensing of wetlands. In *Remote Sensing Change Detection: Environmental Monitoring Methods and Application;* Lunetta, R. S., Elvidge, C.D., Eds.; Sleeping Bear Press, Inc.: Chelsea, MI, 1998; 211–243 pp.

26. Henderson, F.M.; Lewis, A.J. Radar detection of wetland ecosystems: a review. Int. J. Remote Sensing **2008**, *29* (20), 5809–5835.

27. Bourgeau-Chavez, L.L.; Kasischke, E.S.; Brunzell, S.M.; Mudd, J.P.; Smith, K. B.; Frick, A. L. Analysis of space-borne SAR data for wetland mapping in Virginia riparian ecosystems. Int. J. Remote Sensing **2001**, *22* (18), 3665–3687.

28. Lang, M.W.; McCarty, G. W. Lidar intensity for improved detection of inundation below the forest canopy. Wetlands **2009**, *29* (4), 1166–1178.

29. Julian, J.; Young, J.; Jones, J.; Snyder, C.; Wright, C. The use of local indicators of spatial association to improve LiDAR-derived predictions of potential amphibian breeding ponds. J. Geographical Syst. **2009**, *11* (1), 89–106.

30. Lefsky, M.A.; Cohen, W.B.; Parker, G.G; Harding, D. J. Lidar remote sensing for ecosystem studies. BioScience **2002**, *52* (1), 19–30.

31. Elvidge, C.D.; Miura, T.; Jansen, W.T.; Groeneveld, D.P.; Ray, J. Monitoring trends in wetland vegetation using a land-sat MSS time series. In *Remote Sensing Change Detection: Environmental Monitoring Methods and Application;* Lunetta, R.S., Elvidge, C.D., Eds.; Sleeping Bear Press, Inc.: Chelsea, MI, 1998; 191–210 pp.

32. Townsend, P.A.; Walsh, S. J. Remote Sensing of Forested Wetlands: Application of Multitemporal and Multispectral Satellite Imagery to Determine Plant Community Composition and Structure in Southeastern USA. Plant Ecol. **2001**, *157* (2), 129–149.

33. Sørensen, R.; Zinko, U.; Seibert, J. On the calculation of the topographic wetness index: evaluation of different methods based on field observations. Hydrol. Earth Syst. Sci. **2006**, *10* (1), 101–112.

34. Creed, I.F.; Sanford, S.E.; Beall, F.D.; Molot, L.A.; Dillon, P.J. Cryptic wetlands: integrating hidden wetlands in regression models of the export of dissolved organic carbon from forested landscapes. Hydrol. Processes **2003**, *17* (18), 3629–3648.

35. Rodhe, A. & J. Seibert. Wetland occurrence in relation to topography: a test of topographic indices as moisture indicators. Agric. Forest Meteorol. **1999**, *98–99,* 325–340.

36. Curie, F.; Gaillard, S.; Ducharne, A.; Bendjoudi, H. Geomorphological methods to characterise wetlands at the scale of the Seine watershed. Sci. Total Environ. **2007**, *375* (1–3), 59–68.

37. Murphy, P.; Ogilvie, J.; Connor, K.; & Arp, P. Mapping wetlands: A comparison of two different approaches for New Brunswick, Canada. Wetlands **2007**, *27* (4), 846–854.

38. Fan, Y.; Miguez-Macho, G. A simple hydrologic framework for simulating wetlands in climate and earth system models. Climate Dynamics **2011**, *37* (1–2), 253–278.

39. Leavesley, G.A.; Hay, L.E., Viger, R.J.; Markstrom, S. L. Use of a priori parameter-estimation methods to constrain calibration of distributed-parameter models. Water Sci. Appl. **2003**, *6*, 255–266.

40. Oreskes, N.; Shrader-Frechette, K. Belitz, K. Verification, validation, and confirmation of numerical models in the earth sciences. Science **1994**, *263* (5147), 641–646.

41. Wilson, J.P.; Gallant, J. C. *Terrain Analysis: Principles and Applications;* John Wiley & Sons, Inc.: New York, NY, 2000.

42. Bourgeau-Chavez, L.L.; Riordan, K.; Powell, R.B.; Miller, N.; Barada, H. Improving Wetland Characterization with MultiSensor, Multi-Temporal SAR and Optical/Infrared Data Fusion. In *Advances in Geoscience and Remote Sensing;* Jedlovec, G., Ed.; InTech: 2009; 742 p.

43. Li, J. Chen, W. A rule-based method for mapping Canada's wetlands using optical, radar and DEM data. Int. J. Remote Sensing **2005**, *26* (22), 5051–5069.

44. Hart, J.K.; Martinez, K. Environmental Sensor Networks: A revolution in the earth system science? Earth Sci. Rev. **2006**, *78* (3–4), 177–191.

45. Quinn, N.W.T.; Ortega, R.; Rahilly, P. J. A.; Royer, C. W. Use of environmental sensors and sensor networks to develop water and salinity budgets for seasonal wetland real-time water quality management. Environ. Model. Soft. **2010**, *25* (9), 1045–1058.

46. Pitt, A.; Baldwin, R.; Lipscomb, D.; Brown, B.; Hawley, J.; Allard-Keese, C.; Leonard, P. The missing wetlands: using local ecological knowledge to find cryptic ecosystems. Biodivers. Conserv. **2012**, *21* (1), 51–63.

7

Riparian Zones: Groundwater Nitrate (NO_3^-) Cycling

D. Q. Kellogg
University of Rhode Island

Introduction

Groundwater nitrate cycling in riparian zones is an important component in managing nitrogen delivery to surface waters. Humans have had an enormous effect on the global nitrogen (N) cycle, dramatically increasing nitrogen pools in atmospheric, terrestrial and aquatic ecosystems over the last century, largely through the production and use of mineral fertilizers, but also through the burning of fossil fuels and cultivation of nitrogen-fixing crops, such as legumes.[1] One effect has been the over-fertilization of estuarine ecosystems, which are generally nitrogen-limited. When nitrogen is added to these nitrogen-limited systems algae blooms are excessive and are not consumed, drifting to bottom waters where they are decomposed by microbes. Microbial activity depletes dissolved oxygen in the water column, and contributes to hypoxia (low oxygen) in coastal ecosystems, stressing marine life.[2] An example of this phenomenon is the so-called "dead zone" in the Gulf of Mexico, into which the Mississippi River flows, carrying nitrogen from distant reaches in the watershed to the outlet in the Gulf.[3] Other coastal estuaries around the world are also showing symptoms of over-fertilization. For example, in the U.S. the Chesapeake Bay is a large and economically important estuary on the east coast where nitrogen has been identified as a pollutant of concern, prompting aggressive nitrogen management throughout the watershed, i.e., the area of land that contributes water to a given water body.[4] Additionally, nitrate in groundwater contaminates drinking water supplies and is a human health hazard, regulated by the Clean Water Act. Groundwater nitrate is increasing in many areas of the United States and is being actively monitored by the United States Geological Survey.[5]

Sources of Nitrate in Groundwater

Anthropogenic sources of nitrate to groundwater include mineral fertilizers applied to lawns and agricultural lands, on-site septic systems, the cultivation of nitrogen-fixing crops, and animal waste. Atmospheric deposition, as rain or particulates, carries N from the burning of fossil fuels and also contributes nitrogen to ecosystems.[6] About 78% of the earth's atmosphere is made up of nitrogen gas in the form of dinitrogen (N_2), a relatively inert gas due to its triple bond. Nitrogen is an essential nutrient to life, and most ecosystems are relatively N limited.[7] With the development of the Haber-Bosch process that takes N_2 out of the atmosphere to create mineral fertilizers, agricultural systems have expanded in size and production to meet the demand for food for the world's growing population.[8] As a result nitrogen delivery to terrestrial and aquatic ecosystems has increased, frequently overfertilizing these systems. Nitrogen is found in both organic and inorganic forms. Inorganic, or mineral, nitrogen forms include ammonium (NH_4^+), nitrite (NO_2^-) and nitrate (NO_3^-). Ammonium and nitrate are "bioavailable" or accessible to plants and microbes, while nitrite is rarely found in high concentrations in nature and is toxic to plants. Nitrate is the most stable form of N in groundwater because of the presence of oxygen in most groundwaters. Due to its negative charge nitrate is also mobile and easily transported with groundwater. Fine soil particles (i.e., clays and humus) are almost always negatively charged and repel nitrate, except where soils are very acidic.[9,10]

Riparian Zones and Management of Groundwater Nitrate

Riparian zones are the lands next to surface water (e.g., rivers, streams, lakes, reservoirs) and can play an important role in mediating nitrogen delivery to N-limited estuaries.[11] Situated between terrestrial and aquatic ecosystems (Figure 7.1), riparian zones display gradients in water table, vegetation, soil wetness, soil carbon, temperature, oxygen, and many other biogeochemical characteristics, providing opportunities for biogeochemical transformations.[12]

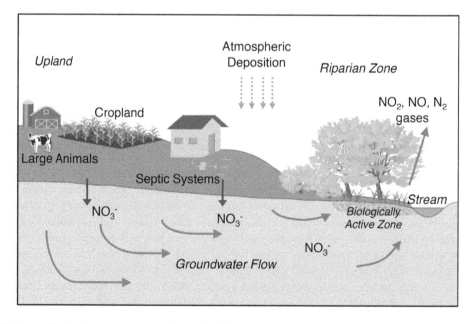

FIGURE 7.1   Idealized riparian zone, with cropland, large animals, septic systems and atmospheric deposition as sources of nitrogen to groundwater. In humid climates groundwater flows toward surface water, carrying nitrate (NO_3^-) to streams and ponds, and eventually to the coast.

Groundwater Nitrate Pathways for Transformation

As groundwater flows toward surface water, interaction with shallow, biologically active soils provides opportunities for nitrate transformations: plant uptake, microbial immobilization, and denitrification (Figure 7.2). These shallow soils are rich in organic matter (mostly decomposing plant debris), providing essential elements for microbial activity and plant growth. Decomposition begins the cycle of converting organic nitrogen back to inorganic nitrogen, returning bioavailable nitrogen to the ecosystem as ammonium and nitrate.

Plant Uptake

Nitrogen is necessary for plants to thrive, providing the building blocks for chlorophyll and essential proteins that drive photosynthesis and growth. With the exception of nitrogen-fixing plants (e.g., legumes) that extract N$_2$ from the atmosphere, plants absorb mineral nitrogen as nitrate and ammonium from soil pore water.[10] During the growing season trees and grasses actively transpire, pulling groundwater up toward the soil surface and taking up nutrients, often limiting nitrate transport towards surface waters.[13] During the dormant season when plants are in senescence and temperatures are cooler, evaporative demand slows dramatically as does plant uptake of nutrients. Nitrogen may also be removed from riparian zones through active riparian zone management such as tree harvesting.

Microbial Immobilization

There are countless types of microbes found in soils and each plays a different role in ecosystem function, but all require nitrogen to grow and thrive. Microbial immobilization refers to the process by which microbes take up mineral nitrogen, transforming it to an organic form as it is used for biological function. When microbes die, they decompose and mineral N is released back to the ecosystem.

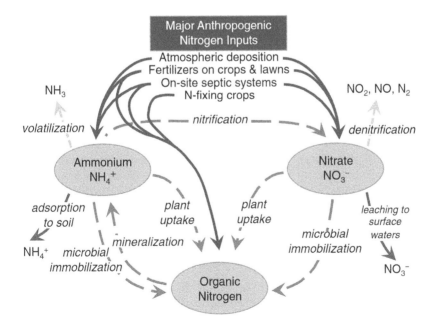

FIGURE 7.2 Simplified nitrogen cycle in the subsurface. Anthropogenic (human) inputs include atmospheric deposition, fertilizers, on-site septic systems and N-fixing crops.

Decomposition

Decomposition of plant debris initiates the active cycling of nitrogen, with plant-based N undergoing transformations by soil invertebrates and microbial communities, resulting in organic matter and microbial biomass. These organic N pools can be stored for long periods in wetter soils. Wetter soils (also called hydric soils and frequently identified as wetlands) tend to be low in oxygen (anaerobic) because oxygen diffusion into water from the atmosphere is slow. Decomposition is slower in anaerobic soils, allowing organic matter to accumulate. When these soils are drained, nitrogen can be released as decomposition increases. Organic N is transformed to inorganic forms through mineralization of organic matter, releasing bioavailable nitrogen (NH_4^+ and NO_3^-) back to the ecosystem.[7] The process begins with the production of ammonium (NH_4^+). Ammonium may be taken up by plants, immobilized by microbes, volatilized as ammonia (NH_3), adsorbed to negatively charged soil particles, or oxidized to NO_3^- (nitrification) in the presence of oxygen. Nitrate may be taken up by plants, immobilized by microbes, denitrified, or leached to surface waters. In some environments nitrate may also be adsorbed to positively charged soil particles.[14]

Denitrification

Denitrification is the subject of much study because it is the only process by which nitrogen is returned to the atmosphere and permanently removed from aquatic or terrestrial ecosystems. Denitrification is largely carried out by microbes that use the oxygen in the NO_3^- for respiration, releasing dinitrogen (N_2), nitrous oxide (N_2O), and nitric oxide (NO) gases. Microbial denitrification requires low oxygen (anaerobic) conditions, a source of electron donors, most frequently in the form of organic carbon such as decomposing plant matter, and a supply of NO_3^-. In general, nitrate is denitrified in a series of steps:

$$2NO_3^- \rightarrow 2NO_3^- \rightarrow 2NO \rightarrow N_2O \rightarrow N_2.$$

The conditions and types of denitrifiers determine the specific mechanisms involved. The process does not always go to completion, which results in the production of NO and N_2O gases, both of which have negative environmental impacts. They contribute to acid rain and ground-level ozone, act as greenhouse gases in the upper atmosphere, and may contribute to the destruction of the protective ozone layer in the stratosphere.[10] Alternative electron donors that have been found to play a significant role in groundwater denitrification include inorganic sulfide (often in the form of pyrite, FeS_2), and reduced iron, $Fe(II)$.[15] Other potentially important microbial transformations of groundwater nitrate are dissimilatory reduction of nitrate to ammonium (DNRA), and anaerobic ammonium oxidation (anammox).[16]

Riparian zones vary spatially in soils, slope, vegetation, and other physical characteristics affecting the extent to which they can remove groundwater nitrate through denitrification. Some characteristics vary seasonally, such as temperature and water table elevation. Some factors directly influence denitrification rates, while others control groundwater flow paths and residence time that influence the extent to which nitrate-laden groundwater interacts with denitrifying microbes in the biologically active zone.

Factors That Directly Influence Denitrification Rates

Laboratory and field studies have explored the limiting factors of microbial denitrification. These include a supply of nitrate, low oxygen (anaerobic) conditions, and the availability of electron donors. Temperature and pH play a less critical role, but also influence denitrification rates.[17] The transformation of ammonium to nitrate requires oxygen, while denitrification requires anaerobic conditions. Riparian zones frequently provide the conditions necessary to support nitrification and denitrification because they are at the interface between terrestrial and aquatic systems and possess a gradient of biogeochemical conditions, such as oxygen status. Therefore these wetland riparian soils are most favorable for groundwater denitrification; however, not all riparian zones have wet soils. If a stream is deeply incised, then riparian soils tend to be dryer, keeping surface soils more aerated and limiting the accumulation of organic matter.[18] Groundwater denitrification tends to be patchy and occurs in hot spots in the subsurface[19] making field measurements of groundwater denitrification rates challenging.[20]

Factors That Influence Groundwater Nitrate Interaction with Denitrifying Microbes

Field studies have focused on the many physical factors that influence the interaction between nitrate-laden groundwater and denitrifying microbes, such as groundwater flow paths[21] and residence time. In humid climates, groundwater generally flows towards streams and ponds and supplies water to these systems when it is not actively raining (baseflow) (Figure 7.1). Total groundwater nitrate removal by denitrification in riparian zones depends on slope, the depth of permeable soils in the upland, and the depth of permeable soils in the riparian area.[22] These factors determine the supply of nitrate to the riparian zone as well as the residence time within the riparian zone. Residence time can also be expressed in terms of riparian width or the distance necessary to remove a given percentage of nitrate. The longer the residence time (or greater the width) the greater the potential for nitrate removal. Highly permeable soils allow for a large volume of groundwater to move through the riparian zone and potentially a large supply of nitrate, but highly permeable soils also reduce the residence time. All of these factors explain why riparian nitrate removal is so variable across the landscape.[23] Regardless of the variability, vegetated riparian zones help to buffer aquatic ecosystems from nitrate loading and are an effective nitrogen management practice.[24] Urbanization and intensive agriculture have degraded riparian buffers in the past, prompting restoration efforts to improve water quality.[25]

Conclusions

Groundwater nitrate cycling in riparian zones affects nitrogen delivery to surface waters. In many environments nitrate is the most mobile form of nitrogen in groundwater and can be transported long distances without transformation. Coastal estuaries are sensitive to excessive nitrogen loading from the watershed, with over-fertilization contributing to eutrophication and hypoxia, stressing marine life. Vegetated riparian zones are a common management practice used to reduce nitrate delivery to coastal waters. Because of their position between terrestrial and aquatic ecosystems, vegetated riparian zones provide opportunities for nitrate transformation before emerging to surface water. Riparian zones vary spatially and temporally and this variation affects the extent to which groundwater nitrate cycling occurs. Plant uptake, microbial assimilation and denitrification are major groundwater nitrate transformation pathways. Denitrification is the only process that returns nitrogen to the atmosphere as N_2, N_2O and NO, with the latter two gases representing incomplete reduction to N_2 and having negative impacts on the environment. Riparian wetlands provide conditions that are particularly favorable to microbial denitrification. Nitrogen removal may also occur through agroforestry management approaches such as selective tree harvesting. Mineralization of decomposing plants and microbes returns bioavailable inorganic nitrogen, as nitrate and ammonium, to the riparian ecosystem.

References

1. Galloway, J.N.; Dentener, F.J.; Capone, D.G.; Boyer, E.W.; Howarth, R.W.; Seitzinger, S.P.; Asner, G.P.; Cleveland, C.C.; Green, P.A.; Holland, E.A.; Karl, D.M.; Michaels, A.F.; Porter, J.H.; Townsend, A.R.; Vorosmarty, C.J. Nitrogen cycles: past, present, and future. Biogeochemistry **2004**, 70(2), 153–226.

2. Diaz, R.J.; Rosenberg, R. Spreading dead zones and consequences for marine ecosystems. Science **2008**, *321*, 926–929.

3. Rabalais, N.N.; Turner, R.E.; Wiseman, Jr., W.J. Hypoxia in the Gulf of Mexico. J. Environ. Qual. **2001**, *30* (2), 320–329.

4. Boesch, D.F.; Brinsfield, R.B.; Magnien, R.E. Chesapeake Bay eutrophication: Scientific under-standing, ecosystem restoration, and challenges for agriculture. J. Environ. Qual. **2001**, *30*, 303–320.

5. Lindsey, B.D.; Rupert, M.G. Methods for evaluating temporal groundwater quality data and results of decadal-scale changes in chloride, dissolved solids, and nitrate concentrations in groundwater in the United States, 1988–2010. U.S. Geological Survey Scientific Investigations Report 20125049.

6. Vitousek, P.M.; Aber, J.; Howarth, R.W.; Likens, G.E.; Matson, P.A.; Schindler, D.W.; Schlesinger, W.H.; Tilman, G.D. *Human Alteration of the Global Nitrogen Cycle: Causes and Consequences.* Issues in Ecology 1997, Issue 1. Ecological Society of America, Washington, D.C.

7. Schlesinger, W.H. *Biogeochemistry: An Analysis of Global Change*; Academic Press: London, 1997.

8. Erisman, J.W.; Sutton, M.A.; Galloway, J.; Klimont, Z.; Winiwarter, W. How a century of ammonia synthesis changed the world. Nat. Geosci. **2008**, *1* (10), 636–639.

9. Freeze, R.A.; Cherry, J.A. *Groundwater*; Prentice-Hall, Inc.: Englewood Cliffs, NJ, 1979.

10. Brady, N.C.; Weil, R.R. *The Nature and Properties of Soils*; 12th edition. Prentice-Hall, Inc.: Upper Saddle River, N.J, 1999.

11. Hill, A.R. Nitrate removal in stream riparian zones. J. Environ. Qual. **1996**, 25 (4), 743–755.

12. McClain, M.E.; Boyer, E.W.; Dent, C.L.; Gergel, S.E.; Grimm, N.B.; Groffman, P.M.; Hart, S.C.; Harvey, J.W.; Johnston, C.A.; Mayorga, E.; McDowell, W.H.; Pinay, G. Biogeochemical hot spots and hot moments at the interface of terrestrial and aquatic ecosystems. Ecosystems **2003**, *6* (4), 301–312.

13. Kellogg, D.Q.; Gold, A.J.; Groffman, P.M.; Stolt, M.H.; Addy, K. Riparian ground-water flow patterns using flownet analysis: Evapotranspiration-induced upwelling and implications for N removal. J. Am. Water Res. Assoc. **2008**, *44* (4), 1024–1034.

14. Korom, S.F.; Seaman, J.C. When "conservative" anionic tracers aren't. Ground Water **2012**, doi: 10.1111/j.1745–6584.2012.00950.x.
    Korom, S.F.; Schuh, W.M.; Tesfay, T.; Spencer, E.J. Aquifer denitrification and in situ mesocosms: Modeling electron donor contributions and measuring rates. J. Hydrol. **2012**, *432–433*, 112–126.

15. Bergin, A.J.; Hamilton, S.K. Have we overemphasized the role of denitrification in aquatic ecosys-tems? A review of nitrate removal pathways. Front. Ecol. Environ. **2007**, *5* (2), 89–96.

16. Korom, S.F. Natural denitrification in the saturated zone: A review. Water Resour. Res. **1992**, *28* (6), 1657–1668.

17. Gold, A.J.; Groffman, P.M.; Addy, K.; Kellogg, D.Q.; Stolt, M.; Rosenblatt, A.E. Landscape attri-butes as controls on ground water nitrate removal capacity of riparian zones. J. Am. Water Resour. Assoc. **2001**, *37* (6), 1457–1464.

18. Parkin, T.B. Soil microsites as a source of denitrification variability. Soil Sci. Soc. Am. J. **1986**, *51* (5), 1194–1199.

19. Addy, K; Kellogg, D.Q.; Gold, A.J.; Groffman, P.M.; Ferendo, G.; Sawyer, C. In situ push-pull method to determine ground water denitrification in riparian zones. J. Environ. Qual. **2003**, *31* (3), 1017–1024.

20. Speiran, G.K. Effects of groundwater-flow paths on nitrate concentrations across two riparian for-est corridors. J. Am. Water Resour. Assoc. **2010**, *46* (2), 246–260.

21. Vidon, P.G.F.; Hill, A.R. Landscape controls on nitrate removal in stream riparian zones. Water Resour. Res. **2004**, *40*, W03201, DOI: 10.1029/2003WR002473.

22. Mayer, P.M.; Reynolds, S.K.; McCutchen, M.D.; Canfield, T.J. Meta-analysis of nitrogen removal in riparian buffers. J. Environ. Qual. **2007**, *36* (4), 1172–1180.

23. Lowrance, R.; Altier, L.S.; Newbold, D.; Schnabel, R.R.; Groffman, P.M.; Denver, J.M.; Correll, D.L.; Gilliam, J.W.; Robinson, J.L.; Brinsfield, R.B.; Staver, K.W.; Lucas, W.; Todd, A.H. Water quality functions of riparian forest buffers in Chesapeake Bay watersheds. Environ. Manage. **1997**, *21* (5), 687–712.

24. Mitsch, W.J.; Day, J.W.; Gilliam, J.W.; Groffman, P.M.; Hey, D.L.; Randall, G.W.; Wang, N. Reducing nitrogen loading to the Gulf of Mexico from the Mississippi River Basin: Strategies to counter a persistent ecological problem. BioScience **2001**, *51* (5), 373–388.

25. Weller, D.E.; Baker, M.E.; Jordan, T.E. Effects of riparian buffers on nitrate concentrations in watershed discharges: new models and management implications. Ecol. Appl. **2011**, *21* (5), 1679–1695.

8

River Delta Processes and Shapes

Douglas A.
Edmonds and
Rebecca L. Caldwell
Indiana University

Introduction

The flow of water and sediment creates environments of breath-taking beauty and complexity and few places exhibit this as clearly as river deltas. In deltas, the processes of water flow and sediment transport give rise to a seemingly ordered network of river channels that construct a semicircular platform from the sediments they carry (Figure 8.1). These channels are the delta's artery system; they transport the water, sediment, and nutrients that promote delta growth and keep ecosystems healthy.

Beyond the aesthetic appeal of river deltas, they represent one of the most important environments on Earth. River deltas have rich, productive ecosystems, abundant coastal waterways for transport, and valuable hydrocarbon reservoirs in the subsurface. Humans have long recognized this importance. For example, many early civilizations emerged around river deltas,[1] and since then deltas have become home to 10–25% of the world's population.[2]

By their nature deltas are flat-lying landforms with little topographic relief. For instance, the average elevation of delta land is ~5 m above sea level.[3] Given their low elevations, they are vulnerable to flooding from sea-level rise and coastal subsidence.[4] To protect deltaic environments, we must understand how they function and how they will respond to environmental changes such as relative sea level change. Here, we focus on the processes and shapes that form the subaerial part of the delta, since that is where most humans live. This entry summarizes what is known about the subaerial part of river deltas by answering the following questions: 1) What is a river delta? 2) How and why do river deltas form? 3) What processes create river deltas? And 4) what determines the shape of a river delta? Finally, we conclude by considering the future of deltaic environments given the stress of changing sea-level, sediment fluxes, and subsidence rates.

What Is a River Delta?

A river delta is an accumulation of sediment on a coastline adjacent to a river. Deltas can form on any coastline, and recognizing them on satellite images is not always straightforward, especially if the "sediment accumulation" of the delta is small or not easily visible. We define two geomorphic criteria for

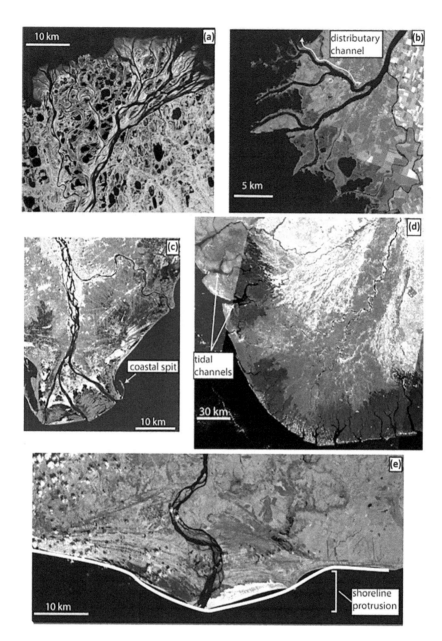

FIGURE 8.1   Satellite imagery showing five examples of river deltas. Black represents water. (**a**) Colville River delta, Alaska, USA; (**b**) St. Clair River delta, Ontario, Canada; (**c**) Krishna River delta, India; (**d**) Niger River delta, Niger; and (**e**) Sao Francisco River delta, Brazil.
Source: Images are from Landsat 7 satellite obtained from the USGS through usgs.glovis.org.

recognizing deltas. These criteria are chosen because they imply substantial sediment accumulation. First, deltas may have a noticeable depositional protrusion (or bump) that causes deviation from the mean shoreline position and is unambiguously linked to the river (see Figure 8.1e for example). The minimum scale of the protrusion is not easy to define, but it should be wider (in a shore-perpendicular direction) than a few multiples of the river width to avoid confusing deltas with prograding river mouths or tie channels that have no delta.[5] The second criterion is clear evidence for a distributary channel network (Figure 8.1a–d). Distributary channels are river channels that branch off from a parent channel and do not rejoin before intersecting the coast (Figure 8.1b). By their nature, these channels facilitate sediment accumulation as they distribute sediment over a broad area. Only one of these criteria is necessary for classifying a geomorphic feature as a delta, and often deltas exhibit both a protrusion and a distributary network (Figure 8.1c).

How and Why Do River Deltas Form?

A delta will form if sediment from a river or other coastal processes accumulates faster than is taken away by sediment compaction, and riverine and coastal erosive forces, while also accounting for a changing volume due to sea-level fluctuations and coastal subsidence. This is effectively a statement of conservation of mass for a control volume at the coastline. There are three main sources that supply sediments to that control volume. Rivers are the largest source of deltaic sediment; and on average, those rivers with larger drainage basins, higher relief, and in warmer climates carry more sediment.[4,6] It is somewhat unsurprising that the largest rivers in the world often have the largest river deltas. But, interestingly, two rivers can have nearly the same sediment flux, such as the Mississippi River and the Amazon River, yet the Amazon delta has an area ten times that of the Mississippi delta.[3] Waves and tides, in some cases, can be a net sediment source to a delta,[7,8] especially during hurricanes when wind blows onshore. Finally, organic sediment created from the production and decay of plants can create substantial amounts of sediment, and this is a mechanism by which deltas aggrade their surfaces in response sea-level rise.[9] For example, in parts of the Mississippi delta, organic sediment makes up one-third of the total sediment volume.[10]

In principle, deltas should form at locations where these sediment sources are maximized, though it also depends on the size of the volume to be filled and the sediment sinks. The volume depends on the offshore depth and changes due to varying sea-level and coastal subsidence. Sediment sinks include fluvial bypass, which is the proportion of incoming sediment that is not deposited within the delta. In some cases, up to 40% of the river sediment brought to the delta is not deposited in the delta, but carried out to sea.[11] Other sediment sinks include sediment compaction, and waves and tides which can disperse river sediment, especially during storms.[8]

This conservation of mass framework helps us understand the origin of most of the world's modern deltas. Modern delta formation started around 7,000 years ago,[12] which is coincident with a deceleration of global sea-level rise. A decelerating global sea level rise reduced the volume to fill with sediment and tipped the balance in favor of delta formation.

What Processes Create River Deltas?

The natural processes that create deltas range in scale from microbial processes that influence sediment mobility to tectonic processes that influence sediment flux. Here, we focus on those mesoscale processes that construct the subaerial, geomorphic expression of the delta and its channel network. We acknowledge that even in this context many processes are not discussed and those processes may become more important depending on the delta and scale of observation.

The key hydrodynamic process that creates deltas is the turbulent expanding jet (Figure 8.2). A turbulent jet forms at the river mouth as the river flow enters a standing body of water and transitions from confined to unconfined. Expansion due to loss of flow confinement and bed friction causes flow deceleration along the jet centerline. At the jet margins, there is strong lateral shear between the fast-moving currents in the jet and the relatively still-standing body of water, which produces turbulent eddies.[13] Flow expansion and deceleration within the jet ultimately cause sediment deposition, leading to embryonic delta formation.

The processes of levee and river mouth bar formation are linked to the dynamics of turbulent jets (Figure 8.2). Subaqueous levees form when advection and diffusion transport sediment to the turbulent jet margins where it is deposited in the relatively still ambient water.[13,14] Concurrent with the process of

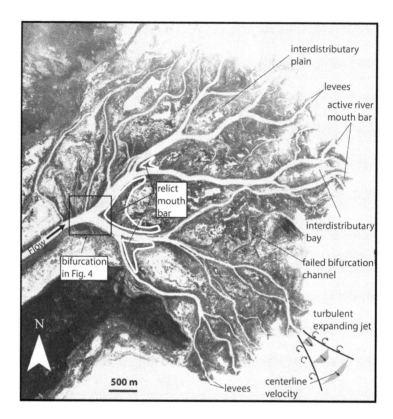

FIGURE 8.2   Aerial image of the Mossy River delta, Saskatchewan, Canada taken in February 2009. Channel network and water are white due to snowfall. The turbulent jet depicted in the lower right is a cartoon, where the gray curves represent the horizontal velocity profile across the jet, and the curved arrows represent turbulence generation at jet margins. See text for a detailed discussion on the other features labeled in this image. Image copyright Digital Globe, Inc., obtained through National Geospatial Intelligence Agency commercial imagery program.

levee formation, a river mouth bar forms at the location of maximum deceleration of velocity along the turbulent jet centerline.[15] River mouth bars are usually triangular in plan view when formed only by fluvial processes,[16] but can be deflected in the presence of obliquely approaching waves.[17] Over time, the river mouth bar aggrades to the surface and flow bifurcates around it, creating two or more channels where there was originally one (Figure 8.2).[15] In this way, the mouth bar sets the point of bifurcation, and relict mouth bars can be identified in the older, upstream parts of the delta (Figure 8.2). As channels prograde past mouth bars, turbulent jets once again form at the river mouths and the process of levee growth and mouth bar deposition starts over.

Muddy, interdistributary plains form in bays between deltaic channels and often behind mouth bars (Figure 8.2). These bays are usually shallow water bodies (1–4 m deep) and the water can be fresh or saline.[18] These interdistributary bays are sheltered from wave and tidal erosion during storms, allowing sediment brought in during floods to accumulate. As sediment slowly accumulates over time an inter-distributary plain composed of coastal marshes will form.[18,19] These are also locations where vegetation is easily established allowing organic sediment to accumulate.

Deltas are also influenced by the process of channel avulsion. Avulsion occurs when flow is diverted out of an existing river channel and establishes a new channel on the adjacent floodplain (Figure 8.3).[20,21] Deltaic avulsions are probable if there is a more attractive flow path down the delta, due to a steeper slope or superelevation of the channel. Superelevation occurs when channels perch themselves above their floodplain by sedimentation on the channel bed and levees. In some sense, avulsions are an inevitable

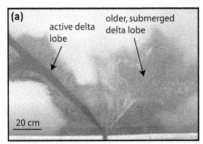
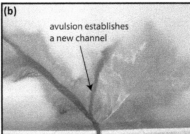

FIGURE 8.3 **(See color insert.)** An example of deltaic avulsion in an experimental delta created in the laboratory. Elapsed time between the images **(a)** and **(b)** is 80 minutes. The water is dyed so that channelized flow is visible. **Source:** Figure modified from Edmonds et al.[26]

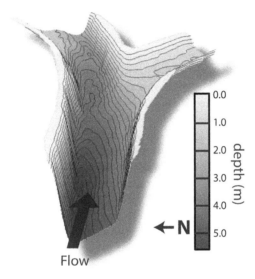

FIGURE 8.4 Perspective view of the bathymetry of a deltaic bifurcation from the Mossy River delta (see Figure 8.2 for location). The downstream bifurcate channels are asymmetric; the northern one is wider and deeper than the southern channel. The contour interval is 0.25 m and the vertical exaggeration is 50 times.

outcome of delta growth. Differential progradation of the delta front will create more attractive, steeper paths on other parts of the delta.[22,23] Also, as deltas prograde their channels must aggrade lest their bed slopes decrease. Maintaining a constant slope requires aggradation which will lead to superelevation of the channel.

Once the conditions for avulsion are met, there are two main scales of deltaic avulsions that occur in different places within the delta system. Avulsions can occur upstream of the channel network, which results in creation of a new delta lobe. For these avulsions, the upstream extent of the backwater zone correlates with avulsion locations[24] because this is an area of higher sedimentation rates leading to superelevation.[25] Avulsions also occur within the channel network, creating new delta sublobes. This tends to occur where there is sustained overbank flow leading to time-dependent weakening of the levee. [26] In either case, once an avulsion occurs, it initiates the construction of new deltaic land as the processes of levee growth, mouth bar deposition, and interdistributary plain formation begin in new places.

As deltas prograde basinward through the deposition of mouth bars, levees, and interdistributary plains, they create channel networks. The basic unit of a channel network is the bifurcation, wherein one channel branches into two or more (Figure 8.4). Bifurcations are created by channel splitting around mouth bars and avulsions. A global survey of modern deltas shows that on average bifurcations are asymmetric where one channel is ~70% wider than the other.[15] For example, the bifurcation at the head of the Mossy River delta (Figure 8.2) is asymmetric—the northern channel is wider and deeper than the southern one (Figure 8.4). Deltaic bifurcations can be stable where both channels receive water and sediment, or unstable where one channel captures all the flow. Stable bifurcations distribute water and sediment asymmetrically between their downstream bifurcate channels because this configuration is stable to perturbations, such as floods. Symmetric bifurcations, on the other hand, are not as common because if they are perturbed they return to the more stable asymmetric configuration.[27–30]

What Determines Delta Shape?

We contend that delta morphology is intimately linked to the relative influences of processes operating on the delta. Those processes, in turn, are affected by the upstream and downstream boundary conditions of the system. Here we provide three generic measures of shape that capture first-order differences in delta morphology. We then describe how these shapes are linked to the different delta building processes, and finally how different boundary conditions affect the balance of delta building processes and thus morphology. We make the distinction between those boundary conditions originating from upstream (e.g., sediment type and flux) and downstream (e.g., marine processes such as waves and tides). This is only meant to serve as a guide to what controls delta shape because ultimately many different delta shapes exist depending on how these processes and boundary conditions interact.

Delta shape can be described by three attributes—bulk external shape, shoreline pattern, and channel network configuration. The bulk external shape of a delta, defined as the delta width divided by the length, ranges from wide flattened protrusions, to semicircles, to elongate shapes. An elongate delta, with a length longer than the width, preferentially constructs land further from the coastline compared to a semicircular delta. Delta shorelines range from smooth (Figure 8.1d) to rough (Figure 8.1b). Roughness can be defined by a sinuosity index, such as the ratio of the actual shoreline length to a spatially averaged shoreline length. For example, a sinuosity index greater than one represents a rough shoreline characterized by large-scale deviations from the average shoreline position. The configuration of a delta channel network has two end-member patterns: single-channel (Figure 8.1e) to multichannel (Figure 8.1a). Each channel can be a single-thread channel or a braided multithread channel, allowing for the possibility of a channel network where each channel is braided with multiple threads.[31]

A given combination of processes operating in a delta should create a distinct morphology. Here, we take a simple view and describe the delta morphology if a single process was dominant. For instance, deltas dominated by mouth bar construction are semicircular because the mouth bars create a well-developed distributary network that delivers sediment to a broad area. The shorelines are rough due to localized progradation at river mouths.[32] The channel networks are fractal in character because the channels reduce in size with successive bifurcations.[33,34] Deltas dominated by levee construction are usually elongate in shape, with smooth shorelines, and few channels.[32,35,36] This occurs because channels with well-developed levees inhibit overbank flow, and focus sedimentation at the river mouth creating localized progradation and an elongate form. Deltas dominated by avulsion have semicircular shapes due to frequent channel movement around the delta, but their shorelines have intermediate roughness, and single-to multi-channel networks. These end-member morphologies may be infrequently encountered in nature and in reality delta morphology depends on the relative influence of these processes and how they interact.

Predicting the morphology of a delta requires knowing the upstream and downstream boundary conditions, because those conditions determine the relative influence of each delta building process. Upstream boundary conditions discussed here include sediment size and total sediment input. Deltas constructed from coarse-grained sediments have steeper topset slopes and the additional shear stress creates channels that are more mobile because the banks are easily eroded. This creates channels that frequently avulse and in turn should limit mouth bar formation since channel mouths rarely stay in one position long. Morphologically, the result is a semicircular delta with a smooth shoreline. Deltas constructed from intermediate grain sizes have channels that are more stable and thus able to produce mouth bars. Mouth bar construction creates semicircular deltas with rougher shorelines. Deltas constructed from muddy sediments, on the other hand, tend to be elongate since mud is easily transported to the margins of a turbulent jet and preferentially deposited in levees.[32] An increase in sediment flux will increase the delta slope. This can trigger braiding in the distributary channels.[37] Higher sediment fluxes will create aggradation and frequent avulsions on the delta resulting in a semicircular planform with a smooth shoreline.[38] Additionally, bed load dominated systems should preferentially construct mouth bars,[15] and suspended load dominated systems should build levees since that material is more easily advected to the turbulent jet margins.[36]

Downstream boundary conditions that influence delta building processes include buoyancy forces, offshore bathymetry, waves, tides, and relative sea-level variations. Buoyancy forces and deep offshore bathymetry cause turbulent jets to detach from the bed, which prevents them from spreading as rapidly. These narrow turbulent jets, in turn, preferentially produce levees resulting in elongate delta shapes.[16,36] In general, strong wave energy creates smooth deltaic shorelines by redistributing sediment laterally via alongshore transport, and suppressing mouth bar formation (Figure 8.1e).[39] The lateral distribution creates morphological features, such as the coastal spits of the Krishna delta (Figure 8.1c). Additionally, waves have the effect of decreasing avulsion frequency because they decrease progradation rates and limit aggradation and superelevation potential.[40] On the contrary, certain kinds of waves can actually enhance delta growth[7] and influence mouth bar formation patterns[17] due to varying degrees of wave amplitude and angle to the shoreline. Tides, on the other hand, cause the remobilization of sediment, eroding mouth bars and maintaining abandoned distributary channels which results in delta morphologies characterized by dendritic channel networks with a large number of distributaries that widen basinward (Figure 8.1d).[41,42] Tidal effects on bifurcation morphology and stability are not well constrained; in some cases, tides increase discharge asymmetry between bifurcate channels and in other cases they minimize it.[43,44] Little is also known about effects of sea-level rise on deltaic processes. Sea-level rise is hypothesized to limit mouth bar formation and increase avulsion frequency, since the rising elevation of the downstream boundary will drown mouth bars and force aggradation (and thus avulsion) upstream,[45] but this has never been tested.

What Is the Future of Deltaic Systems?

The future of deltaic systems is bleak because global environmental changes have tipped conditions toward delta destruction, rather than creation. This is not altogether surprising; deltas occupy the shoreline where changes in upstream and downstream boundary conditions converge. For example, the Mississippi delta feels the combined effects of dam building and other engineering structures that have reduced the sediment flux of the Mississippi River by 50% compared to prehuman levels.[46] Alone this would be problematic, but it is further compounded by human withdrawal of subsurface natural resources (e.g., water and hydrocarbons) that has created accelerated sediment compaction, and by sea-level rise in the Gulf of Mexico at rates of 2 mm/year.[47] These observations suggest that there is not enough sediment reaching the Mississippi delta to offset these changes, and the delta will inevitably drown in the next 100 years.[47,48] Unfortunately, similar estimates have been made for the many of world's largest deltas, and the reality is that if nothing is done deltas could drown.[4]

In light of this, there is a broad call for developing restoration schemes that will save deltaic landforms.[11,49] Local restoration schemes cannot slow or reverse global trends of sea-level rise, but they can reverse some of the negative impacts of humans and restore the processes that effectively created the original delta. To start, this requires a predictive understanding of delta processes and shapes so that restoration schemes can be tuned for success. It is our contention that we know enough about delta processes to make scientifically defensible restoration decisions. Evidence of this includes recent advances in simulating delta growth (Figure 8.5).[26,32,50] These tools simulate the complex interactions between delta processes and changing boundary conditions, and tests show that delta growth and shapes predicted by these methods are statistically similar to field-scale deltas.[33,34]

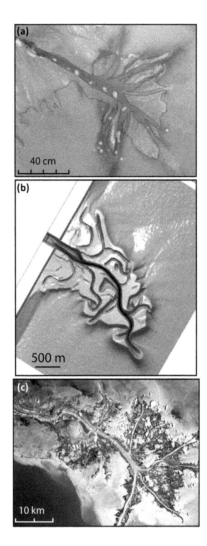

FIGURE 8.5 Experimental (**a**) and numerical models (**b**) realistically simulate the basic morphology and shape of the Mississippi River delta (**c**). In the experimental delta, the water is dyed so that channelized flow is visible, the white dots on the surface are foam, and the gray line is a trace of the shoreline. For more details on experimental setup see Edmonds et al.[26] The numerical delta depicts bed elevation as shaded relief. For more details on numerical setup see Edmonds and Slingerland.[32] Image of the Mississippi delta is 2001 Advanced spaceborne thermal emission and reflection radiometer (ASTER) data. (Courtesy of U.S. Geological Survey National Center for Earth Resources Observation and Science, and National Aeronautics and Space Administration Landsat Project Science Office).

There are three promising areas of future research on deltaic systems. First, there are individual processes and boundary conditions left unexplored. For instance, there is little empirical data or theory that predicts how vegetation growth alters delta form. Second, we have only a minimal understanding of how delta processes interact to produce their morphology. We presented simple models for the effects of delta processes (i.e., mouth bar growth, avulsion, and levee growth) that operate in isolation. But, in most cases, many processes are interacting, possibly in nonlinear feedback loops. Experiments designed to understand these complex interactions and feedback loops will be important breakthroughs. Third, we must understand how deltas will respond to changing boundary conditions. Numerical and experimental models are valuable tools in this regard because they can simulate different scenarios of changing sea-level, coastal subsidence, and sediment flux.

References

1. Day Jr, J.W.; Gunn, J.D.; Folan, W.J.; Yanez-Arancibia, A.; Horton, B.P. Emergence of complex societies after sea level stabilized. Eos, Trans. Am. Geophys. Union **2007**, *88* (15), 169.

2. Ericson, J.P.; Vorosmarty, C.J.; Dingman, S.L.; Ward, L.G.; Meybeck, M. Effective sea-level rise and deltas: Causes of change and human dimension implications. Global Planet. Change **2006**, *50*, 63–82.

3. Syvitski, J.; Saito, Y. Morphodynamics of deltas under the influence of humans. Global Planet. Change **2007**, 57, 261–282.

4. Syvitski, J.P.M.; Kettner, A.J.; Overeem, I.; Hutton, E.W.H.; Hannon, M.T.; Brakenridge, G.R.; Day, J.; Vorosmarty, C.; Saito, Y.; Giosan, L.; Nicholls, R.J. Sinking deltas due to human activities. Nat. Geosci **2009**, *2* (10), 681–686.

5. Rowland, J.C.; Dietrich, W.E.; Day, G.; Parker, G. Formation and maintenance of single-thread tie channels entering floodplain lakes: Observations from three diverse river systems. J. Geophys. Res.-Earth Surf. **2009**, *114*.

6. Orton, G.J.; Reading, H.G. Variability of deltaic processes in terms of sediment supply, with particular emphasis on grain size. Sedimentology **1993**, *40*, 475–512.

7. Ashton, A.D.; Giosan, L. Wave-angle control of delta evolution. Geophys. Res. Lett. **2011**, *38* (13), L13405.

8. Cahoon, D.R. A review of major storm impacts on coastal wetland elevations. Estuar. Coast. **2006**, *29* (6), 889–898.

9. Lorenzo-Trueba, J.; Voller, V.R.; Paola, C.; Twilley, R.R.; Bevington, A.E. Exploring the role of organic matter accumulation on delta evolution. J. Geophys. Res. **2011**, *117*, F00A02.

10. Wilson, C.A.; Allison, M.A. An equilibrium profile model for retreating marsh shorelines in southeast Louisiana. Estuar. Coast. Shelf Sci. **2008**, *80* (4), 483–494.

11. Paola, C.; Twilley, R.; Edmonds, D.; Kim, W.; Mohrig, D.; Parker, G.; Viparelli, E.; Voller, V. Natural processes in delta restoration: Application to the Mississippi delta. Annu. Rev. Mar. Sci **2011**, 3, 3.1–3.25.

12. Stanley, D.J.; Warne, A.G. Worldwide initiation of Holocene marine deltas by deceleration of sea-level rise. Science **1994**, *265* (5169), 228–231.

13. Rowland, J.; Stacey, M.; Dietrich, W. Turbulent characteristics of a shallow wall-bounded plane jet: Experimental implications for river mouth hydrodynamics. J. Fluid Mech.**2009**, 627, 423–449.

14. Rowland, J.C.; Dietrich, W.E.; Stacey, M.T. Morphodynamics of subaqueous levee formation: Insights into river mouth morphologies arising from experiments. J. Geophys. Res.-Earth Surf. **2010**, *115* (F4), F04007.

15. Edmonds, D.A.; Slingerland, R.L. Mechanics of river mouth bar formation: Implications for the morphodynamics of delta distributary networks. J. Geophys. Res. **2007**, *112* (F2).

16. Wright, L.D. Sediment transport and deposition at river mouths: A synthesis. Geol. Soc. Am. Bull. **1977**, 88, 857–868.

17. Nardin, W.; Fagherazzi, S. The effect of wind waves on the development of river mouth bars. Geophys. Res. Lett. **2012**, *39* (12), L12607.

18. Elliott, T. Interdistributary bay sequences and their genesis. Sedimentology **1974**, *21* (4), 611–622.

19. Tamura, T.; Saito, Y.; Nguyen, V.L.; Ta, T.K.O.; Bateman, M.D.; Matsumoto, D.; Yamashita, S. Origin and evolution of interdistributary delta plains; insights from Mekong River delta. Geology **2012**, *40* (4), 303–306.

20. Slingerland, R.; Smith, N.D. River avulsions and their deposits. Annu Rev. Earth Planet. Sci. **2004**, *32*, 257–285.

21. Hajek, E.A.; Wolinsky, M.A. Simplified process modeling of river avulsion and alluvial architecture: Connecting models and field data. Sediment. Geol. **2012**, *257*, 1–30.

22. Reitz, M.; Jerolmack, D.; Swenson, J. Flooding and flow path selection on alluvial fans and deltas. Geophys. Res. Lett. **2010**, 37 (6), L06401.

23. Reitz, M.D.; Jerolmack, D.J. Experimental alluvial fan evolution: Channel dynamics, slope controls, and shoreline growth. J. Geophys. Res. **2012**, *117* (F2), F02021.

24. Jerolmack, D.J.; Swenson, J. Scaling relationships and evolution of distributary networks on wave-influenced deltas. Geophys. Res. Lett. **2007**, *34*.

25. Chatanantavet, P.; Lamb, M.P.;Nittrouer, J.A. Backwater controls of avulsion location on deltas. Geophys. Res. Lett. **2012**, *39* (1), L01402.

26. Edmonds, D.A.; Hoyal, D.C.J.D.; Sheets, B.A.; Slingerland, R.L. Predicting delta avulsions: Implications for coastal wetland restoration. Geology **2009**, *37* (8), 759–762.

27. Edmonds, D.; Slingerland, R.; Best, J.; Parsons, D.; Smith, N. Response of river-dominated delta channel networks to permanent changes in river discharge. Geophys. Res. Lett., *37* (12), L12404.

28. Edmonds, D.A.; Slingerland, R.L. Stability of delta distributary networks and their bifurcations. Water Resour. Res. **2008**, *44* (W09426).

29. Bolla Pittaluga, M.; Repetto, R.; Tubino, M. Channel bifurcations in braided rivers: Equilibrium configurations and stability. Water Resour. Res. **2003**, *39*, 10–46.

30. Miori, S.; Repetto, R.; Tubino, M. A one-dimensional model of bifurcations in gravel bed channels with erodible banks. Water Resour. Res. **2006**, 42, W11413.

31. Jerolmack, D.J.; Mohrig, D. Conditions for branching in depositional rivers. Geology **2007**, *35* (5), 463–466.

32. Edmonds, D.A.; Slingerland, R.L. Significant effect of sediment cohesion on delta morphology. Nat. Geosci. **2010**, *3* (2), 105–109.

33. Edmonds, D.A.; Paola, C.; Hoyal, D.C.J.D.; Sheets, B.A. Quantitative metrics that describe river deltas and their channel networks. J. Geophys. Res. **2011**, *116* (F4), F04022.

34. Wolinsky, M.; Edmonds, D.A.; Martin, J.M.; Paola, C. Delta allometry: Growth laws for river deltas. Geophys. Res. Lett. **2010**, *37* (21), L21403.

35. Kim, W.; Dai, A.; Muto, T.; Parker, G. Delta progradation driven by an advancing sediment source: Coupled theory and experiment describing the evolution of elongated deltas. Water Resour. Res. **2009**, *45*.

36. Falcini, F.; Jerolmack, D. A potential vorticity theory for the formation of elongate channels in river deltas and lakes. J. Geophys. Res.-Earth Surf. **2010**, *115*.

37. McPherson, J.G.; Shanmugam, G.; Moiola, R.J. Fan-deltas and braid deltas: Varieties of coarse-grained deltas. Geol. Soc. Am. Bull. **1987**, *99* (1), 331–340.

38. Sun, T.; Paola, C.; Parker, G.; Meakin, P. Fluvial fan deltas: Linking channel processes with large-scale morphodynamics. Water Resour. Res. **2002**, *38* (8), 1151.

39. Galloway, W.E. *Process Framework for Describing the Morphologic and Stratigraphic Evolution of Deltaic Depositional Systems*; Deltas: Models for Exploration. M.L. Broussard. Houston, TX, Houston Geological Society: 1975; 87–98.

40. Swenson, J.B. Relative importance of fluvial input and wave energy in controlling the timescale for distributary channel avulsion. Geophys. Res. Lett. **2005**, *32*.

41. Dalrymple, R.W.; Choi, K. Morphologic and facies trends through the fluvial-marine transition in tide-dominated depositional systems: A schematic framework for environmental and sequence-stratigraphic interpretation. Earth-Sci. Rev. **2007**, *81* (3), 135–174.

42. Fagherazzi, S. Self-organization of tidal deltas. Proc. Nat. Acad. Sci. **2008**, *105* (48), 18692.

43. Buschman, F.A.; Hoitink, A.J.F.; van der Vegt, M.; Hoekstra, P. Subtidal flow division at a shallow tidal junction. Water Resour. Res. **2010**, *46* (12), W12515.

44. Sassi, M.G.; Hoitink, A.J.F.; de Brye, B.; Vermeulen, B.; Deleersnijder, E. Tidal impact on the division of river discharge over distributary channels in the Mahakam Delta. Ocean Dyn. **2011**, *61* (12), 2211–2228.

45. Jerolmack, D.J. Conceptual framework for assessing the response of delta channel networks to Holocene sea level rise. Quarter. Sci. Rev. **2009**, *28* (17–18), 1786–1800.

46. Meade, R.H.; Moody, J.A. Causes for the decline of suspended-sediment discharge in the Mississippi River system, 1940–2007. Hydrol. Process. **2010**, *24* (1), 35–49.

47. Blum, M.D.; Roberts, H.H. The Mississippi Delta region: Past, present, and future. Annu. Rev. Earth Planet. Sci. **2012**, *40*, 655–683.
48. Blum, M.D.; Roberts, H.H. Drowning of the Mississippi Delta due to insufficient sediment supply and global sea-level rise. Nat. Geosci. **2009**, 2 (7), 488–491.
49. Edmonds, D.A. Restoration sedimentology. Nat. Geosci. **2012**, 5 (11), 758–759.
50. Hoyal, D.; Sheets, B.A. Morphodynamic evolution of experimental cohesive deltas. J. Geophys. Res.-Earth Surf. **2009**, *114*.

9

Streams: Perennial and Seasonal

Daniel von Schiller,
Vicenç Acuña,
and Sergi Sabater
*Catalan Institute for
Water Research*

Introduction

A stream is defined as a body of running water confined within a channel. Streams and rivers are analogous except that the latter are larger. Although they contain only a minor fraction of the total amount of water in the world (0.0001%), streams and rivers play a key role in the hydrologic cycle by connecting the atmosphere with the land and the ocean.[1]

Most streams originate when groundwater flow surfaces, usually at the lowest topographic areas of a valley.[2] Several connected streams build drainage or stream networks. Each stream has a drainage area, also referred to as watershed or catchment, which is the topographical region from which rainfall drains into a stream. The climate, geology, and vegetation of the drainage area exert a strong influence on the structure and function of streams.[3] For instance, stream flow or discharge (i.e., the volume of water passing through a channel cross-section per unit time) is mainly driven by the balance between precipitation and evapotranspiration within the drainage area, the second depending in turn on the vegetation, solar radiation, and temperature.

The aim of this entry is to provide an overview of perennial and seasonal streams considering their classification, spatial zonation, role as ecosystems, the ecosystem services they provide, as well as their management and conservation.

Classification

Perennial streams have permanent flow year round, whereas seasonal streams only flow at certain times (Figure 9.1). The former are usually marked on topographic maps with a solid blue line, whereas a dashed blue line is used to indicate the latter. Depending on the length of the dry (flow cessation) period, seasonal streams are often referred to as temporary, intermittent, or ephemeral streams.[4] Seasonal streams are not restricted to headwaters, as they can also be found in the mid-reaches and lowlands of stream networks.[5]

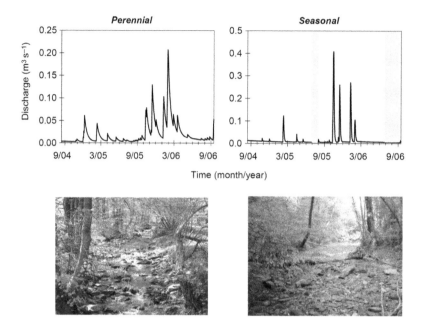

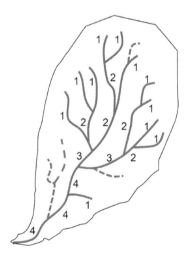

FIGURE 9.1 Biannual hydrographs and pictures of a perennial and a seasonal stream from the Mediterranean region (NE Iberian Peninsula) during the dry summer period. The gray bars indicate the dry (flow cessation) periods in the seasonal stream.

In natural landscapes, the perennial or seasonal nature of a stream is mainly determined by the local climate and geology. Thus, seasonal streams tend to occur in areas where evapotranspiration exceeds precipitation for at least part of the year and where water infiltration rates through soils are high (e.g., areas with calcareous geology). In humanized landscapes, seasonal streams can also be artificially created through excessive water abstraction. In addition, because of climate change many naturally perennial streams are shifting to a seasonal hydrologic regime.[5]

In addition to classifying streams by the permanence of their flow, they are often classified by stream order. The most common method for ordering streams is the Strahler classification system (Figure 9.2).[6]

FIGURE 9.2 Stream network illustrating the Strahler stream order within a fourth-order drainage area. Perennial streams are marked as solid lines. Seasonal streams are marked as dashed lines.

In this hierarchical method, the smallest streams are assigned first order. The order of the streams increases when two streams of the same order join. Streams of lower order joining a higher order stream do not change the order of the higher stream. Based on their order, streams are typically classified as small size (first to third order), medium size (fourth to sixth order), or large size (higher than seventh order). Within a drainage area, low-order streams are more frequent than high-order streams and comprise the longest part of the total stream network. Stream order is positively related to physical characteristics such as stream size, discharge, and drainage area. Therefore, it is a simple and informative classification system. Its major drawback is that it is often difficult to determine the perennial first-order streams on maps.

Streams are also classified by the landscape, predominant soil uses, or the vegetation of their drainage areas.[1] Accordingly, there are desert streams, forest streams, arctic streams, prairie streams, agricultural streams, urban streams, etc. This classification system is useful because the drainage area greatly affects the discharge, morphology, and chemistry of the draining streams. Other characteristics used to classify streams include their discharge patterns, geology, profile, gradient, water color, riparian vegetation, flora, fauna, etc.[7]

Spatial Zonation

Stream networks can be divided into three zones on the basis of the dominant geomorphological processes dominating within each zone:[8] 1) erosion zone, characterized by steep v-shaped valleys, rapidly flowing water, and a high export of sediments; 2) transfer zone, characterized by gradual slope valleys, incipient meandering, and quick reception and delivery of sediments; and 3) deposition zone, with broad and flat valleys, developed stream meandering, and a high deposition of sediments. Nonetheless, rivers form part of a longitudinal continuum defined by changes in geomorphological, physicochemical, and hydrological conditions, where materials distribution and organisms play a significant role.

Streams are hierarchically organized systems that can be partitioned in progressively smaller and overlapping spatial units, i.e., drainage area, drainage network, segment, reach, habitat, and microhabitat.[9] Each spatial unit shows different patterns and processes that influence lower hierarchical units, but not vice versa. This hierarchy implies that many processes are unidirectional and that upstream sections always, at least partially, influence downstream sections. Humans usually view streams at the reach scale perspective, which has many characteristic components or parts (Figure 9.3).

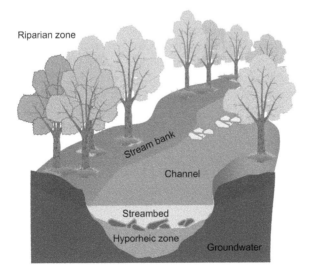

FIGURE 9.3 Cross-section illustrating the main parts of a flowing stream reach.

The Fluvial Ecosystem

Streams have distinct physical, chemical, and biological characteristics when compared with other aquatic or terrestrial ecosystems. Generally, streams are defined by unidirectional flow, high biogeochemical reactivity, and unique biological communities.[10] Moreover, streams are especially open and dynamic ecosystems that have been described as having four dimensions.[11] The upstream-downstream connection constitutes the longitudinal dimension. The lateral dimension includes interactions between the channel and riparian/floodplain systems. The vertical dimension involves the relation between the channel and hyporheic/groundwater. The fourth dimension is time.

As any other ecosystem, streams integrate a variety of interactions between abiotic (non-living) factors and living organisms.[7] Spatial and temporal variation in flow shapes the morphology of the streambed and creates habitats for aquatic biota. Light is the main source of energy for primary production and its availability is primarily controlled by the riparian vegetation. Temperature is a key determinant of the functioning of many organisms and its variation is mainly driven by shading, climate, and water source. The substrata of streams are composed by organic and inorganic materials. Organic materials (e.g., leaves, wood) constitute a key source of energy for the heterotrophic organisms at the base of headwater stream food webs, and their amount is mostly driven by the presence of riparian vegetation. The size of the inorganic substrata is mainly controlled by the stream gradient and typically decreases downstream from headwaters to larger streams. Stream water chemistry is largely determined by the geology and land use of the drainage area and can vary dramatically along the stream network.

The biota inhabiting streams is diverse and adapted to the unique physical and chemical environment.[12] Microorganisms such as algae, bacteria, and fungi constitute the base of the stream food web, often forming biofilms on benthic substrata and/or plankton in the water column of larger streams. Larger algae, mosses, liverworts, and vascular plants contribute to the primary production of the stream and form microhabitats for fauna. Invertebrates (mostly insects) show a high diversity and act as key consumers and prey in streams. Vertebrates (mostly fishes) are at the top of the food web in most fluvial ecosystems, though they may be absent in seasonal streams. Both invertebrates and vertebrates show a large variety of feeding strategies (e.g., shredders, collector-gatherers, grazers, suspension feeders, predators) and occupy myriad habitats. In seasonal streams, the aquatic flora and fauna have developed adaptations (e.g., desiccation-resistant stages, propagules) to the frequent absence of flow.[4–5] Less is known, however, about the terrestrial biota that inhabits the dry streambeds of seasonal streams and the processes that occur in the absence of surface water.[13]

Energy sources in lotic ecosystems can be autochthonous (i.e., originated within the stream) or allochthonous (i.e., originated in the terrestrial environment and later transported to the stream). Autochthonous energy sources include the organic carbon compounds formed by the stream's primary producers through photosynthesis. Allochthonous energy sources include coarse (i.e., leafs, wood, and fruits), fine, or dissolved organic matter introduced from the terrestrial environment through surface or subsurface pathways. The changes in energy sources and the metabolic balance along the stream network are nicely illustrated by the river continuum concept:[14] In low-order streams, production is lower than ecosystem respiration (P/R ratio <1) and allochthonous coarse organic matter is the main source of energy. In mid-order streams, the P/R ratio approaches 1, indicating that much more energy is supplied by autochthonous primary production. In high-order streams, photosynthetic rates by phytoplankton become limited due to turbidity; thus, the P/R ratio decreases again. Headwaters and downstream sections are linked through the export of fine organic matter. The generality of the river continuum concept has been criticized mainly because it mostly applies for temperate climate regions, it does not explicitly account for seasonal streams, and it is based on pristine streams which are rarely found in our highly humanized landscapes. Nonetheless, it remains a useful conceptual model, which has been expanded by other conceptual models such as the serial discontinuity concept,[15] the flood pulse concept[16,17] or the riverine productivity model.[18]

On the whole, streams are not just transport conduits connecting land with the oceans. These ecosystems possess a high capacity to store, transform, and remove biologically reactive elements. Due to the presence of a strong unidirectional flow, in streams we speak of material spiraling instead of material cycling.[19] The study of the contribution of material spiraling in stream networks to the material budgets of whole drainage areas and global elemental budgets as well as the role of flow cessation periods in seasonal streams on these processes are considered important issues in current research.[13,20]

Ecosystem Services

Resources and processes supplied by stream ecosystems provide valuable benefits to humankind. The ecosystem services offered by streams can be divided into three categories:[21] 1) water supply; 2) supply of goods other than water; and 3) non-extractive or in-stream benefits (Table 9.1). In many regions, streams constitute a critical source of freshwater for irrigation, households, aquaculture, and industries. Goods other than water include important food resources such as fish, waterfowl, and shellfish. The natural functioning and geomorphology of streams helps mitigate the energy of floods, dilute and attenuate pollution, and fertilize floodplain soils, estuaries, and coastal areas near the river deltas. In addition, water from streams is used for hydroelectric power generation. Streams can also be used as transportation routes or as recreational sites for swimming, boating, fishing, etc. Moreover, streams provide a crucial habitat for the conservation of aquatic biodiversity and a series of important non-use values. Despite the absence of water during some periods, seasonal streams and their dry streambeds also have important human and ecological values mainly in terms of culture, recreation, and biodiversity conservation.[13] For instance, seasonal streams host a unique mixture of aquatic, terrestrial, and amphibious communities as a result of their dry and wet phases.

Management and Conservation

Perennial and seasonal streams are highly susceptible to anthropogenic pressures.[7] Pollution (e.g., nutrients, heavy metals, and pharmaceuticals), originated from point (e.g., waste water treatment plant effluents) or diffuse (e.g., agricultural fertilizers) sources, represents a major impact that has deleterious effects on stream structure and function. Hydro-morphological modifications including damming, channelization, water abstraction, and removal of riparian vegetation strongly influence the flow, temperature, and sediment regime of streams among other factors. In addition, dams fragment stream networks, thus breaking the longitudinal connectivity characteristic of these ecosystems. Furthermore, humans have intentionally or unintentionally introduced a high number of non-native species into streams. These invasive organisms predate native species, compete with them for prey or habitat, alter stream habitats, and introduce harmful diseases.[12] Furthermore, changes in climate together with land use modifications are expected to cause an increase in the temporal and spatial extent of seasonal

TABLE 9.1 Ecosystem Services Provided by Streams

Water Supply	Supply of Goods Other than Water	Non-Extractive or In-Stream Benefits
Irrigation	Fish	Flood control
Household	Waterfowl	Transportation
Industry	Shellfish	Hydroelectric power
Aquaculture		Pollution, dilution, and attenuation
		Soil fertilization
		Recreation
		Biodiversity
		Non-use values

Source: Based on Postel & Carpenter.[21]

streams in many regions of the world.[5,13] Seasonal streams, mostly if water interruption is not natural, favor the disappearance of sensitive taxa and offer opportunities for invasive species.

In view of these major threads, the management and conservation of stream ecosystems and their associated values remain a major challenge. With this purpose, many countries have developed bodies of legislation such as the Clean Water Act in the U.S.A., the Water Framework Directive in the European Union, or the Water Act in Australia. These represent landmarks in sustainable management, but water security is often brought about by economic investment in water treatment rather than the prevention of impacts on freshwater ecosystems.[22] Unfortunately, many of these legislations fail to protect small low-order streams and are not yet well defined with seasonal streams.[10]

Conclusion

Streams are complex and diverse ecosystems. In a wide sense, streams can be divided into perennial (permanently flowing) or seasonal (temporarily flowing) streams, which show important commonalities but also differences in their structure, function, and dynamics. Moreover, streams are often characterized by their stream order as well as the type of landscape or geology of their drainage areas. Stream networks form a continuum from headwaters to the river mouth, which can be divided into different zones and hierarchical units on the basis of geomorphological and ecological attributes. Fluvial ecosystems are primarily characterized by unidirectional flow, high connectivity, unique biological assemblages, and high biogeochemical reactivity. They have the ability to retain, transform, and remove important amounts of material during downstream transport and provide a series of highly valuable ecosystem services to humankind. Nonetheless, perennial and seasonal streams are highly susceptible to human impacts and their management and conservation remains a major challenge.

Acknowledgments

We thank E. Marti for providing drawings of Figure 9.3. This research was funded through the projects SCARCE (CON-SOLIDER-INGENIO CSD2009-00065) and CARBONET (CGL2011-30474-C02-01) from the Spanish Ministry of Science and Innovation. D. von Schiller was additionally supported by a Juan de la Cierva postdoctoral fellowship from the Spanish Ministry of Economy and Competitiveness (JCI-2010-06397).

References

1. Dodds, W.K.; Whiles, M.R. *Freshwater Ecology: Concepts and Environmental Applications*; Academic Press: San Diego, CA, USA, 2010.
2. Charlton, R. *Fundamentals of Fluvial Geomorphology*; Routledge: London, UK, 2008.
3. Hynes, H.B.N. The stream and its valley. Verh. Internat. Verein. Limnol. **1975**, *19*, 1–15.
4. Williams, D.D. *The Biology of Temporary Waters*; Oxford University Press: New York, NY, USA, 2006.
5. Lake, P.S. *Drought and Aquatic Ecosystems: Effects and Responses*; Wiley-Blackwell: Oxford, UK, 2011.
6. Strahler, A.N. Hypsometric (area-altitude) analysis of erosional topography. Bull. Geol. Soc. Am. **1952**, *63* (11), 1117–1142.
7. Allan, J.D.; Castillo, M.M. *Stream Ecology: Structure and Function of Running Waters*; Springer: Dordrecht, the Netherlands, 2007.
8. Schumm, S.A. *The Fluvial System;* John Wiley & Sons: New York, USA, 1977.
9. Frissell, C.A.; Liss, W.L.; Warren, C.E.; Hurley, M.D. A hierarchical framework for stream habitat classification: Viewing streams in a watershed context. Environ. Manag. **1986**, *10* (2), 199–214.
10. Doyle, M.W.; Bernhardt, E.H. What is a stream? Environ. Sci. Technol. **2011**, *45* (2), 354–359.

11. Ward, J.V. The four-dimensional nature of lotic ecosystems. J. North Am. Benthol. Soc. **1989**, *8* (1), 2–8.

12. Giller, P.S.; Malmqvist, B. *The Biology of Streams and Rivers;* Oxford University Press: New York, 1998.

13. Steward, A.L.; von Schiller, D.; Tockner, K.; Marshall, J.C.; Bunn, S.E. When the river runs dry: Human and ecological values of dry riverbeds. Front. Ecol. Environ. **2012**, *10* (4), 202–209.

14. Vannote, R.L.; Minshall, G.W.; Cummins, K.W.; Sedell, J.R.; Cushing, C.E. The river continuum concept. Can. J. Fish. Aquat. Sci. **1980**, 37 (1), 130–137.

15. Ward, J.V.; Stanford, J.A. The serial discontinuity model: Extending the model to floodplain rivers. Reg. Rivers Res. Manag. **1995**, *10* (2–4), 159–168.

16. Junk, W.J.; Bayley, P.B.; Sparks, R.E. The flood pulse concept in river-floodplain systems. In *Proceedings of the International Large River Symposium;* Dodge, D.P., Ed.; Canadian Special Publications of Fisheries and Aquatic Sciences, 1989; 110–127.

17. Tockner, K.; Malard, F.; Ward, J.V. An extension of the flood pulse concept. Hydrol. Process. **2000**, *14* (16–17), 2861–2883.

18. Thorp, J.H.; Delong, M.D. The riverine productivity model: An heuristic view of carbon sources and organic processing in large river ecosystems. Oikos **1994**, *70* (2), 305–308.

19. Newbold, J.D. Cycles and spirals of nutrients. In *The Rivers Handbook*; Calow, P., Petts, G.E., Eds.; Blackwell Scientific: Oxford, UK, 1992; pp. 379–408.

20. Cole, J.J.; Prairie, Y.T.; Caraco, N.F.; McDowell, W.H.; Tranvik, L.J.; Striegl, R.G.; Duarte, C.M.; Kortelainen, P.; Downing, J.A.; Middelburg, J.J.; Melack, J. Plumbing the global carbon cycle: Integrating inland waters into the terrestrial carbon budget. Ecosystems **2007**, *10* (1), 171–184.

21. Postel, S.; Carpenter, S. Freshwater ecosystem services. In *Nature's Services;* Daily, G.C., Ed.; Island Press: Washington D.C., USA, 1997; pp. 195–214.

22. Vörösmarty, C.J.; McIntyre, P.B.; Gessner, M.O.; Dudgeon, D.; Prusevich, A. Green, P.; Glidden, S.; Bunn, S.E.; Sullivan, C.A.; Reidy Liermann, C.; Davies, P.M. Global threats to human water security and river biodiversity **2010**, *467* (7315), 555–561.

10

Vernal Pool

Peter W. C. Paton
University of Rhode Island

Introduction

Broadly defined, vernal pools are generally hydrologically isolated from open water bodies, temporary to semipermanent wetlands that form in small, shallow permanent basins during the coolest part of the year, and dry annually or during drought years.[1–3] In the United States, these ephemeral wetlands are found in Pacific Coast states, in the North, and Northeast,[1] however they can be found globally.[4,5] They are called vernal pools because in many areas, particularly Mediterranean climates such as California, these wetlands are inundated with water in the spring; vernal is from the Latin word "vernalis" which means "pertaining to spring" and "pool" is often used for a small water body.[3] A somewhat more restrictive definition for vernal pools in California is "precipitation-filled seasonal wetlands inundated during periods when temperature is sufficient for plant growth, followed by a brief waterlogged-terrestrial stage and culminating in extreme desiccating soil conditions of extended duration."[4] In the widely used Cowardin wetland classification system,[6] vernal pools with vegetation are classified as seasonally flooded emergent wetlands, while ponds with no vegetation are open water or unconsolidated bottom habitats in the palustrine system, although some ponds would be better classified as intermittently flooded or semi-permanently flooded. Other terms that biologists have used for this wetland type include spring pools, ephemeral wetlands, semipermanent pond, fishless ponds, seasonal astatic waters, vernal ponds, vernal marshes, buffalo wallows, seasonal pools, seasonal ponds, vleis, or temporary waters.[2,4]

Four phases have been identified in the annual cycle of vernal pools: (i) desiccated phase: pond substrate is completely dry and desert-like, which often occurs in the summer or fall; (ii) preinundation phase: initial precipitation makes the soil moist, allowing germination of flora and hatching of some vertebrates and invertebrates; (iii) pool phase: the pond basin is completely full of water from precipitation; and (iv) drying phase: evapotranspiration directly or indirectly causes the pond and soil substrate to dry.[4] To classify vernal pools, biologists suggest that one needs to understand three elements: first, the water source to fill pools, which may be rain-fed or groundwater or a combination; second, the hydroperiod, which is the duration of inundation and the waterlogged phase; and third, the hydroregime, which is the timing of inundation.[1,4] The geomorphologic setting also helps one to understand the function of vernal pools.[7]

Vernal pools primarily occur in confined, permanent basins with no permanent inflow or outflow outlet; thus, there is no continuous surface water connection with permanently inundated water bodies. Vernal pools in Mediterranean climates are usually found in small, shallow basins, where surface area usually covers 50–5000 m², [8] and depths ranging from 0.1 to 1 m.[9,10] In California, vernal pools occur in small depressions that are underlain by an impervious substrate that limits drainage.[11,12] Therefore, they tend to be clustered on the landscape because they are associated with specific geological formations and soils.[13] In the glaciated, northeastern North America, vernal pools are found in a broad array of geomorphic settings.[7] The surface area of pond basins in the glaciated Northeast averages about 0.1 ha in size, but ranges from 0.01 to 2.25 ha.[2] Pond basins typically are often about 1 m depth at maximum water depth, with a range from 0.2 to >3.8 m deep.[2]

Hydroperiod, the number of days with surface inundation by water, is a function of climate and water balance, where changes in surface water in a pool is equal to the sum of inputs from precipitation, groundwater, and surface water, minus losses from evapotranspiration.[10] Water storage will vary as a function of climatic conditions and hydrogeologic settings, with considerable variation among pond basins. In ponds in the glaciated Northeast, the hydro-period for most vernal pools averages about 190–270 days inundation annually and can range from 70 days to permanently flooded in some wet years.[14] One set of vernal pools in California has average hydroperiods of 115 days,[15] while the hydroperiod at other pools ranged from 10 to 200 days.[16,17] Precipitation is the primary water input for most vernal pools, with weekly precipitation accounting for over 50% of the variation in water depth at one site in Massachusetts.[18] However, in certain geomorphic settings, vernal pools can be fed by groundwater.[7,10,15]

Vernal pools are unique ecosystems that include many regionally endemic (e.g., in California) or obligate species (e.g., glaciated Northeast).[1] However, many genera and species that are found in vernal pools are distributed across the continent. Vernal pools are important for local, regional, and national biodiversity because they provide habitat for unique flora,[19] invertebrates[2,20] and vertebrates.[21] In particular, pool-breeding amphibians have their greatest reproductive success in fishless vernal pools, although they will use other aquatic habitats for breeding.[22]

Flora

Hydroperiod and hydroregime, coupled with soil and canopy cover, are probably the most influential factors that affect plant community composition within vernal pools. One of the unique attributes of vernal pools is the flora adapted to these ephemeral wetlands, particularly in California. Vernal pools are among the last ecosystems in California dominated by native flora.[13] Along the Pacific Coast, over 100 plant taxa commonly occur in vernal pools, with over half endemic to California and the other species dispersed from Washington to Baja California. Although vernal pools have a global distribution, California is the only region that evolved an extensive flora endemic to vernal pools.

At least 15 taxa are federally listed as either endangered (10 species) or threatened (5 species) and 37 other species are either state-listed or state species of concern.[23] In the glaciated Northeast, there are at least 20 at-risk plant taxa that occur in vernal pools, with 1 federally listed species (Northern bulrush *Scirpus ancistrochaetus*) and 4 threatened species.[24]

Plants occurring in vernal pools are generally dispersed along a hydrological gradient, with obligate wetlands plants found in deeper parts of a pond and facultative wetland plants occupying the shallow edges of vernal pools.[24] Vegetation has been used successfully in the Northeast to predict the hydroperiod of vernal pools. Researchers in Rhode Island were able to predict the hydroperiod class of 72% of 65 ponds they studied, but many ponds had no vascular vegetation which made it impossible to predict the hydroperiod of those ponds.[25]

Invertebrates

Vernal pools with different hydroperiods and hydroregimes usually are occupied by different community assemblages, which can include hundreds of species.[2,20] Invertebrate community structure and food webs within vernal pools can be complex, with invertebrate species occupying a range of trophic levels. Lower trophic levels include decomposers, detritivores, and photosynthesizers; intermediate levels include grazers, shredders, filter feeders, and deposit feeders; while upper tropic levels include predators and parasites.[2] Two primary ecological forces structuring vernal pool communities are competition and predation. Predatory dragonfly nymphs and diving-beetle larvae can affect the composition of the entire community, including vertebrates breeding in the pools.[20]

Fairy shrimp (Order Anostraca) rank among the most charismatic invertebrates that are closely linked to vernal pools. In the northeastern United States, most species are the genus *Eubranchipus,* which are filter feeders that eat algae, zooplankton and bacteria; several species in the region are rare or endangered. On the Pacific Coast, several species of endemic fairy shrimp are among the flagship species leading to conservation efforts for this imperiled wetland because several are either federally listed as endangered (four species: conservancy, longhorn, san diego, and riverside fairy shrimp) or threatened (one species: vernal pool fairy shrimp).[26] These species are threatened by habitat loss and the introduction of alien species, such as predatory fish (*Gambusia affinis*).[27]

Vertebrates

Among the vertebrates that use vernal pools, pool-breeding amphibians are the taxa that many biologists associate with vernal pools. In California, at least two species of amphibians (California Tiger Salamander (*Ambystoma californiense*) and Western Spadefoot Toad (*Spea hammondii*)) regularly use vernal pools, but these two species also use other types of wetlands.[19] In the glaciated Northeast, at least 10 species of salamanders and three species of frogs use vernal pools as their primary breeding habitat, and one other species of salamander and 14 species of frogs use vernal pools in addition to other types of wetlands for breeding.[22] Pool-breeding amphibians have complex life histories, with aquatic larvae developing through metamorphosis in vernal pools, while juveniles and adults spend most of their life in the terrestrial realm surrounding vernal pools. Vernal pool characteristics that are important in determining amphibian use of the pool include hydro-period, canopy cover over the pool, potential predator composition, and vegetation composition in area surrounding the pond.[22]

There are additional vertebrates that often use vernal pools for various aspects of their life history. Although reptiles, birds, and mammal may not often breed in vernal pools, they often use these ephemeral wetlands as foraging and resting habitat.[25,28] In the glaciated Northeast, of the 24 species of turtles in the region, at least 64% obtain food, shelter, or other resources from vernal pools, with at least three species considered as vernal pool specialists.[25] Of the 43 snakes in the Northeast, over 50% use vernal pools. Over 80% of the 232 birds in the region occasionally use vernal pools. In the western United States, vernal pools provide important habitat for resident and migratory birds, with many species of waterfowl and shorebirds often using these wetlands as foraging habitat.[29]

Conclusion

In North America, vernal pools are abundant on the Pacific Coast, in the glaciated Northeast, and in parts of the Southeast. Due to their hydroperiod and hydroregime, they provide important habitat for many species of regionally endemic invertebrates and vertebrates that are adapted to this unique type of wetland. Many species of endangered plants, invertebrates, and vertebrates occur in vernal pools, thus vernal pools are a global conservation concern for biologists. Therefore, land planners and biologists have developed a variety of strategies to conserve vernal pools.[30] However, for many species, it is just as important to protect habitat surrounding the vernal pool, as some species may disperse hundreds of meters from vernal pools during other parts of their annual cycle.

References

1. Zedler, P.H. Vernal pools and the concept of isolated wetlands. Wetlands **2003**, *2* (3), 597–607.
2. Colburn, E. A. *Vernal pools: Natural History and Conservation*; McDonald and Woodward Publishing Co.: Blacksburg, Virginia, 2004.

3. Calhoun, A.J.K.; deMaynadier, P.O. *Science and Conservation of Vernal Pools in Northeastern North America;* CRC Press: Boca Raton, Florida. 2008.

4. Keeley, J.E.; Zedler. P.H. Characterization and global distribution of vernal pools. In *Ecology, Conservation, And Management of Vernal Pool Ecosystems – Proceedings from a 1996 Conference;* Witham, C.W., Bauder, E.T., Belk, D., Ferren W.R., Jr., R. Ornduff, R, Eds.; California Native Plant Society: Sacramento, CA, 1998; 1–14.

5. Vanschoenwinkel, B.; Hulsmans, A.; De Roeck, E.; De Vries, C; Seaman, M; Brendonck, L. Community structure in temporary freshwater pools: disentangling the effects of habitat size and hydroregime. Freshwater Biol. **2009**, *54* (7), 1476–1500.

6. Cowardin, L.M.; Carter, V.; Oolet, F.C.; LaRoe, E.T. *Classification of Wetlands and Deepwater Habitats of the United States.* U.S. Fish and Wildlife Service, Washington, D.C. 1979.

7. Rheinhardt, R.D.; Hollands, O.O. Classification of vernal pools: geomorphic setting and distribution. In *Science and Conservation of Vernal Pools in Northeastern North America;* Calhoun, A.J.K., deMaynadier, P.O. Eds.; CRC Press: Boca Raton, FL, 2008; 31–54.

8. Mitsch, W.J.; Oosselink, J.O. *Wetlands;* 3rd Ed. John Wiley and Sons Inc: New York, 2000.

9. Hanes, T.; Stromberg, L. Hydrology of vernal pools on non-volcanic soils in the Sacramento Valley. In *Ecology, Conservation, and Management of Vernal Pool Ecosystems;* Witham, C.W., Bauder, E.T., Belk, D., Ferren, W.R. Jr, Ornduf, R. Eds.; California Native Plant Society: Sacramento, CA; 1998; 38–49.

10. Brooks, R.T.; Hayashi, M. Depth-area-volume and hydroperiod relationships of ephemeral (vernal) forest pools in southern New England. Wetlands **2002**, *22* (2), 247–255.

11. Leibowitz, S.O.; Brooks, R.T. Hydrology and landscape connectivity of vernal pools. In *Science and Conservation of Vernal Pools in Northeastern North America*; Calhoun, A.J.K., deMaynadier, P.O., Eds.; CRC Press: Boca Raton, FL; 2008; 31–53.

12. Holland, R.F. The vegetation of vernal pools: a survey. In *Vernal Pools: Their Ecology and Conservation;* Jain, S. Ed.; Institute of Ecology Publication No. 9. University of California: Davis, CA; 1976; 11–15.

13. Rains, M.C.; Fogg, O.E.; Harter, T.; Dahlgren, R.A.; Williamson, R.J. The role of perched aquifers in hydrological connectivity and biogeochemical processes in vernal pool landscapes, Central Valley, California. Hydrol. Proces. **2006**, *20* (5), 1157–1175.

14. Smith, D.W.; Verrill, W.L. Vernal pool-soil-landform relationships in the Central Valley, California. In *Ecology, Conservation, and Management of Vernal Pool Ecosystems*, Witham, C.W., Bauder, E.T., Belk, D., Ferren, W.R. Jr, Ornduf, R. Eds.; California Native Plant Society: Sacramento, CA. 1998; 15–23.

15. Skidds, D.E.; Oolet, F.C. Estimating hydroperiod suitability for breeding amphibians in southern Rhode Island seasonal forest ponds. Wetland Ecol. Manag. **2005**, *13* (3), 349–366.

16. Marty, J.T. Effects of cattle grazing on diversity in ephemeral wetlands. Conserv. Biol. **2005**, *19* (5), 1626–1632.

17. Black, C.; Zedler, P.H. An overview of 15 years of vernal pool restoration and construction activities in San Diego County, California. In *Ecology, Conservation, and Management of Vernal Pool Ecosystems,* Witham, C.W., Bauder, E.T., Belk, D., Ferren, W.R. Jr, Ornduf, R. Eds.; California Native Plant Society: Sacramento, CA; 1998; 195–205.

18. Brooks, R.T. Weather-related effects on woodland vernal pool hydrology and hydroperiod. Wetlands **2004**, *24* (1), 104–114.

19. Zedler, P.H. The ecology of southern California vernal pools: A community profile. Biology Report 85, U.S. Fish and Wildlife Service: Washington, D.C. 1987.

20. Colburn, E.A.; Weeks, S.C.; Reed, S.K. Diversity and ecology of vernal pool invertebrates. In *Science and Conservation of Vernal Pools in Northeastern North America*; Calhoun, A.J.K., deMaynadier, P.O., Eds.; CRC Press: Boca Raton, FL, 2008; 105–126.

21. Mitchell, J.C.; Paton, P.W.C.; Raithel, C.J. The importance of vernal pools to reptiles, birds, and amphibians. In *Science and Conservation of Vernal Pools in Northeastern North America;* Calhoun, A.J.K.; deMaynadier, P.O., Eds.; CRC Press: Boca Raton, FL, 2008; 169–192.

22. Semlitsch, R.D.; Skelly, D.K. Ecology and conservation of pond-breeding amphibians. In *Science and Conservation of Vernal Pools in Northeastern North America*; Calhoun, A.J.K., deMaynadier, P.O., Eds.; CRC Press: Boca Raton, FL, 2008; 127–148.

23. Lazar, K.A. Characterization of rare plant species in the vernal pools of California. MS thesis, Plant Biology Graduate Group, University of California: Davis, CA, 2006.

24. Cutka, A.; Rawinski, T.J. Flora of Northeastern vernal pools. In *Science and Conservation of Vernal Pools in Northeastern North America;* Calhoun, A.J.K.; deMaynadier, P.O., Eds.; CRC Press: Boca Raton, FL, 2008; 71–104.

25. Mitchell, J. Using plants as indicators of hydroperiod class and amphibian habitat suitability in Rhode Island seasonal pond. M.S. thesis. University of Rhode Island: Kingston, RI, 2005. http://www.fws.gov/endangered (accessed October 2011).

26. Leyse, K.E.; Lawler, S.P.; Strange, T. Effects of an alien fish, *Gambusia affinis,* on an endemic California fairy shrimp, *Linderiella occidentalis*: implications for conservation of diversity in fish-less waters. Biol. Conserv. **2003**, *118* (1), 57–65.

27. Paton, P.W.C. A review of vertebrate community composition in seasonal forest pools of the northeastern United States. Wetlands Ecol. Manag. **2005**, *13* (3), 235–246.

28. Silveira, J.O. Avian uses of vernal pools and implications for conservation practice. In *Ecology, Conservation, and Management of Vernal Pool Ecosystems;* Witham, C.W., Bauder, E.T., Belk, D., Ferren, W.R. Jr, Ornduf, R. Eds.; California Native Plant Society: Sacramento, CA, 1998; 195–205.

29. Calhoun, A.J.K; Reilly, P. Conserving vernal pool habitat through community based conservation. In *Science and Conservation of Vernal Pools in Northeastern North America;* Calhoun, A.J.K.; deMaynadier, P.O., Eds.; CRC Press: Boca Raton, FL, 2008; 319–341.

30. deMaynadier, P.O.; Houlahan, J.E. Conserving vernal pool amphibians in managed forests. In *Science and Conservation of Vernal Pools in Northeastern North America*; Calhoun, A.J.K.; deMaynadier, P.O., Eds.; CRC Press: Boca Raton, FL, 2008; 253–280.

11

GIS Analysis of Groundwater Salinity

Yuming Wen
University of Guam

Introduction

Guam, an unincorporated U.S. territory in the Western Pacific, is the largest (about 541 km^2) and southernmost member of the Mariana Islands chain. It currently supports a local population of about 170,000 and receives over 1.5 million tourists annually [1,2]. About 5,000 marines will be relocated to Guam from Okinawa in the following years. With the marines and their dependents moving to Guam, and the importation of foreign labor to assist with the military buildup activities, the island's population is expected to soar in the next few years. Groundwater from a karst limestone aquifer in the northern Guam currently supplies local residents and tourists with approximately 90% of their daily water needs. While estimates of the aquifer's sustainable yield remain adequate for the current population and tourists, there is widespread concern that the projected increase of local population and tourists will severely compromise the island's drinking water supplies and have an unprecedented impact on water quality and quantity. The high porosity and rapid recharge characteristics of the Northern Guam Lens Aquifer (NGLA) make it especially susceptible to contamination from seawater intrusion, urban runoff, chemical spills, effluents for septic tanks, and sewage overflows [3].

Guam Waterworks Authority (GWA) is the custodian of Guam's public water supply and is responsible for ensuring that it meets all appropriate standards as mandated under the Safe Drinking Water Act of U.S. Environmental Protection Agency (USEPA). To this end, GWA regularly evaluates the physical, chemical, and biological integrity of the island's drinking water in accordance with USEPA requirements. While the agency has obtained a considerable amount of monitoring data over the years, only values that approach or exceed the water quality standards, including salinity, are paid attention.

Since high level of salinity in drinking water poses risks to human's health, it is crucial to locate wells with problems of salinity, and archive those wells in a format that permits a rapid visualization of spatial and temporal trends of salinity in the wells in the NGLA using the state-of-the-art GIS technology. GIS is especially useful for identifying and predicting subtle long-term trends (increases or decreases) in salinity levels within the aquifer visually and analytically. Some recent related research on groundwater salinity in NGLA is available in [3] and [4]. History and trends of chloride concentration based on average decadal chloride data from 1973 to 1999 are discussed in [3], and [4] extends the research in [4] to figure out the relationships between chloride and factors, including precipitation, mean sea level, southern oscillation index, and well production rate. Most of the results and discussions are based on Excel spreadsheet data, graphs, and charts. This research aims to focus on GIS-based analysis of groundwater salinity in Guam.

Methodology

Study Area

Guam, a tropical island in the Western Pacific, is a strategic place for the United States. It is close to major Asian cities, including Tokyo, Seoul, Beijing, Shanghai, Hong Kong, and Manila (Figure 11.1). The north half of the island of Guam is mainly composed of a limestone plateau bounded by cliffs, and the south half of the island consists primarily of lowlands and uplands fringed with volcanic deposits (Figure 11.2). There are no streams in the porous limestone plateau in the north, whereas numerous

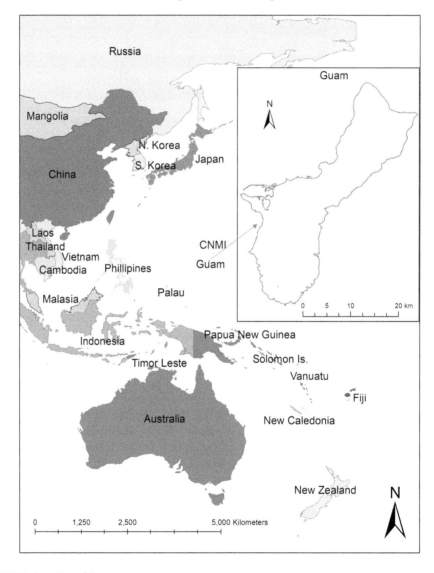

FIGURE 11.1 Location of Guam.

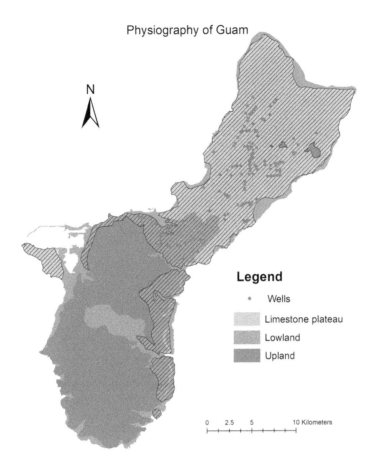

FIGURE 11.2 Physiography of Guam.

streams can be found in the south part of the island. The study area focuses on the NGLA located in the northern part of Guam. Since about 90% of drinking water comes from groundwater, and all wells located in NGLA are fringed with limestone, groundwater quality is a significant concern in Guam. Because of its size and location between the interactive area of the Philippine Sea in the west and the Pacific Ocean in the east, and the physiographic feature of limestone in the NGLA, seawater intrusion into the groundwater resources, and therefore risk of groundwater salinity to the public's health are an emphasis on stakeholders' agendas.

Data Sources and Data Processing

Data of subbasin boundaries and well locations are available. There are six subbasins located in the northern Guam, most of which is composed of limestones. There are 148 wells available in this research (Figure 11.3). Chloride data from 2001 to 2009 collected quarterly by GWA is employed. GWA collects groundwater samples from wells quarterly. The concentration of chloride is used as an indicator for salinity level. The recommended maximum level of chloride in U.S. drinking water suggested by USEPA is 250 mg/L [5]. The local benchmark maximum level of salinity in Guam is 150 ppm or 150 mg/L of chloride. The chloride data from GWA is provided in MS Excel spreadsheet format. Since the unique identification information for each well is used in both the spreadsheet data and GIS data of well locations, the chloride data in MS Excel spreadsheet format is geo-coded and saved in a GIS data format, in this case, a shapefile so that it can be further processed and analyzed to figure out salinity problems

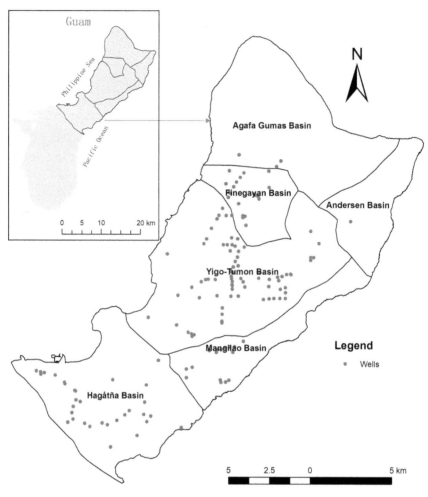

FIGURE 11.3   Subbasins and wells in the NGLA.

with some drinking wells in Guam. A research about the impacts of sinkholes on groundwater salinity concentration in the Finegayan area of Guam is discussed in [6]. GIS analysis is applied to locate wells with salinity problems, and delineate sinkholes in the Finegayan area, and therefore, the relationship between salinity and sinkholes is evaluated. The similar procedures of GIS analysis are utilized to deal with salinity problems in this research.

Results

Based on the suggested maximum contaminant level (MCL) of chloride at 250 mg/L, an indicator for salinity, and the local benchmark MCL at 150 mg/L, the results of spatial analysis of the chloride data categorize problematic wells with high levels of salinity into two groups: wells with water quality deficiency of salinity if the chloride concentration from a well is equal to or greater than 250 mg/L (Figure 11.4) and wells with potential water quality deficiency of salinity if the chloride concentration from a well is less than 250 mg/L and equal to or greater than 150 mg/L (Figure 11.5). The table in Figure 11.4 indicates that there are 20 wells with salinity problems in some times during the study period. Based on

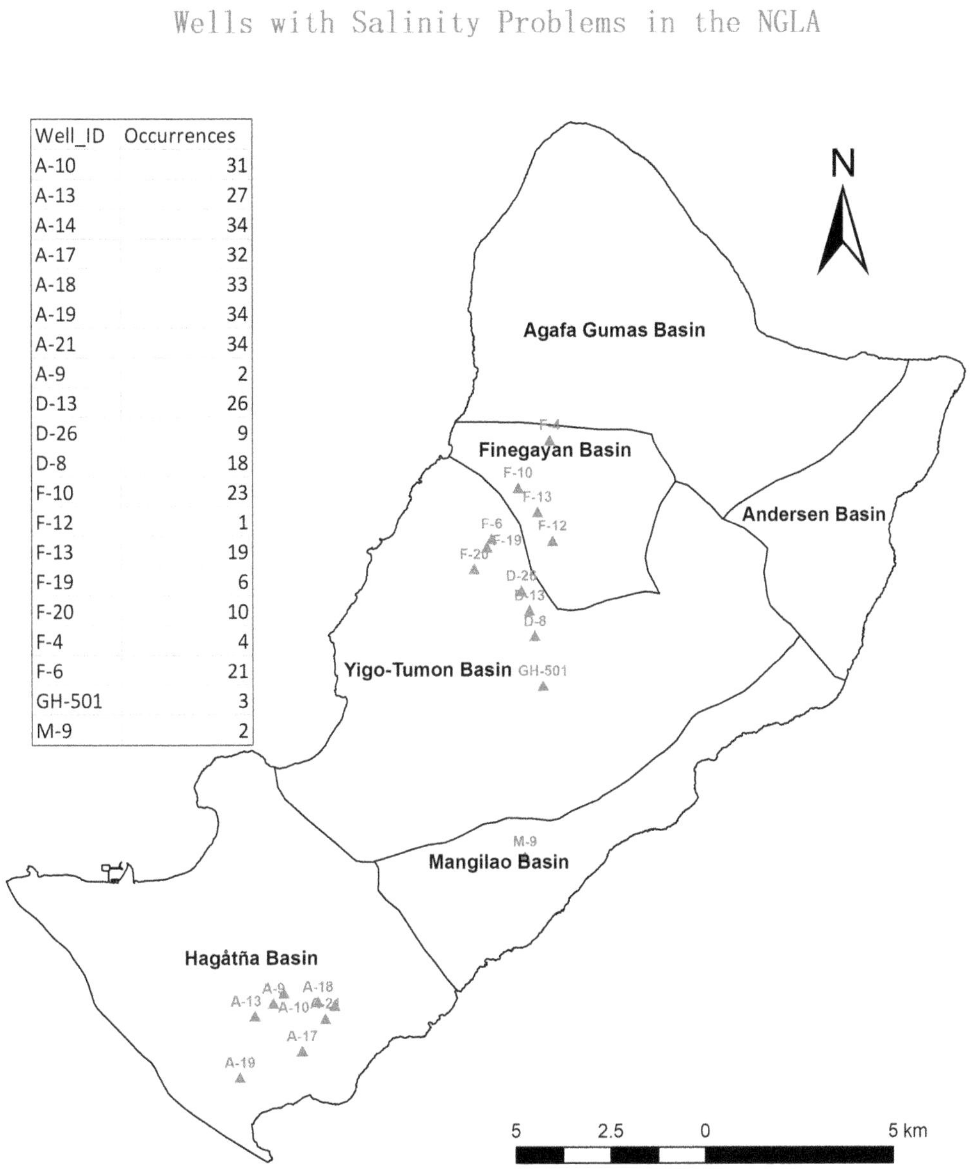

FIGURE 11.4 Wells with salinity problems in the NGLA.

the occurrences of problematic wells, six wells have serious salinity problems. All of those six wells (A-10, A-14, A-17, A-18, A-19, and A-21) are observed more than 30 occurrences out of 36 samples with chloride levels over the MCL of 250 mg/L. Another six wells (A-13, D-13, F-10, F-6, F-13, and D-8) are found to have at least 18 occurrences out of 36 samples with chloride levels over the MCL. Most of the rest of the wells just have a few occurrences of samples with chloride levels over MCL. Well F-12 has only one occurrence with chloride level over MCL, and Wells A-9 and M-9 have been detected only twice with chloride levels over MCL. The table in Figure 11.5 manifests that 39 wells have potential salinity problems over 9 years from 2001 to 2009. Six wells (A-9, D-9, F-11, H-1, M-1, and M-9) are detected with at least 20 occurrences with chloride levels over the local benchmark MCL of 150 mg/L. One-third of those 39 wells (A-15, A-28, D-8, F-1, F-2, F-4, F-5, F-6, F-13, F-19, F-20, M-2, and M-6) have at least 10 but

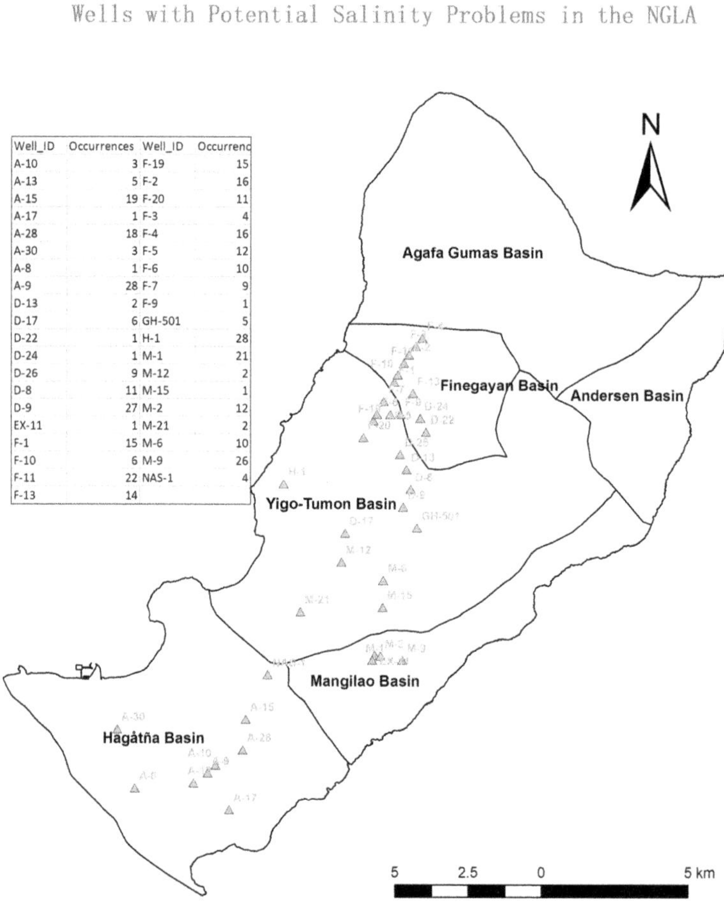

Well_ID	Occurrences	Well_ID	Occurrences
A-10	3	F-19	15
A-13	5	F-2	16
A-15	19	F-20	11
A-17	1	F-3	4
A-28	18	F-4	16
A-30	3	F-5	12
A-8	1	F-6	10
A-9	28	F-7	9
D-13	2	F-9	1
D-17	6	GH-501	5
D-22	1	H-1	28
D-24	1	M-1	21
D-26	9	M-12	2
D-8	11	M-15	1
D-9	27	M-2	12
EX-11	1	M-21	2
F-1	15	M-6	10
F-10	6	M-9	26
F-11	22	NAS-1	4
F-13	14		

FIGURE 11.5   Wells with potential salinity problems in the NGLA.

less than 20 occurrences with chloride levels over 150 mg/L. The rest of the wells (A-8, A-10, A-13, A-17, A-30, D-13, D-17, D-22, D-24, D-26, EX-11, F-3, F-7, F-9, F-10, GH-501, M-12, M-15, M-21, and NAS-1) are observed just a few occurrences with chloride levels over the local benchmark maximum level. Among the wells with insignificant salinity problems, 7 wells have just one occurrence with chloride level over 150 mg/L, and another 3 wells are detected with chloride levels over the local benchmark maximum level only twice over the study period.

Conclusions and Discussions

Guam is an island located in the Western Pacific. It is surrounded by the Philippine Sea in the west and the Pacific Ocean in the east. The physiographic characteristics indicate that the northern part of Guam is mostly composed of limestone, and the southern part is mainly composed of volcanic deposits. Since the wells in the NGLA provide about 90% of drinking water to around 170,000 local residents and over 1.5 million tourists, and the porosity of limestone poses risks for groundwater to be contaminated by salinity caused by seawater intrusion into the NGLA, it is important to conduct research on salinity levels from drinking wells, and locate wells with water quality deficiency of salinity. Based on the analysis of salinity data for 148 wells in the NGLA over 9 years, 20 wells are detected with salinity problems—of which, 6 wells are serious with salinity levels and 6 wells are significant in salinity levels. Based on the map in Figure 11.5, the wells with water quality problems of salinity are located in two clusters: one located in the south of Hagatna Basin and the other located between Finegayan Basin and Yigo-Tumon Basin and in the southwest of Finegayan. Only one well with salinity problems is far away from those two cluster areas and located in Mangilao Basin. Based on the local benchmark maximum level, 39 wells have problems with salinity. Technically, the local benchmark maximum level of salinity at 150 ppm or chloride at 150 mg/L is recommended for new well drills. If samples from potential new well locations are detected with chloride concentration over 150 mg/L, new wells will not be designed in such locations and nearby.

The research does not address the factors affecting the salinity levels in the groundwater resources in Guam. Wen and Jenson have discussed the impacts of sinkholes on salinity levels [6]. Other data, including southern oscillation index (SOI), precipitation, anthropogenic activities, land-cover/land-use change, well pumping rates, vegetation types, sea-level change, just to name some, may be combined with salinity data to evaluate whether those factors affect salinity levels, and if so, how they may affect salinity levels.

Acknowledgment

The research is funded by the USGS 104b Program and Guam Hydrological Survey.

References

1. Guam Bureau of Statistics and Plans. 2018. *2017 Guam Statistical Yearbook*. Guam Bureau of Statistics and Plans, Office of the Governor, Guam.
2. RNZ. 2019. Guam Visitor Arrivals Reach Record High. RNZ News, 15 January 2019.
3. Wen, Y.; Jenson, J. 2019. Impacts of Sinkholes on Salinity Level of Groundwater in Finegayan Area, Guam, USA. *Proceedings of the 29th International Laser Radar Conference, Hefei, China*, 24–28 June 2019.
4. McDonald, M. Q.; Jenson, J. W. 2003. Chloride History and Trends of Water in Production Wells in the Northern Guam Lens Aquifer. WERI Technical Report No. 98, Water & Environmental Research Institute of the Western Pacific, University of Guam.
5. USEPA. 2009. National Primary Drinking Water Regulations. United States Environmental Protection Agency EPA 816-F-09-004, May 2009.
6. Wen, Y.; Jenson, J. 2019. Impacts of Sinkholes on Salinity Level of Groundwater in Finegayan Area, Guam, USA. *The 29th International Laser Radar Conference, Hefei, China*, June 24–28, 2019.

12

Basic Evaluation Units and Physical Structural Integrality in Riparian Zone Evaluation

Bolin Fu
*Guilin University
of Technology*

Yeqiao Wang
University of Rhode Island

Ying Li
*Northeast Institute of
Geography and Agroecology*

Introduction

Riparian zones are narrow strips of land located along the banks of rivers, streams, and water networks. Riparian zones are widely acknowledged as an ecological transition zone of material and energy exchange between terrestrial and aquatic ecosystems (USDI Bureau of Land Management. Riparian Area Management, 1998; Tang et al., 2014). Riparian zones can provide a range of ecosystem functions and services, for example, bank stabilization and protection, water purification, reservoirs of biodiversity, wetland products, as well as recreation and tourism (Bennett & Simon, 2004; Ghermandi, et al., 2009; Hruby, 2009). Riparian zones are also a focus of human activities, such as urban expansion, agriculture, mining, grazing, erosion, and point- and non-point-source pollutions (Dixon et al., 2006; Ivits et al., 2009; Ranalli & Macalady, 2010). It is essential that riparian zones are managed appropriately to avoid degradation and damage that have become increasingly evident (Munné et al., 2003; Jansen et al., 2005; Ministry of Water Resources of the People's Republic of China, 2010; Chen et al., 2012; Fernández et al., 2014).

The physical structural integrality (PSI) characterizes riparian ecological conditions using indicators, for example, bank condition, riparian vegetation condition, and human intervention. Methodologies for assessing PSI of riparian zones have been developed to provide different evaluating indicators (Munné et al., 2003; Jansen et al., 2005; Innis et al., 2000; Barquin et al., 2011; González del Tánago & García de Jalón, 2011). Most of the methods are based on expert knowledge in selection of measuring sections and monitoring sites, then evaluating the PSI through field measurements. However, those methods are challenged in assessing long stretches of riparian zone, in particular vast regions and remote locations.

The evaluating methods have usually been concentrated on field measurements of a few hundred meters, which could be very laborious or even unpractical when attempting to evaluate an entire catchment or a long river corridor (Johansen et al., 2007). In addition, selection of measuring sections and sites might not be able to take into account simultaneously the representativeness, accessibility, and security, which made it difficult to fully characterize a riparian zone with site-based field data alone. Remote sensing techniques have been utilized to map indicators of riparian condition because of advantages in spatial extensiveness, noninvasiveness, and repetitive capability (Congalton et al., 2002; Johansen & Phinn, 2006; Johansen et al., 2010a, 2010b). Those studies demonstrated that indicators such as streambed width, riparian zone width, riparian vegetation, and bank stability were important and feasible to extract from remote sensing data for condition assessment of riparian zones. However, evaluate riparian conditions using those indicators still remain to be examined.

This study evaluated the riparian condition of the Songhua River across the Northeastern Plain of China, using a series of indicators developed from 2,081 basic evaluation units (BEUs) (Fu et al., 2017). The specific objectives of this paper are to demonstrate the feasibility of the multimetric approach through comparisons with field measurements.

Materials and Methods

Study Area

Songhua River is the largest tributary of the Heilongjiang (Amur River) in Northeast China. With the length of 1,897 km long and the drainage area of 545,600 km², Songhua is ranked the fifth longest river in China. It is originated from Changbai Mountain, passing through the Northeastern Plain, and injected into Amur River at about 48° north latitude. This study focused on the 1,679 km riparian zone of Songhua River started from the Fengman hydrologic dam passing through major cities such as Jilin, Harbin, and Jiamusi, and ended at the Amur River, the border river between China and Russia. The elevation of the river basin is from 54 to 2,735 m above mean sea level. The basin has a temperate continental climate with four clearly distinct seasons. The mean annual rainfall in the area is about 550–800 mm. The river reaches its maximum flow in summer and is frozen from November to April next year.

Songhua River Basin is a major agricultural center and commodity grain production base of China. Large-scale operations of state farms were established after the establishment of the People's Republic of China in 1949 and land reclamation projects begun. Songhua River is an important transportation artery for agricultural products of the region. The river flows through major port cities and serves as the source of drinking water for millions of people. The ecological functions and integrality of riparian zones of the Songhua River have experienced much degradation under the influence of anthropogenic activities such as overgrazing, construction of transportation infrastructure, urban sprawl, sand mining, tourism development, reclaimed wetland, and other human activities.

Land-Cover Data and Ancillary Data

This study adopted the land-use data at 1:100,000 scale for 1976, 1986, 1995, 2000, 2005, and 2013 from the Resources and Environment Science Data Center, Chinese Academy of Sciences. The land-use datasets included information on forest, grassland, farmland, wetland, water body, barren land, and buildup (Table 12.1). The land-use datasets of 2013 and 2015 are derived from updating the land-use map of 2005 with interpretation keys from field measurement sites using Landsat eight Operational Land Imager (OLI) images acquired in 2013 and 2015, respectively. Table 12.2 describes selected multispectral OLI images that covered the study area with 30 m spatial resolution. Other datasets adopted, including 1:500,000 geomorphic map and 1:100,000 topographic map developed by the Institute of Geographic Sciences and Natural Resources Research, Chinese Academy of Sciences, and Shuttle Radar

TABLE 12.1 The Categories of 1:100,000 Land-Use Data in Riparian Zone

Categories	Subcategories
Forest	Arboreal forest, sparse woodland, and shrub woodland
Grassland	High-, mid-, and low-cover grassland
Farmland	Paddy field, glebe field
Wetland	Marsh, riverine wetland
Water body	Rivers, reservoirs, fishery, and lakes
Barren land	Lands unused or difficult for using, saline-alkaline land
Buildup	Industrial and commercial, residential, transportation

Source: Fu et al., 2017.

TABLE 12.2 Multispectral Landsat 8 OLI Images

Years	Path/Row	Acquisition Date	Sensor
2013	118/29	Sep. 17, 2014	OLI
	119/29	Sep. 24, 2014	
2015	115/27	Sep. 15, 2015	OLI
	116/27	Sep. 19, 2014	
	116/28	Sep. 06, 2015	
	117/28	Sep. 13, 2015	
	118/28	Sep. 04, 2015	
	119/28	Sep. 24, 2014	

Source: Fu et al., 2017.

Topographic Mission (SRTM), generated Digital Elevation Model (DEM) data at 30 m spatial resolution. Referencing the 1:100,000 topographic map and the 1:500,000 geomorphic map, the elevation range of the riparian zone was between 115 and 370 m.

Field Data Collection

The studied riparian zone was divided into 13 measurement sections according to the Chinese Ministry of Water Resources technical protocol. Within each section, three or four measurement sites were chosen, and each site consisted of three transections (Figure 12.1). Each transection was located in an area with a consistent management regime and with similar vegetation and stream characteristics. The entire riparian zone consisted of 45 measurement sites and 135 transections. Each transection laid two 30 m width × 50 m length measurement areas on both sides of banks. The positions of measurement sites and transections were guided by Global Positioning System (GPS). The subindicators of field-based PSI include bank slope and slope length of river bank, canopy cover, bank sediment size (clay, sand, gravel, cobbles, or bedrock), bank erosion, and anthropogenic activities (Table 12.3). The bank slope and slope length were directly measured by a laser rangefinder and used to calculate the slope height. Canopy cover was measured by 1 × 1 m quadrat within the 30 × 50 m evaluation area. The 1 × 1 m vegetation quadrat was photographed with a camera mounted on a vertical shooting platform. These photos were subsequently processed to calculate vegetation coverage in every quadrat with the binary segmentation method by image processing, and then to calculate all canopy cover in the evaluation area. Anthropogenic activities, bank sediment size, erosion, and river connectivity were recorded by field photos. All field measurements were executed during the normal river flow period, in September of 2013 and 2015, respectively.

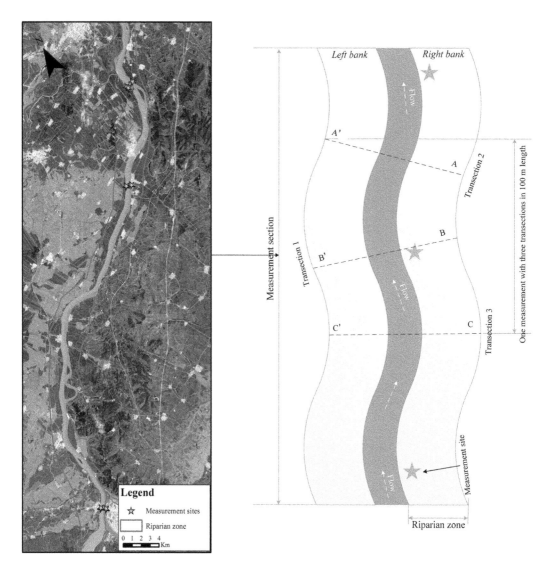

FIGURE 12.1 (See color insert.) Landsat eight OLI data (left) acquired on September 06, 2015, illustrated example locations of the measurement sites and the layout of field measurements (right). One measurement section had 3–4 measurement sites, and one measurement site had three transections with 30 m width × 50 m length measurement area on both sides of banks.
Source: Fu et al., 2017.

Description of BEUs

The BEUs are defined as the basic evaluation units with various sizes in the measurement sections of the riparian zone. To obtain the BEUs, the riparian zone was retrieved as a single featured GIS polygon using the 1:500,000 geomorphic map. The river centerline was extracted from the river mouth to the source for the main channel based on the 1:100,000 topographic map and the DEM. The 600 m length was adopted following the Chinese Ministry of Water Resources technical protocol (Ministry of Water Resources of the People's Republic of China, 2010) as a splitting criterion to divide the river centerline. Then, the polygon of the riparian zone was separated using splitting lines perpendicular to the river

TABLE 12.3   The Field-Based Indicators and Weights of PSI

Indicators		Subindicators
Riparian stability (RST) (0.5)	Bank stability (BKS) (0.25)	Bank slope (BS) (0.2)
		Slope height (SH) (0.2)
		Vegetation percent cover (VPC) (0.2)
		Bank sediment size (BSS) (0.2)
		Erosion status (ES) (0.2)
	Vegetation canopy coverage (BVC) (0.5)	Canopy cover
	Human disturbance (RD) (0.25)	Human-built structures, agriculture, transportations, sand mining, and other activities
River connectivity (RC) (0.25)		Human-built instream structures, including bridges, dams, culverts
Natural wetland conservation (NWC) (0.25)		Wetland area in the evaluating year (A_C) and historical year (1976) (A_R)

Source: Fu et al., 2017.

centerline. This process generated 2,081 BEUs in the study area. The minimum, average, and maximum size of the BEUs were 0.28, 2.89, and 19.81 km², respectively. All indicators and subindicators were calculated and scored based on the grid of BEUs.

PSI Indicators Derived from Remote Sensing Data

Adapting field-based indicators to remote sensing has been done through a combination of direct conversion and substitution (Table 12.4). *Bank slope* was calculated by DEM. *Vegetation percent cover* (VPC) and *human disturbance* (RD) were calculated by Equations (12.1) and (12.2) using the updated 1:100,000 land-use data. The width of riparian zone is affected synthetically by bank sediment size, flow erosion, and deposition (Bren, 1993). Previous studies utilized *water-level width* and *riparian zone area* as subindicators to assess *bank stability* (Barbour et al., 1999; Fryirs, 2003; Dixon et al., 2006). In addition, these subindicators could be accurately calculated by remote sensing data. In this paper, *the area of riparian zone* was calculated based on the 1:500,000 geomorphic map and 1:100,000 topographic map. The *water-level width* of transection was extracted using OLI images. The two subindicators were normalized for converting the value to 0–100 (Equation (12.3)). *Natural wetland conservation* (NWC) was calculated using the 1976, 2013, and 2015 1:100,000 land-use data (Equation (12.4)). *Vegetation canopy coverage* (BVC) was calculated by Equation (12.5) (Gutman & Ignatov, 1998).

TABLE 12.4   The Evaluating Indicators and Weights of PSI Based on Remote Sensing Method

Indicators		Subindicators
Riparian stability (RST) (0.5)	Bank stability (BKS) (0.25)	Bank slope (BS) (0.25)
		Water-level width (WL) (0.25)
		Vegetation percent cover (VPC) (0.25)
		Riparian zone area (RA) (0.25)
	Vegetation canopy coverage (BVC) (0.5)	Percent canopy cover
	Human disturbance (RD) (0.25)	Buildup and farmland percent cover
River connectivity (RC) (0.25)		Human-built instream structures, including bridges, dams, culverts
Natural wetland conservation (NWC) (0.25)		Wetland area in the evaluating year (A_C) and historical year (1976) (A_R)

Source: Fu et al., 2017.

$$\text{VPC} = 100\% \times \frac{\sum_{i=1}^{6} \text{Area}_i}{\text{Area}_{\text{BEU}}} i = 1,2,\ldots,6 \tag{12.1}$$

where Area_{BEU} is the area of a BEU; Area_i is the area of each vegetation type in the BEU; and i is the vegetation types, including arboreal forest land, open forest land, and shrubland, high-, mid-, and low-cover grassland.

$$\text{RD} = 100\% \times \frac{\sum_{i=1}^{5} \text{Area}_i}{\text{Area}_{\text{BEU}}} i = 1,2,\ldots,5 \tag{12.2}$$

where Area_i is the area of each kind of human activity in the BEU, and i denotes the types of human activities, including farmland, urban, buildup land, rural residential land, and pasture.

$$\text{Range Standardization}(X) = 100 \times \frac{X - X_{\min}}{X_{\max} - X_{\min}} \tag{12.3}$$

$$\text{NWC} = 100\% \times \left(\sum_{i=1}^{Ns} \text{Area}_C \middle/ \sum_{i=1}^{Ns} \text{Area}_R \right) \tag{12.4}$$

where Area is the area of natural wetland with close hydraulic connection with the river, C is the year of evaluating PSI, R is the historical year, and Ns is the number of wetlands.

$$\text{BVC} = 100 \times \left[\left(\text{NDVI} - \text{NDVI}_{\text{soil}} \right) \middle/ \left(\text{NDVI}_{\text{veg}} - \text{NDVI}_{\text{soil}} \right) \right] \tag{12.5}$$

where normalized difference vegetation index (NDVI) was calculated using the September OLI images; $\text{NDVI}_{\text{soil}}$ and NDVI_{veg} represent the NDVI in the vegetated and nonvegetated areas, respectively. In this study, $\text{NDVI}_{\text{soil}}$ was the NDVI of 5% cumulative frequency, and NDVI_{veg} was the NDVI of 95% cumulative frequency in the study area (Jiang et al., 2006; Kallel et al., 2007).

Calculated PSI of the Riparian Zone

Subindicators and indicators of PSI based on remote sensing method were scored by the modification of the score sheets of the Chinese Ministry of Water Resources technical protocol. Each indicator was given a score between 0 and 100, with higher number implying better conditions. The indicators were calculated with associated subindicators that were scored multiplying by their corresponding weights. The indicator of bank stability was calculated by Equation (12.6). The PSI of riparian zone was calculated by Equation (12.7).

$$\text{BKS} = \left(\text{BS} + \text{WL} + \text{VPC} + \text{RA} \right) / 4 \tag{12.6}$$

$$\text{PSI} = \text{RST} \times 1/2 + \text{RC} \times 1/4 + \text{NWC} \times 1/4 \text{ or } \text{PSI} = \text{RST} \times 2/3 + \text{RC} \times 1/3 \tag{12.7}$$

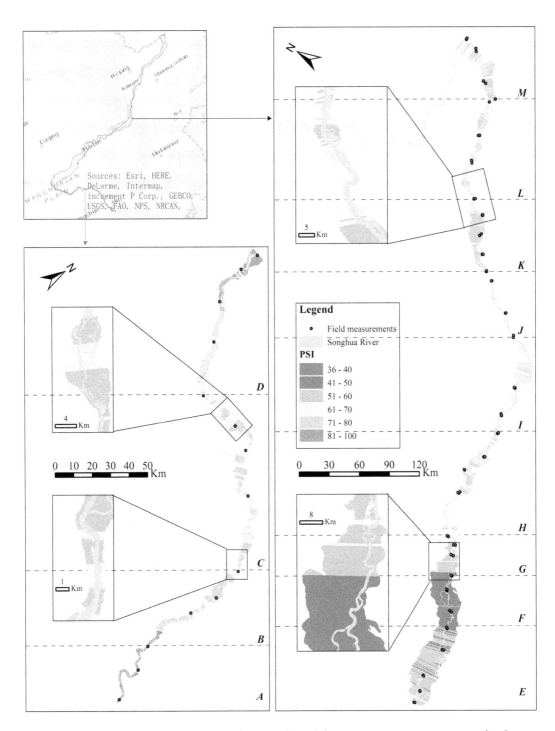

FIGURE 12.2 Spatial distribution of PSI values was derived from 2,081 BEUs. A–M represent the thirteen measurement sections.
Source: Fu et al., 2017.

TABLE 12.5   The Mean, Standard Deviation of Riparian PSI, and Number of BEUs in Each Measurement Section

Measurement Sections	Mean	Standard Deviation	Number of BEUs
A	41.29	1.78	83
B	56.74	5.20	108
C	60.85	4.65	187
D	74.27	10.62	141
E	52.11	2.68	212
F	44.33	1.70	139
G	59.41	2.80	101
H	63.49	5.93	338
I	66.31	2.51	254
J	65.95	2.69	114
K	57.67	2.77	101
L	62.99	4.42	189
M	60.76	3.06	116

Source: Fu et al., 2017.

Results

Riparian PSI

The BEUs-based PSI results of the entire riparian zone are illustrated in Figure 12.2. The mean and standard deviation of PSI, and the number of BEUs for each measurement section are summarized in Table 12.5. The riparian condition had a significant difference among 13 measurement sections due to the range of PSI values. The dominated land-cover type of riparian zones in sections A and F is the buildup category, and the riparian zones lost its ecological functions, resulting in PSI values below 45, which were lower than those of other measurement sections. In addition, the two sections with the smallest standard deviation (i.e., less than 2) indicated that the riparian zones were completely developed by human-induced buildup structures. Section D has been less disturbed by human activities and resulted in a better riparian condition, with the highest PSI value of over 70. However, the section also possessed the highest standard deviation up to 10.62, indicating a wide difference within the section. The PSI value in sections I and J was over 65 and the standard deviation was less than 3, indicating that the riparian zones have been in a homogeneous and stable condition. Section B possessed the PSI value of 56.74, but a high standard deviation (5.20), indicating a weak stability of the riparian zone.

Accuracy Validation

The comparison between PSI value derived from field measurements and that derived from remote sensing observation was made both in measurement sites and in measurement sections to validate the evaluating results (Figure 12.3). The trend of the riparian PSI value was consistent in both measurement sites and measurement sections, and the riparian PSI values were between 38 and 85. Specifically, there were a little differences in some measurement sites and sections. The PSI differences between field measurements and remote sensing observations for all sites were mainly distributed between 0 and 10, except for eleventh measurement site, which produced the highest difference (15.84). Measurement sites with a difference less than 5 account for over 44% of all measurement sites. The PSI differences between field measurements and remote sensing observations were less than 8 for all measurement sections. The measurement sections with a difference less than 3 account for over 46.2%. The tenth measurement section produced the highest difference of 8.52.

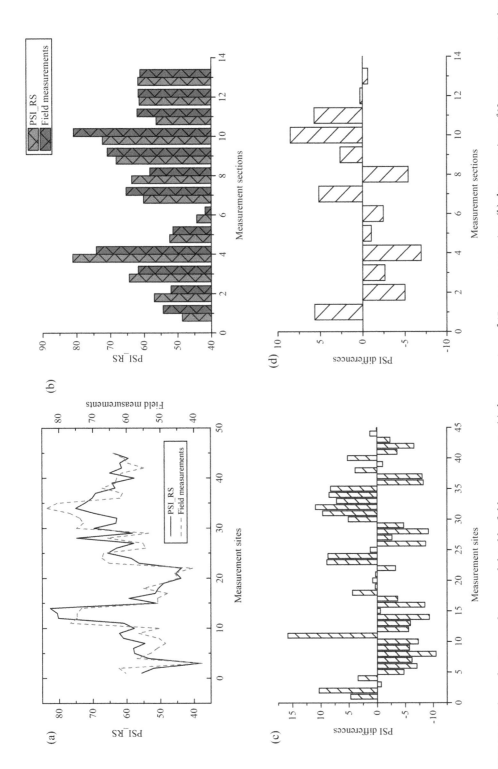

FIGURE 12.3 The evaluation results were validated by field measurements: (a) the comparison of 45 measurement sites; (b) the comparison of 13 measurement sections; (c) and (d) the differences in PSI values derived from measurement sites and measurement sections.

Source: Fu et al., 2017.

Conclusions

A BEUs method of remote sensing assessment was used to calculate the riparian PSI of Songhua River, Northeast China. The study confirmed that the approach was accurate and remote sensing method was practical to evaluate the riparian PSI. The NS model coefficient quantified the similarity of PSI values between the two approaches, which demonstrates that remote sensing-derived PSI values are promising to facilitate a repeated monitoring PSI of a riparian zone. Evaluation results derived from 13 measurement sections indicated that over 60% of riparian zone was disturbed by human activities with the PSI values below 65. Land-use patterns of riparian zone had an important effect on the riparian condition. The dominated buildup and farmland area had a poor riparian condition, and the dominated wetland and forest area had a good riparian condition. The buildup and farmland had been the main human disturbance to the riparian condition.

References

Barbour, M. T., Gerritsen, J., Snyder, B. D. & Stribling, J. B. *Rapid Bioassessment Protocols for Use in Streams and Wadeable Rivers: Periphyton, Benthic Macroinvertebrates and Fish*, Second Edition. EPA 841-B-99-002. U.S. Environmental Protection Agency; Office of Water, Washington, DC, Vol. 5, 1999, pp. 10–30.

Barquin, J., Fernández, D., Alvarez-Cabria, M. & Penas, F. Riparian quality and habitat heterogeneity assessment in Cantabrian Rivers. *Limnetica* 2011, 30, 329–346.

Bennett, S. J. & Simon, A. Riparian vegetation and fluvial geomorphology. *American Geophysical Union* 2004, 8, 1–10.

Bren, L. J. Riparian zone, stream, and floodplain issues: A review. *Journal of Hydrology* 1993, 150, 277–299.

Chen, Q., Liu, J., Ho, K. C. & Yang, Z. Development of a relative risk model for evaluating ecological risk of water environment in the Haihe River Basin estuary area. *Science of the Total Environment* 2012, 420, 79–89.

Congalton, R. G., Birch, K., Jones, R. & Schriever, J. Evaluating remotely sensed techniques for mapping riparian vegetation. *Computers and Electronics in Agriculture* 2002, 37, 113–126.

Dixon, I., Douglas, M., Dowe, J. & Burrows, D. *Tropical rapid Appraisal of Riparian Condition Version 1 (for use in tropical savannas)*. River Management Technical Guideline No. 7. Land & Water Australia, Canberra, 2006, pp. 1–31.

Fernández, D., Barquín, J., Álvarez-Cabria, M. & Peñas, F. J. Land-use coverage as an indicator of riparian quality. *Ecological Indicators* 2014, 41, 165–174.

Fryirs, K. Guiding principles for assessing geomorphic river condition: Application of a framework in the Bega catchment, South Coast, New South Wales, Australia. *Catena* 2003, 53, 17–52.

Fu, B., Li, Y., Wang, Y., Campbell, A., Zhang, B., Yin, S. & Jin, X. Evaluation of riparian condition of Songhua river by integration of remote sensing and field measurements. *Scientific Reports* 2017, 7(1), 2565.

Ghermandi, A., Vandenberghe, V., Benedetti, L., Bauwens, W. & Vanrolleghem, P. A. Model-based assessment of shading effect by riparian vegetation on river water quality. *Ecological Engineering* 2009, 35(1), 92–104.

González del Tánago, M. & García de Jalón, D. Riparian quality index (RQI): A methodology for characterising and assessing the environmental conditions of riparian zones. *Limnetica* 2011, 30, 235–251.

Gutman, G. & Ignatov, A. The derivation of the green vegetation fraction from NOAA/AVHRR data for use in numerical weather prediction models. *International Journal of Remote Sensing* 1998, 19, 1533–1543.

Hruby, T. Developing rapid methods for analyzing upland riparian functions and values. *Environmental Management* 2009, 43, 1219–1243.

Innis, S. A., Naiman, R. J. & Elliott, S. R. Indicators and assessment methods for measuring the ecological integrity of semi-aquatic terrestrial environments. *Hydrobiologia* 2000, 422, 111–131.

Ivits, E., Cherlet, M., Mehl, W. & Sommer, S. Estimating the ecological status and change of riparian zones in Andalusia assessed by multi-temporal AVHHR datasets. *Ecological Indicators* 2009, 9, 422–431.

Jansen, A., Robertson, A., Thompson, L., Wilson, A. *Rapid Appraisal of Riparian Condition, Version 2.* Land &Water Australia, Canberra, 2005.

Jiang, Z., Huete, A. R., Chen, J., Chen, Y., Li, J., Yan, G. & Zhang, X. Analysis of NDVI and scaled difference vegetation index retrievals of vegetation fraction. *Remote Sensing of Environment* 2006, 101(3), 366–378.

Johansen, K., Phinn, S., Dixon, I., Douglas, M. & Lowry, J. Comparison of image and rapid field assessments of riparian zone condition in Australian tropical savannas. *Forest Ecology and Management* 2007, 240(1–3), 42–60.

Johansen, K. & Phinn, S. Mapping structural parameters and species composition of riparian vegetation using IKONOS and Landsat ETM+ data in Australian tropical savannahs. *Photogrammetric Engineering & Remote Sensing* 2006, 72, 71–80.

Johansen, K., Arroyo, L. A., Armston, J., Phinn, S., & Witte, C. Mapping riparian condition indicators in a sub-tropical savanna environment from discrete return LiDAR data using object-based image analysis. *Ecological Indicators* 2010a, 10(4), 796–807.

Johansen, K., Phinn, S. & Witte, C. Mapping of riparian zone attributes using discrete return LiDAR, QuickBird and SPOT-5 imagery: Assessing accuracy and costs. *Remote Sensing of Environment* 2010b, 114, 2679–2691.

Kallel, A., Le Hégarat-Mascle, S., Ottlé, C. & Hubert-Moy, L. Determination of vegetation cover fraction by inversion of a four-parameter model based on isoline parametrization. *Remote Sensing of Environment* 2007, 111, 553–566.

Munné, A., Prat, N., Solà, C., Bonada, N. & Rieradevall, M. J. A. C. M. A simple field method for assessing the ecological quality of riparian habitat in rivers and streams: QBR index. *Aquatic Conservation: Marine and Freshwater Ecosystems* 2003, 13(2), 147–163..

Ministry of Water Resources of the People's Republic of China. River (Lake) Health Indicators, Standards and Methods V1.0. The Ministry of Water Resources of the People's Republic of China, Beijing, China. 2010, pp. 27–42 (In Chinese).

Ranalli, A. J. & Macalady, D. L. The importance of the riparian zone and in-stream processes in nitrate attenuation in undisturbed and agricultural watersheds–A review of the scientific literature. *Journal of Hydrology* 2010, 389, 406–415.

Tang, Q., Bao, Y., He, X., Zhou, H., Cao, Z., Gao, P. & Zhang, X. Sedimentation and associated trace metal enrichment in the riparian zone of the three gorges reservoir, China. *Science of the Total Environment* 2014, 479, 258–266.

USDI Bureau of Land Management. Riparian Area Management: A User Guide to Assessing Proper Functioning Condition and the Supporting Science for Lotic Areas. Technical Reference TR 1737-15. 1998, pp, 4–7.

13

Riparian Zones Evaluation: Remote Sensing

Bolin Fu
*Guilin University
of Technology*

Yeqiao Wang
University of Rhode Island

Ying Li
*Northeast Institute of
Geography and Agroecology*

Introduction

Riparian zones are the transition zones between aquatic and terrestrial systems along inland watercourses. Riparian zones have been identified as vital components of the landscape because of their provision of ecosystem functions and benefits, for example, erosion control and bank stabilization, water purification, biodiversity maintenance, wetland products, as well as recreation (Hruby, 2009; Barquın et al., 2011). Nevertheless, threats to riparian zones are compounded by increased anthropogenic development and disturbances, such as urban expansion, agriculture, mining, overgrazing, erosion, and point- and nonpoint-source pollutions (Ranalli & Macalady, 2010; Fernández et al., 2014; González et al., 2017). Maintaining and improving the condition of riparian zones is likely to be identified as a high priority in most of the natural resource management plans. Consequently, there is a growing need for practical techniques for assessing and monitoring the condition of riparian environments.

Numerous field-based methods have been developed for evaluation of riparian zone condition in specific areas, such as Riparian Quality Index (RQI) (Del & De, 2011), Tropical Rapid Appraisal of Riparian Condition (TRARC) method (Dixon et al., 2006), Rapid Appraisal of Riparian Condition (RARC) (Jansen et al., 2005), visual assessment methods (Ward et al., 2003), Biological Quality of the Riparian zones (QBR) index method (González Del Tánago & García de Jalón, 2006), measurement of river health using gross primary production (Bunn et al., 1999), multivariate statistics, and predictive model method (Reynoldson et al., 1997; Chen et al., 2012). Most of those methods are based on expert knowledge in selection of measuring sections and monitoring sites, then evaluating the riparian condition through field measurements. However, those methods could be very laborious or even unpractical, in particular vast regions and remote locations. Remote sensing techniques have been utilized to map indicators of riparian condition (Johansen et al., 2010), and compare the measurement of riparian condition between the TRARC method and remote sensing observation (Johansen et al., 2007). Fu et al. (2017) present a multimetric and basic evaluation unit (BEU)-based method of remote sensing observations to assess the riparian zone of Songhua River, Northeast China, using the physical structural integrality (PSI) indicator. The evaluation of riparian condition derived from remote sensing observations was consistent with field measurement in both measurement sites and measurement sections.

This study evaluated the riparian condition of the Songhua River, Northeast China, using BEU-based method of remote sensing observations. The specific objectives of this paper are to (i) compare evaluation results of riparian condition between remote sensing observations and field measurements using the independent-samples *t*-test and hydrological modeling coefficient of Nashe and Sutcliffe (NS); (ii) discriminate the variation and clusters of riparian conditions to identify the vulnerability and stability of the riparian zones; and (iii) explore the change of landscape of the riparian zone using the land-use data from 1976 to 2013 to understand the effects of human impacts on riparian conditions.

Materials and Methods

Study Area

Songhua River is the largest tributary of the Heilongjiang (Amur River) in Northeast China. With the length of 1,897 kilometers long and the drainage area of 545,600 square kilometers, Songhua is ranked the fifth longest river in China. This study focused on the 1,679-kilometer riparian zone of Songhua River that started from the Fengman hydrologic dam passing through major cities such as Jilin, Harbin, and Jiamusi, and ended at the Amur River, the border river between China and Russia (Figure 13.1). Songhua River Basin is a major agricultural center and commodity grain production base of China. Large-scale operations of state farms were established after the establishment of the People's Republic of China in 1949 and land reclamation projects begun. Songhua River is an important transportation artery for agricultural products of the region. The river flows through major port cities and serves as the source of drinking water for millions of people. The ecological functions and integrality of riparian zones of the Songhua River have experienced much degradation under the influence of anthropogenic activities such as overgrazing, construction of transportation infrastructure, urban sprawl, sand mining, tourism development, reclaimed wetland, and other human activities.

Remote Sensing Images and Field Data

This study adopted the land-use dataset at 1:100,000 scale (Figure 13.2) for 1976–2015, including information of forest, grassland, farmland, wetland, water body, barren land, and buildup (Table 13.1). Other datasets adopted, including the 1:500,000 geomorphic map and the 1:100,000 topographic map developed by Shuttle Radar Topographic Mission (SRTM), generated Digital Elevation Model (DEM) data at 30 m spatial resolution.

The studied riparian zone was divided into 13 measurement sections. Within each section, three or four measurement sites were chosen, and each site consisted of three transections. The entire riparian zone consisted of 45 measurement sites (Figure 13.1) and 135 transections. Each transection laid two 30 m width × 50 m length measurement areas on both sides of banks. The subindicators of field measurement include bank slope and slope length of riverbank, canopy cover, bank sediment size (clay, sand, gravel, cobbles, or bedrock), bank erosion, and anthropogenic activities. Anthropogenic activities, bank sediment size, bank erosion, and river connectivity were recorded by field photos. All field measurements were executed during the normal river flow period, in September of 2013 and 2015, respectively.

Calculated BEU-Based PSI of the Riparian Zone

The BEUs are defined as the basic evaluation units with various sizes in the measurement sections of the riparian zone (Figure 13.1). The 600 m length was adopted as a splitting criterion to divide the river centerline, and this process generated 2,081 BEUs in the study area. All indicators and subindicators were calculated and scored based on the grid of BEUs.

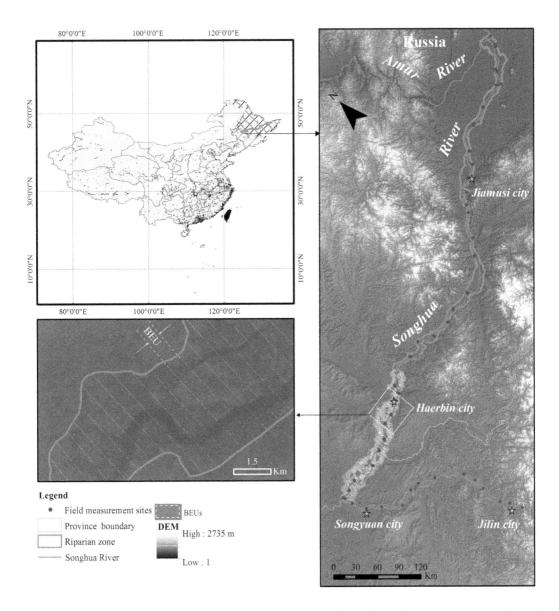

FIGURE 13.1 The study area and field measurement sites in Northeast China. The riparian zone was divided into 13 measurement sections, and each measurement section was partitioned into BEUs.
Source: Fu et al., 2017.

Subindicators and indicators (Table 13.2) of PSI based on remote sensing method were calculated following the paper of Fu et al. (2017). The indicators were calculated with associated subindicators that were scored multiplying by their corresponding weights. The indicator of bank stability was calculated by Equation (13.1). The PSI of the riparian zone was calculated by Equation (13.2):

$$BKS = (BS + WL + VPC + RA)/4 \qquad (13.1)$$

$$PSI = RST \times 1/2 + RC \times 1/4 + NWC \times 1/4 \text{ or } PSI = RST \times 2/3 + RC \times 1/3 \qquad (13.2)$$

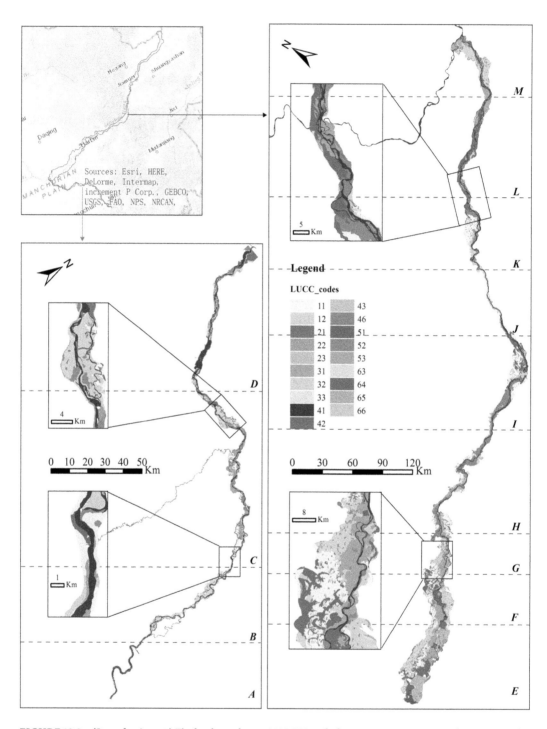

FIGURE 13.2   (See color insert.) The land-use data at 1:100,000 scale for riparian zone in 2012. The LUCC_codes represent different land-use types, that is, 11–12: paddy field, glebe field; 21–23: arboreal forest, sparse woodland, and shrub woodland; 31–33: high-, mid-, and low-cover grassland; 41–43: rivers, lakes, and reservoirs; 46: beach; 51–53: residential, rural, and transportation; 63: saline-alkali soil land; 64: wetland; 65–66: barren land.
Source: Fu et al., 2017.

TABLE 13.1   The Categories of 1:100,000 Land-Use Data in Riparian Zone

Categories	Subcategories
Forest	Arboreal forest, sparse woodland, and shrub woodland
Grassland	High-, mid-, and low-cover grassland
Farmland	Paddy field, glebe field
Wetland	Marsh, riverine wetland
Water body	Rivers, reservoirs, fishery, and lakes
Barren land	Lands unused or difficult for using, saline-alkaline land
Buildup	Industrial and commercial, residential, transportation

Source: Fu et al., 2017.

TABLE 13.2   The Evaluating Indicators and Weights of PSI Based on Remote Sensing Method

Indicators		Subindicators
Riparian stability (RST) (0.5)	Bank stability (BKS) (0.25)	Bank slope (BS) (0.25)
		Water-level width (WL) (0.25)
		Vegetation percent cover (VPC) (0.25)
		Riparian zone area (RA) (0.25)
	Vegetation canopy coverage (BVC) (0.5)	Percent canopy cover
	Human disturbance (RD) (0.25)	Buildup and farmland percent cover
River connectivity (RC) (0.25)		Human-built instream structures, including bridges, dams, culverts
Natural wetland conservation (NWC) (0.25)		Wetland area in the evaluating year (A_C) and historical year (1976) (A_R)

Source: Fu et al., 2017.

Statistical Validation of Evaluation Results

To validate the evaluation results, an independent-samples t-test and F-test were performed to examine the statistically significant difference in PSI scores between field evaluation and remote sensing observation. A test of normality was first performed using the Kolmogorov-Smirnov statistic and Shapiro-Wilk statistic to examine the possibility that the PSI scores derived from 45 measurement sites conform to a normal distribution.

Under the circumstance of no significant difference, the study sought to understand the similarity between the remote sensing- and the field-derived PSI, which led to the use of coefficients for understanding model outputs. We used the common hydrological modeling coefficient of NS to evaluate the similarity of riparian PSI between the field and remote sensing observation. NS is a method for quantitatively analyzing the modeled outputs as they were compared to the observed values (Nash & Sutcliffe, 1970; Lebecherel et al., 2016). The method was appropriate as we compared the observed field data and the modeled remote sensing approach.

Spatial Statistic and Analysis of Riparian PSI

The Getis-Ord General Gi and the Anselin local Moran's I statistics were calculated to analyze the spatial distribution of riparian PSI. The General Gi indicator was used to identify concentrations of high or low PSI values, while Moran's I indicator was applied for calculating spatial autocorrelation of PSI between each BEU and its neighboring BEUs, and identifying spatial outliers of PSI value (García-Palomares et al., 2015; Swetnam et al., 2015; Peerbhay et al., 2016). A search radius or threshold distance is essential for the two local statistics. In this study, the "Incremental Spatial Autocorrelation" Toolbox in ArcGIS was used to determine the optimal distance radius for spatial statistics analysis. This tool measures spatial autocorrelation for a series of distances and optionally creates a line graph of those distances and their corresponding z-score. The statistically significant peak z-scores indicate distance where spatial processes promoting clustering are most pronounced. The distance corresponding to the first peak z-score was adopted for spatial statistics analysis by the default recommendation. The Getis-Ord General G statistic was calculated using the fixed Euclidean distance method with an optimal threshold distance by the "Hot-Spot Analysis" Toolbox in ArcGIS. The Anselin local Moran's I was calculated using the inverse distance method with the "Cluster and Outlier Analysis" Toolbox. The existence of a dispersed or a clustered distribution of riparian PSI is based on a z-score (the standard deviation about the *mean*) where the p-value of statistical significance is typically set at 0.05 or 0.01 (a 95% or 99% confidence interval (CI)). Positive z-scores indicated clustering of large PSI values, and negative z-scores indicated clustering of small PSI values (e.g., a z-score of 2.58 indicated at the 99% CI BEUs with large neighbors, and a z-score of −2.58 indicated BEUs with small neighbors).

Results

Statistic Validation and Analysis of Riparian PSI

The comparison between PSI values derived from field measurements and those derived from remote sensing observation was made in both measurement sites and measurement sections to validate the evaluating results. The normality test indicated that the PSI scores derived from 45 measurement sites represented a normal distribution (Shapiro-Wilk statistic, $p = 0.51$). The field- and remote sensing-derived measurements were compared with Student's t-test and found that there was no significant difference between the two measurements ($p = 0.16$) (Table 13.3).

The Anselin local Moran's I and Getis-Ord General Gi statistics were calculated to discriminate structural differences and look for evidence of clustering throughout the riparian zone (Figures 13.3 and 13.4). The Anselin local Moran's I statistic found that the high-high clusters, that is, high PSI values with high neighbors, appeared in sections C, D, F, I, and L. However, the PSI value in the high cluster of section F was below 50, which was significantly less than the other sections. Sections A and B with dominated urban and farmland areas had the lowest PSI values and did not present strong high-high clusters. The low-low clusters, that is, low PSI values with low neighbors, presented in all measurement sections. A few cases presented outliers, in the form of low-high or high-low clusters for each section. Within section D, the riparian PSI value changed from low-low clusters to high-high clusters. For the Getis-Ord Gi statistic, negative z-scores ($z < -1.65$, $p < 0.1$) represent statistically significant clusters of low-low PSI values and become cold spots at the 90%–95% confidence interval. Positive z-scores ($z > 1.65$, $p < 0.1$) indicate statistically significant clusters of high-high PSI values and become hot spots at the 90%–95% confidence interval. Clusters were nonsignificant ($-1.65 < z < 1.65$) in neighborhoods with a wide variation in PSI values, indicating those riparian conditions were randomly distributed.

TABLE 13.3   The Independent-Samples t-Test for Field- and Remote Sensing-Derived Measurements

	Levene's Test for Equality of Variances		t-test for Equality of Means						95% Confidence Interval of the Difference	
	F	Sig.	t	df	Sig. (2-Tailed)	Mean Difference	Std. Error Difference		Lower	Upper
Equal variances assumed	1.65	0.20	0.16	88	0.87	0.36	2.20		−4.02	4.74
Equal variances not assumed			0.16	86.51	0.87	0.36	2.20		−4.02	4.74

Source: Fu et al., 2017.

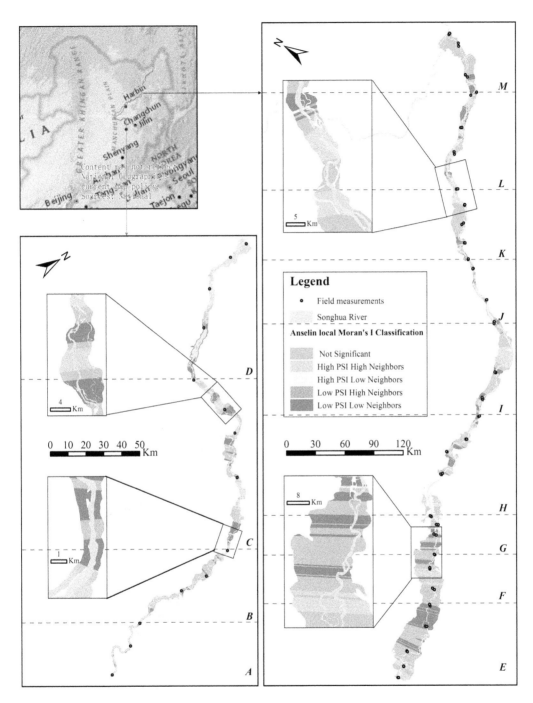

FIGURE 13.3 Anselin local Moran's *I* statistics for the riparian PSI values was calculated and mapped. A–M were the thirteen measurement sections. "High PSI High Neighbors" is that the riparian zone with a high PSI value is surrounded by the neighbors with high PSI values; "High PSI Low neighbors" is that the riparian zone with a high PSI value is surrounded by the neighbors with low PSI values; "Low PSI Low Neighbors" is that the riparian zone with a low PSI value is surrounded by the neighbors with low PSI values. A–M were the thirteen measurement sections.

Source: Fu et al., 2017.

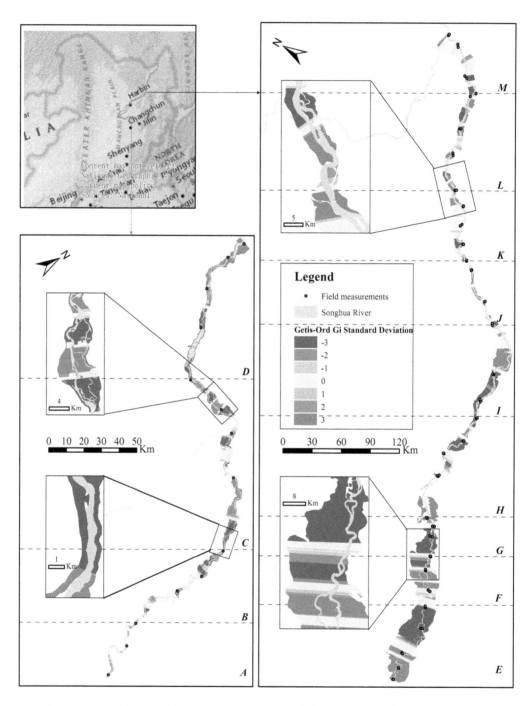

FIGURE 13.4 Getis-Ord *Gi* spatial statistics for the riparian PSI value was calculated and mapped. Negative *z*-scores ($z < -1.65$, $p < 0.1$) represent statistically significant clusters of small PSI values and become cold spots at the 90%–95% confidence interval. Positive *z*-scores ($z > 1.65$, $p < 0.1$) indicate statistically significant clusters of high PSI values and become hot spots at the 90%–95% confidence interval. Clusters were nonsignificant ($-1.65 < z < 1.65$) in neighborhoods with a wide variation in PSI values, indicating those riparian conditions were randomly distributed. A–M were the thirteen measurement sections.

Source: Fu et al., 2017.

Land-Use Patterns in Riparian Zone

According to the evaluation results of the riparian PSI and its spatial statistics, the land-use pattern was analyzed and found that the riparian landscape in the study area experienced a significant change from 1976 to 2013 (Tables 13.4 and 13.5). The areas of buildup and farmland categories accounted for the total section had increased in every measurement section from 1976 to 2013. In addition, the average annual change rate of buildup and farmland in each section had a positive growth, except for section *K*, in which the average annual change rate was −1.4% per year. The combination of riparian PSI values in each measurement section found that for sections A and F with low PSI values, the buildup and farmland

TABLE 13.4 The Main Land-Use Types Comprise Percent of the Total Area of the Section in the Riparian Zone from 1976 to 2013. "*a*": Buildup and Farmland; "*b*": Forest and Grassland; "*c*": Wetland

	1976			1986			1995			2000			2005			2013		
	Percentage (%)			Percentage (%)			Percentage (%)			Percentage (%)			Percentage (%)			Percentage (%)		
	a	*b*	*c*	*a*	*b*	*c*	*A*	*b*	*c*	*a*	*b*	*c*	*a*	*b*	*c*	*a*	*b*	*c*
A	83.7	11.4	5.0	80.5	14.9	4.6	86.5	7.8	5.7	87.3	6.8	5.9	86.9	8.2	4.9	89.4	9.4	1.2
B	86.0	10.4	3.6	81.1	16.8	2.1	84.4	7.0	8.6	87.4	8.6	4.1	81.7	12.4	5.9	92.9	3.6	3.5
C	33.8	31.8	33.5	44.9	44.8	5.4	29.7	37.7	31.3	34.3	46.7	17.7	43.2	35.5	20.6	58.5	29.4	11.4
D	25.7	21.9	51.6	14.4	53.2	27.9	14.8	37.1	48.0	15.7	38.4	45.6	25.0	32.8	41.9	35.8	33.6	30.3
E	45.8	10.4	42.5	44.4	8.5	44.7	56.9	6.2	3.3	46.9	16.2	34.5	47.8	20.0	29.8	57.5	16.8	23.2
F	88.0	1.8	10.0	88.1	3.1	8.5	90.0	1.9	7.9	89.3	1.7	8.8	89.1	1.8	8.9	90.7	1.7	7.3
G	51.7	22.0	26.2	48.1	2.7	49.2	53.6	40.3	6.1	50.9	39.5	9.6	51.0	30.5	18.3	55.8	29.3	14.8
H	33.1	24.4	42.5	39.3	15.7	45.0	37.9	26.1	36.0	36.7	22.9	40.4	40.8	19.0	40.2	42.5	18.3	39.2
I	43.7	6.9	49.4	43.5	5.7	50.8	44.2	5.5	50.3	49.3	5.6	45.1	49.8	5.6	44.5	51.7	4.7	43.8
J	6.6	28.3	65.0	5.2	25.6	69.2	8.8	62.3	28.9	8.0	37.5	54.5	25.0	34.3	40.7	43.4	33.3	23.3
K	72.8	2.6	24.6	72.0	5.9	22.0	75.1	18.0	6.8	65.5	1.8	32.7	66.8	1.9	31.3	69.1	1.9	28.7
L	47.2	7.9	43.3	56.5	8.5	35.0	60.6	6.8	32.2	50.9	20.2	27.4	50.9	20.9	26.7	52.6	20.1	25.7
M	26.7	56.7	16.6	37.5	3.7	58.8	44.3	3.5	52.3	60.7	22.7	16.6	60.9	22.1	17.0	62.1	21.8	16.1

Source: Fu et al., 2017.

TABLE 13.5 The Average Annual Change Rate of Buildup and Farmland in Each Section

	Average Annual Change Rate of Buildup and Farmland (‰)					
Sections	1976–1986	1986–1995	1995–2000	2000–2005	2005–2013	1976–2013
A	1.1	3.0	3.4	−5.8	5.4	1.9
B	−5.1	4.1	7.7	−14.6	15.8	2.2
C	24.8	−37.3	32.0	41.9	40.6	14.9
D	−61.5	4.7	16.5	88.8	20.9	2.3
E	−5.2	24.1	−32.3	6.4	20.1	5.8
F	−2.9	1.9	4.6	0.8	2.0	0.9
G	−8.0	12.1	−11.5	2.0	10.3	2.0
H	15.9	−2.8	−4.7	22.9	4.1	6.9
I	−3.8	1.7	27.6	3.4	3.6	4.4
J	−34.5	50.0	10.1	46.2	16.3	14.8
K	−4.2	3.9	−20.7	5.6	3.2	−1.4
L	18.3	7.5	−35.5	1.5	3.2	2.9
M	31.7	18.2	73.1	1.3	1.7	23.6

Source: Fu et al., 2017.

TABLE 13.6   The Main Land-Use Types Comprise Percent of the Total Area of the BEUs in the Hot Spots and Cold Spots. "*a*": Buildup and Farmland; "*b*": Forest and Grassland; "*c*": Wetland.

Sections	Hot Spots			Cold Spots		
	a	*b*	*c*	*a*	*b*	*c*
A	73.20	16.52	4.75	93.17	2.72	4.11
B	84.59	14.14	1.27	95.69	1.81	2.50
C	27.09	47.15	24.69	56.51	27.78	15.02
D	23.09	26.93	49.98	19.32	50.83	23.84
E	47.45	2.52	47.68	48.34	33.39	13.03
F	89.35	1.51	7.11	93.85	2.67	3.15
G	69.82	7.25	21.89	59.33	30.85	7.17
H	15.43	21.53	62.74	45.26	7.69	46.73
I	61.61	3.83	27.43	36.14	3.12	60.38
J	10.19	22.80	65.85	9.66	48.19	40.79
K	50.08	1.48	47.34	71.88	1.71	23.72
L	46.20	17.56	30.19	50.68	21.08	24.82
M	38.87	36.38	15.18	54.48	19.53	16.39

Source: Fu et al., 2017.

categories comprised over 80% of the total area from 1976 to 2013. However, in section *D* with the highest PSI value, forest, grassland, and wetland accounted for over 60% of the total area, especially over 80% from 1976 to 1986. For sections H, I, J, and L with the moderate PSI values, forest, grassland, and wetland comprised over 45% of the total area. The average annual change rate of buildup and farmland of sections C, J, and M was over 14%, more than other sections, which indicated that these sections were seriously disturbed for 37 years. Sections A and F with the lowest average annual change rate of buildup and farmland indicated that those riparian zones were changed into the urban and farmland area.

In addition, quantification of land-use pattern of the BEUs in the high-high (hot spots) and low-low (cold spots) clusters found that buildup and farmland areas in sections A, B, and F both accounted for over 70% of the total area in the hot spots and cold spots, while wetland, forest, and grassland of section *D* comprised over 70% of the total area (Table 13.6). In BEUs of the hot spots, wetland of sections E, H, and J accounted for over 47% of the total area, and forest and grassland of section *C* also accounted for over 47%. Buildup and farmland of sections *G*, *I*, and *K* comprised over 50% of the total area. In BEUs of the cold spots, buildup and farmland accounted for over 50% of the total area in all sections, except for sections *D*, *E*, *H*, *I*, and *J*. Wetland, forest, and grassland of sections *H*, *I*, and *J* accounted for over 50% of the total area.

Conclusions

The multimetric method derived a single indicator of riparian PSI from the combination of a number of subindicators to evaluate riparian conditions. The consistency of BEU-based evaluation results with field measurements was validated by the statistical *t*-test and the NS model coefficient. No significant difference between two evaluation results and the NS value of 0.63 demonstrated that remote sensing-derived riparian PSI and evaluation of riparian conditions were efficient, reliable, and comparable. Anselin local Moran's *I* and Getis-Ord *Gi* statistics were able to identify the variation and clusters of riparian PSI. The low-low clusters of riparian PSI discriminated the vulnerable area of the riparian zone. The combination of spatial cluster analysis and BEU-based evaluation results was able to make a correct judgment for riparian conditions. Land-use types and patterns in a riparian zone have a great influence on riparian conditions, and it was essential to find a balance between riparian development and fluvial ecosystem preservation. The two land-cover types increased over the past 37 years. The area of buildup and farmland had increased by 787.9 km^2 from 3476.0 km^2 in 1976 to 4263.9 km^2 in 2013.

References

Barquın, J., Fernández, D., Alvarez-Cabria, M., & Penas, F. Riparian quality and habitat heterogeneity assessment in Cantabrian Rivers. *Limnetica*, 2011, 30, 329–346.

Bunn, S. E., Davies, P. M., & Mosisch, T. D. Ecosystem measures of river health and their response to riparian and catchment degradation. *Freshwater Biology*, 1999, 41(2), 333–345.

Chen, Q., Liu, J., Ho, K. C., & Yang, Z. Development of a relative risk model for evaluating ecological risk of water environment in the Haihe River Basin estuary area. *Science of the Total Environment*, 2012, 420, 79–89.

Del Tánago, M. G., & De Jalón, D. G. Riparian quality index (RQI): A methodology for characterising and assessing the environmental conditions of riparian zones. *Limnetica*, 2011, 30(2), 235–254.

Dixon, I., Douglas, M., Dowe, J. & Burrows, D. *Tropical Rapid Appraisal of Riparian Condition Version 1 (For Use in Tropical Savannas)*, River Management Technical Guideline No. 7, Land & Water Australia, Canberra. 2006, pp. 1–31.

Fernández, D., Barquín, J., Álvarez-Cabria, M., & Peñas, F. J. Land-use coverage as an indicator of riparian quality. *Ecological Indicators*, 2014, 41, 165–174.

Fu, B., Li, Y., Wang, Y., Campbell, A., Zhang, B., Yin, S., & Jin, X. Evaluation of riparian condition of Songhua river by integration of remote sensing and field measurements. *Scientific Reports*, 2017, 7(1), 2565.

García-Palomares, J. C., Gutiérrez, J., & Mínguez, C. Identification of tourist hot spots based on social networks: A comparative analysis of European metropolises using photo-sharing services and GIS. *Applied Geography*, 2015, 63, 408–417.

González, E., Felipe-Lucia, M. R., Bourgeois, B., Boz, B., Nilsson, C., Palmer, G., & Sher, A. A. Integrative conservation of riparian zones. *Biological Conservation*, 2017, 211, 20–29.

González Del Tánago, M., & García de Jalón, D. Attributes for assessing the environmental quality of riparian zones. *Limnetica*, 2006, 25(1–2), 389–402.

Hruby, T. Developing rapid methods for analyzing upland riparian functions and values. *Environmental Management*, 2009, 43, 1219–1243.

Jansen, A., Robertson, A., Thompson, L., Wilson, A. *Rapid Appraisal of Riparian Condition, Version 2.* Land &Water Australia, Canberra, 2005.

Johansen, K., Phinn, S., Dixon, I., Douglas, M., & Lowry, J. Comparison of image and rapid field assessments of riparian zone condition in Australian tropical savannas. *Forest Ecology and Management*, 2007, 240(1–3), 42–60.

Johansen, K., Phinn, S., & Witte, C. Mapping of riparian zone attributes using discrete return LiDAR, QuickBird and SPOT-5 imagery: Assessing accuracy and costs. *Remote Sensing of Environment*, 2010, 114(11), 2679–2691.

Lebecherel, L., Andréassian, V., & Perrin, C. On evaluating the robustness of spatial-proximity-based regionalization methods. *Journal of Hydrology*, 2016, 539, 196–203.

Nash, J. E., & Sutcliffe, J. V. River flow forecasting through conceptual models part I-A discussion of principles. *Journal of Hydrology*, 1970, 10, 282–290.

Peerbhay, K., Mutanga, O., Lottering, R., & Ismail, R. Mapping Solanum mauritianum plant invasions using WorldView-2 imagery and unsupervised random forests. *Remote Sensing of Environment*, 2016, 182, 39–48.

Ranalli, A. J., & Macalady, D. L. The importance of the riparian zone and in-stream processes in nitrate attenuation in undisturbed and agricultural watersheds–A review of the scientific literature. *Journal of Hydrology*, 2010, 389, 406–415.

Reynoldson, T. B., Norris, R. H., Resh, V. H., Day, K. E., & Rosenberg, D. M. The reference condition: A comparison of multimetric and multivariate approaches to assess water-quality impairment using benthic macroinvertebrates. *Journal of the North American Benthological Society*, 1997, 16(4), 833–852.

Swetnam, T. L., Lynch, A. M., Falk, D. A., Yool, S. R., & Guertin, D. P. Discriminating disturbance from natural variation with LiDAR in semi-arid forests in the southwestern USA. *Ecosphere*, 2015, 6, 1–22.

Ward, T. A., Tate, K. W., & Atwill, E. R. *Visual assessment of riparian health. Rangeland Monitoring Series*. UC ANR, Davis, CA, 2003.

II

Wetland Ecosystem

14

Cooling Effects of Urban and Peri-Urban Wetlands: Remote Sensing

Zhenshan Xue and
Zhongsheng Zhang
*Northeast Institute of
Geography and Agroecology,
Chinese Academy of Science*

Caifeng Cheng
*Shandong University of
Science and Technology*

Tingting Zhang
Jilin Normal University

Introduction

Over the past three decades, the extent of cities, particularly in developing countries, grew dramatically across the globe. In 2016, over 54% of the world's population lived in urban areas (United Nations, 2015). By 2030, global urban areas are projected to increase by 1.2 million km². Nearly half of the increase is forecasted to occur in Asia, with China and India absorbing over half of the regional total.[1,2] Also, China's urban population is projected to nearly double from 2000 to 2030.[3] Rapid urban expansion could directly alter climate and hydrosystems locally to regionally.[4] A well-known climatic phenomenon caused by urbanization is the urban heat island (UHI) effect: Replacing natural landscapes with man-made surfaces causes excess heat storage, which is slowly released at night.[5–7] UHI has both direct and indirect impacts on energy consumption, environmental quality, and human health.[8]

Compared with traditional air temperature measurement techniques, remote sensing and GIS techniques are increasingly used in recent studies of the UHI effect.[9,10] By using land surface temperatures (LST) data detected by thermal infrared sensors (TIRSs), such as NOAA/AVHRR, TERRA/MODIS, and Landsat, the distribution, dynamicity, and heterogeneity of UHI effect could be explicitly revealed.[11–13] Unlike Landsat TM/ETM sensors with only one thermal band, Landsat-8 TIRS contains two thermal bands at 100-m resolution.[14,15] This makes it is possible to apply split-window (SW) algorithms and get more accurate LST data.[16]

Urban and peri-urban wetlands are those that are located in or adjacent to the boundaries of a city and provide a variety of benefits and services.[17,18] Remarkably, urban and peri-urban wetlands can provide a significant cooling effect (CE) to surrounding areas to weaken increasing UHI effects.[11,19] However, most UHI studies generally focus on the CE of parks and forest cover. To fully explain the effect of urban and peri-urban wetlands on the temperature of cities, more research on the design, distribution, and type of urban wetlands is necessary. Here, we quantitatively assessed the CE of wetlands in Changchun and Jilin, two major cities of Northeast China, that have experienced rapid development in recent decades. By using LST data derived from Landsat-8 TIRS and land-cover data, we determined the boundary threshold for CE and calculated the normalized cooling capability index (NCCI) and normalized cooling efficiency index (NCEI) of each wetland. Specifically, our aims were (i) to measure the efficiency and capability of urban wetlands' CE and (ii) to quantify the relationship between CE and wetland characteristics, such as size, hydrologic connectivity, landscape, and height of surrounding buildings. The results from this study have important implications for many other inland cities.

Remote Sensing Monitoring of UHI and the Cooling Effects of Urban and Peri-Urban Wetlands

Study Area and Remote Sensing Image Processing

Changchun City (43° 52.8′ N, 125° 21′ E) is the capital and largest city of the Jilin Province of China and has a total area of approximately 7,557 km². Jilin City (43° 52′ N, 126° 33′ E) covers an area of 3,636 km², second to Changchun City in Jilin Province. As seen in Figure 14.1, these two cities are at approximately the same latitude in the central part of the Northeast Plain of China and are about 120 km apart. The average elevations of Changchun City and Jilin City are 236.8 and 183.4 m above sea level, respectively. Both cities are under the influence of semi-humid and continental monsoon climates, characterized by hot and humid summers and cold winters. The builtup area (BUA) of Changchun City has increased from 90 km² in 1980 to 506.3 km² in 2015, an increase of 462.6%. The BUA of Jilin City also increased within this period from 72 to 185 km², an increase of around 156.9%. Historical levels of urbanization and hydrological connectivity of wetlands in Changchun City were digitized from historical city maps published in 1958. The current urban land-cover data and digital

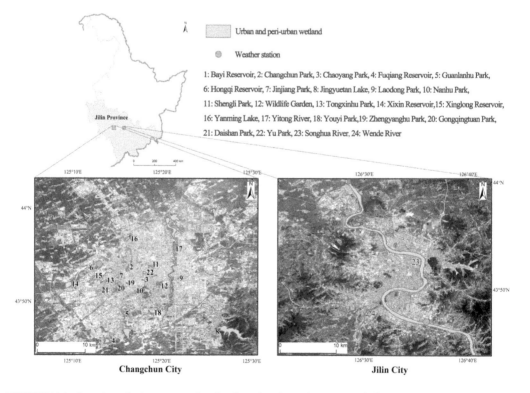

FIGURE 14.1 Location of urban area and wetlands in the Changchun City and Jilin City.

surface model (DSM) of the two cities used in this study were extracted from urban digital datasets provided by Data Center of the Provincial Department of Land and Resources. Different levels of urbanization and nearly identical climatic conditions make the two cities a suitable study area for a contrastive study of urban wetlands' CE.

Before LST conversion, the two Landsat-8 OLI/TIRS images were radiometrically calibrated and geometrically corrected with an accuracy of less than 0.5 pixels. Preprocessing of all images was performed in ENVI 5.1 software. We employed the visual interpretation method to extract wetland information from Landsat-8 OLI images. In this study, 21 wetlands (3 rivers, 5 reservoirs, and 11 lakes), 19 in Changchun City and 2 in Jilin City, with a total area of 36.6 km², were selected for investigation (Figure 14.1). Three urban green spaces without wetlands in Changchun City (Gongqingtuan Park, Daishan Park, and Yu Park) were selected for comparative analysis.

Extent and Intensity of the UHI Effect

As seen in Figure 14.2a, meteorological data from 1981 to 2011 of the two weather stations collected in this study illustrates the rising trend of LST in Changchun City and Jilin City. In the past 30 years, the annual mean ground temperature of Changchun City has risen 4°C, from 5.9°C to 9.9°C, and that of Jilin City has risen 4.6°C, from 5.7°C to 10.3°C. After 2001, the trend of warming in both cities has significantly accelerated. Under urban sprawl and climate warming, the UHI effect in Changchun City and Jilin City has become serious in recent years. Figure 14.2b illustrates that LST is highly correlated with the building volume with R^2 of 0.733.

Average hourly temperature on July 4, 2016, recorded by weather stations shows that the LST in Changchun reached a peak at 13:00 with the maximum LST 56.1°C, while the LST in Jilin reached a peak at 12:00 with the maximum LST 40.4°C (Figure 14.3). Peak values of UHI are mainly located in these open spaces surrounded by tall buildings with high building density. Spatial distribution of LST over Changchun City and Jilin City is shown in Figure 14.4. For Changchun City (Figure 14.4a), the LST ranged between 20.7°C and 52.8°C. For Jilin City (Figure 14.4b), the LST ranged between 16.4°C and 47.2°C. The lowest temperature was recorded from the surface of water bodies in wetlands, while the highest LST was recorded from BUA. An average temperature of 32°C in Changchun City and 29.3°C in Jilin City was recorded in the vegetated areas. The mean LST of 21 urban wetlands was 26.6°C, lower than that of three urban green spaces (31.6°C). For wetland type, the mean LST of rivers (23.3°C) was lower than that of reservoirs (27.3°C) and lakes (29.3°C).

The NCCI and NCEI of Urban Wetlands' CE

Figure 14.5 shows the edge points of urban wetlands and green spaces. The distance between edge points to wetland or green space shows the extent of CE. For the 24 wetlands and green spaces in this study, the extent of CE ranges from 1,000 m of Jingyuetan Lake to 150 m of Hongqi Reservoir and Xixin Reservoir, with an average distance of 371.1 m. The mean distance of lakes (434 m) was the furthest, followed by that of rivers (400 m), green spaces (300 m), and reservoirs (280 m). The mean cooling ability of urban wetlands was 2.74°C, ranging from 6.36°C in Songhua River to 1.27°C in Jinjiang Park. The mean cooling ability of three green spaces was 2.54°C.

Figure 14.6 represents NCCI (normalized volume values of the 24 wetlands and green spaces) in Changchun City and Jilin City. The NCCI of Songhua River is the highest (1.0), followed by Jingyuetan Lake (0.24) and Yitong River (0.18), and Gongqingtuan Park is the lowest (0). The NCCI of Songhua River is nearly 1,128 times higher than that of Gongqingtuan Park. The mean NCCI of rivers is highest (0.4), followed by lakes (0.027), reservoirs (0.019), and green spaces (0.0018). Figure 14.7 shows NCEI of all 24 wetlands and green spaces. The top three most efficient wetlands are Hongqi Reservoir (1.0), Xixin Reservoir (0.99), and Tongxinhu Park (0.82). The mean NCEI of reservoirs is 0.65, higher than that of green spaces (0.58), rivers (0.44), and lakes (0.31).

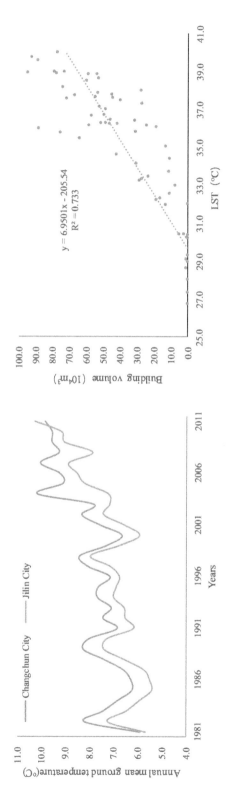

FIGURE 14.2 Annual mean ground temperature of Changchun City and Jilin City during the period from 1981 to 2011.

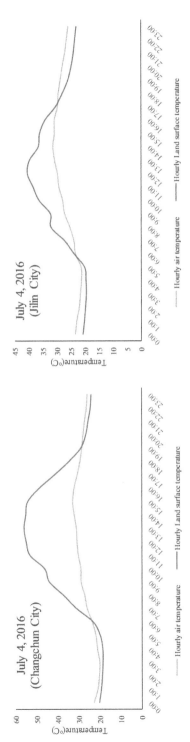

FIGURE 14.3 Average hourly temperature on July 4, 2016, in Changchun and Jilin.

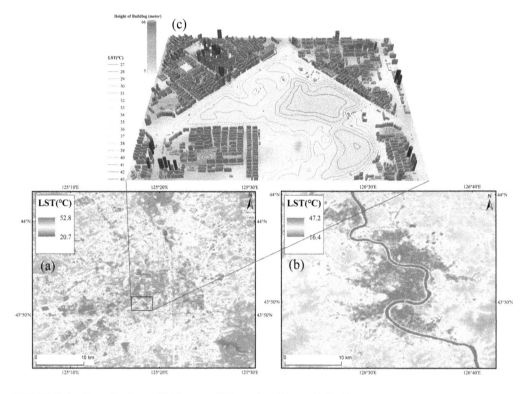

FIGURE 14.4 **(See color insert.)** LST maps of Changchun City and Jilin City.

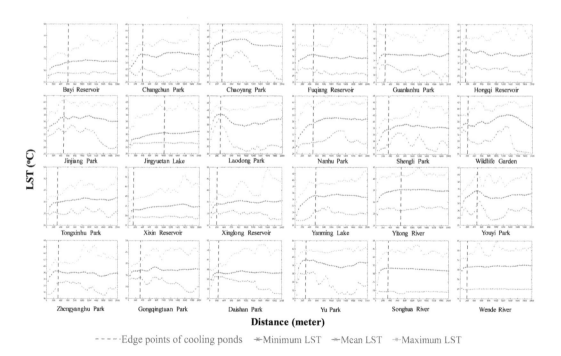

FIGURE 14.5 Min, mean, and max LST in each buffer outside the 24 wetlands and green spaces and edge points.

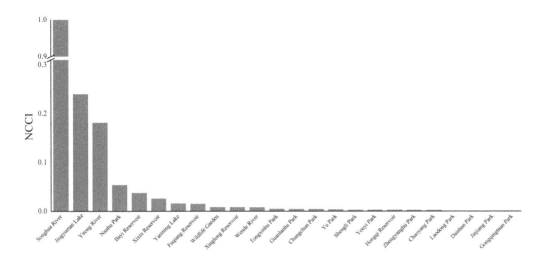

FIGURE 14.6   NCCI of 24 wetlands and green spaces.

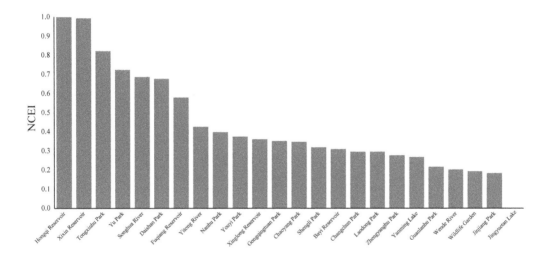

FIGURE 14.7   NCEI of 24 wetlands and green spaces.

Correlation Analysis of the NCCI and NCEI with the Wetland Characteristics

Table 14.1 shows that larger wetlands are cooler than smaller ones, while the CE of urban wetlands is more related to the characteristics of urban wetlands. For example, the area of Bayi Reservoir (143 ha) is larger than that of Nanhu Park (85.9 ha), but the NCCI of Bayi Reservoir (0.038) is lower than that of Nanhu Park (0.055). Still, the Spearman rho is 0.947 ($p < 0.01$) and 0.457 ($p < 0.05$) between NCCI and area of water and vegetation, respectively. There is a positive relationship between NCCI and Shape index, and the Spearman rho is 0.533 ($p < 0.05$). Strong negative correlation has been observed between NCCI and building characteristics, −0.416 ($p < 0.05$) with an average height of the building and −0.625 ($p < 0.01$) with a density of the building. Wetland connectivity is positively correlated with NCCI and NCEI of wetlands, and the Spearman Rho is 0.536 ($p < 0.01$) and 0.393($p < 0.05$), respectively.

The general conviction that air temperature in urban wetlands can be cooler than non-wetlands has been confirmed by many studies.[11,20,21] Intensive evaporation from wetlands could increase air

TABLE 14.1   Spearman's Rho Correlations between Indexes of CE and Wetland Characteristics

Index	Area of Water	Edge Temperature	Lowest Temperature	Wetland Connectivity	Shape Index	Area of Vegetation	Average Height of Building	Area of Building	Density of Building
NCCI	.947[b]	−0.182	−0.158	.536[b]	.553[b]	.457[a]	−416[a]	0.168	−625[b]
NCEI	−0.081	0.044	−0.105	.393[a]	−0.239	0.067	−0.215	−0.129	−0.038

[a] Correlation is significant at the 0.05 level (2-tailed).
[b] Correlation is significant at the 0.01 level (2-tailed).

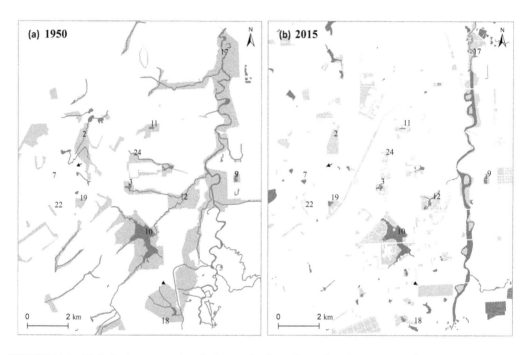

FIGURE 14.8   Hydrologic connectivity of urban wetlands in Changchun City in 1950 and 2015.

humidity and then reduce the temperature of the surrounding areas.[22,23] The results from this study indicate that temperatures close to wetlands are getting reduced about by 2.74°C in comparison with surrounding areas; the effect of urban wetlands on the temperature of cities was determined by urban wetland area, shape, connectivity, and type. The size of wetlands is the most important element for explaining reducing air temperature. In addition, more complex shape is preferable for CE of wetlands. Furthermore, the results of this study also indicated that the hydrologic connectivity of wetlands is another important factor for CE. Hydrologic connectivity of wetlands determines in part their water renewal rate and hydrologic functions.[24,25] The average NCCI of wetlands connected to other surface waters is 0.11, nearly six times higher than that of isolated wetlands without surface inlets or outlets. The NCCI of Yitong River is 0.18, only one-sixth that of Songhua River, for it has been in no-flow condition for years due to upstream reservoir construction. Figure 14.8 shows the hydrologic connectivity of wetlands of Changchun City in 1950 and 2015.

Conclusions

In this study, 21 urban wetlands and 3 green spaces in Changchun City and Jilin City were selected to measure the capability of mitigating UHI by using Landsat-8 TIRS data. The results show that CE of urban wetlands was significantly correlated with wetland area, shape, and hydrologic connectivity.

In contrast, the characteristics of surrounding buildings such as volume, height, and density could limit the CE of urban wetlands. Our findings have important implications for wetland design and urban planning. Reconnection of urban wetlands to rivers could be a cost-efficient way to mitigate UHI effects. High volume, height, and density buildings surrounding urban wetlands should be avoided in urban planning.

References

1. Seto KC, Güneralp B, Hutyra LR. Global forecasts of urban expansion to 2030 and direct impacts on biodiversity and carbon pools. *Proceedings of the National Academy of Sciences of the United States of America*. 2012, 109(40):16083–16088.
2. Angel S, Parent J, Civco DL, Blei A, Potere D. The dimensions of global urban expansion: Estimates and projections for all countries, 2000–2050. *Progress in Planning*. 2011, 75(2):53–107.
3. Cao GY, Chen G, Pang LH, Zheng XY, Nilsson S. Urban growth in China: Past, prospect, and its impacts. *Population and Environment*. 2012, 33(2/3):137–160.
4. Grimm NB, Faeth SH, Golubiewski NE, Redman CL, Wu J, Bai X, Briggs JM. Global change and the ecology of cities. *Science*. 2008, 319(5864):756–760.
5. Arnfield AJ. Two decades of urban climate research: A review of turbulence, exchanges of energy and water, and the urban heat island. *International Journal of Climatology: A Journal of the Royal Meteorological Society*. 2003, 23(1):1–26.
6. Kim HH. Urban heat island. *International Journal of Remote Sensing*. 1992, 13(12):2319–2336.
7. Coutts AM, Tapper NJ, Beringer J, Loughnan M, Demuzere M. Watering our cities: The capacity for water sensitive urban design to support urban cooling and improve human thermal comfort in the Australian context. *Progress in Physical Geography*. 2013, 37(1):2–28.
8. O'Malley C, Piroozfarb PAE, Farr ERP, Gates J, O'Malley C, Farr ERP, Gates J. An investigation into minimizing urban heat island (UHI) effects: A UK perspective. *Energy Procedia*. 2014, 62(1):72–80.
9. Rasul A, Balzter H, Smith C, Remedios J, Adamu B, Sobrino JA, Srivanit M, Weng Q. A review on remote sensing of urban heat and cool islands. *Land*. 2017, 6(2):38–48.
10. Yusuf YA, Pradhan B, Idrees MO. Spatio-temporal assessment of urban heat island effects in Kuala Lumpur metropolitan city using landsat images. *Journal of the Indian Society of Remote Sensing*. 2014, 42(4):829–837.
11. Sun R, Chen A, Chen L, Lü Y. Cooling effects of wetlands in an urban region: The case of Beijing. *Ecological Indicators*. 2012, 20(9):57–64.
12. Clinton N, Peng G. MODIS detected surface urban heat islands and sinks: Global locations and controls. *Remote Sensing of Environment*. 2013, 134(5):294–304.
13. Li YY, Zhang H, Kainz W. Monitoring patterns of urban heat islands of the fast-growing Shanghai metropolis, China: Using time-series of Landsat TM/ETM+ data. *International Journal of Applied Earth Observation & Geoinformation*. 2012, 19(10):127–138.
14. Rozenstein O, Qin Z, Derimian Y, Karnieli A. Derivation of land surface temperature for landsat-8 TIRS using a split window algorithm. *Sensors-Basel*. 2014, 14(4):5768–5781.
15. Barsi J, Schott J, Hook S, Raqueno N, Markham B, Radocinski R. Landsat-8 thermal infrared sensor (TIRS) vicarious radiometric calibration. *Remote Sensing*. 2014, 6(11):11607–11626.
16. Yu XL, Guo XL, Wu ZC. Land surface temperature retrieval from landsat 8 TIRS-comparison between radiative transfer equation-based method, split window algorithm and single channel method. *Remote Sensing*. 2014, 6(10):9829–9852.
17. Wang X, Ning L, Yu J, Xiao R, Li T. Changes of urban wetland landscape pattern and impacts of urbanization on wetland in Wuhan city. *Chinese Geographical Science*. 2008, 18(1):47–53.
18. Lannas KSM, Turpie JK. Valuing the provisioning services of wetlands: Contrasting a rural wetland in Lesotho with a peri-urban wetland in South Africa. *Ecology and Society*. 2009, 14(2):544–544.

19. Manteghi G, Limit HB, Remaz D. Water bodies an urban microclimate: A review. *Modern Applied Science*. 2015, 9(6):1–12.
20. Li C, Yu CW. Mitigation of urban heat development by cool island effect of green space and water body. *Lecture Notes in Electrical Engineering*. 2014, 261(1):551–561.
21. Wu H, Ye LP, Shi WZ, Clarke KC. Assessing the effects of land use spatial structure on urban heat islands using HJ-1B remote sensing imagery in Wuhan, China. *International Journal of Applied Earth Observation and Geoinformation*. 2014, 32(1):67–78.
22. Gunawardena KR, Wells MJ, Kershaw T. Utilising green and bluespace to mitigate urban heat island intensity. *Science of the Total Environment*. 2017, 584(1):1040–1055.
23. Zhao L, Lee X, Smith RB, Oleson K. Strong contributions of local background climate to urban heat islands. *Nature*. 2014, 511(7508):216–219.
24. Golden HE, Lane CR, Amatya DM, Bandilla KW, Kiperwas HR, Knightes CD, Ssegane H. Hydrologic connectivity between geographically isolated wetlands and surface water systems: A review of select modeling methods. *Environmental Modelling and Software*. 2014, 53(3):190–206.
25. Ameli AA, Creed IF. Quantifying hydrologic connectivity of wetlands to surface water systems. *Hydrology & Earth System Sciences*. 2017, 21(3):1791–1808.

15

Ecosystem Service Decline in Response to Wetland Loss

Fengqin Yan
Institute of Geographic Sciences and Natural Resources Research, Chinese Academy of Sciences

Shuwen Zhang
Northeast Institute of Geography and Agroecology, Chinese Academy of Sciences

Introduction

Ecosystem services (ES) represent the direct or indirect contributions of ecosystems to human lives and are necessary for human health, livelihood, and survival. ES that support human livelihoods include climate regulation, clean water provision, and soil fertility maintenance. Assessing ES plays an important role in linking human activities and natural systems. ES have been estimated at multiple spatial scales since the publication of the Millennium Ecosystem Assessment [1,2] all over the world. The Millennium Ecosystem Assessment classified ES into four classes: provisioning function, regulating function, cultural function, and supporting function [3]. While some studies have estimated the loss of regulating functions, such as carbon sequestration, in response to wetland conversion [4], knowledge of ES changes in other functions in response to wetland conversion is currently lacking.

Wetlands, "the kidneys of the landscape," play a vital role in providing ES to humans [4,5]. Studies indicate that wetlands contain around 20%–30% of global carbon storage with about 5%–8% of the world's land surface [6,7]. Despite these great benefits, wetland loss has occurred globally at an alarming rate due to agricultural reclamation, urban expansion, aquaculture, housing developments, golf course development, and climate warming [8]. Wetland loss can reduce biodiversity and water availability, and can increase greenhouse gases (GHGs) emissions and soil erosion [4,9], leading to the diminishing provision of ES. Therefore, it is necessary to estimate ES loss in response to wetland loss quantitatively at local and larger scales.

Land-Use Change Data as Data Sources

The land-use data for different years in this study were obtained from previous studies [10,11]. The land-use map for 1954 was mainly based on topographic maps, and the map for 1976 was generated from Landsat MSS (Multispectral Scanner) data. The land-use data for 1986/2000 and 2015 were reconstructed based on Landsat TM (Thematic Mapper) and Landsat OLI (Operational Land Imager) images, respectively [11,12].

The land-use map for 1954 was reconstructed using the Cellular Automata model based on topographic maps and other auxiliary data such as soil data, vegetation maps, and statistical data. Land-use maps for other years were reconstructed via visual interpretation of remote sensing images. Extensive field surveys were taken to check the accuracy of the land-use data. Additionally, historical data including aerial photos, statistical data, field site data, and unmanned aerial vehicles images were applied to check the accuracy of the land-use data. The mean interpretation accuracies for land-use classification and land-use change detection were both larger than 90% [12].

Ecosystem Service Value Estimation

The ecosystem service value (ESV) (Equation 15.1) in our study was calculated as follows [13]:

$$ESV = \sum A_i \times VC_i \qquad (15.1)$$

where A_i and VC_i stand for the area and ESV per unit area of ith ecosystem, respectively. Zhang et al. amended the coefficients of ESV per unit area by crop yield data in the Sanjiang Plain and calculated the ESV of the Sanjiang Plain in 2010 [14]. Previous studies usually overlooked the difference of ESV per unit area between paddy and dry farmland [1,2]. We applied different coefficients to estimate the ESVs of paddy and dry farmland with reference to study by Zhang et al. [14] (Table 15.1). The correction coefficients of ESV per unit area were calculated by Equation 15.2:

$$\lambda = Q/Q_0 \qquad (15.2)$$

where λ is the correction coefficient of the Sanjiang Plain; Q and Q_0 are the yield per unit of the study area and the yield per unit of China, respectively (Table 15.1).

Spatiotemporal Changes in Wetlands

Figure 15.1 illustrates wetland distributions for the past six decades. Wetland area decreased drastically by 2.8 million ha, approximately 73.3% of wetland area in 1954. Wetland area decreased significantly during 1954–1976 and 1976–1986. The decline in wetland area has slowed down significantly since 1986. Wetland area decreased by 1.27 million ha, approximately 57.5 thousand ha/year during 1954–1976. During 1976–1986, wetland area decreased by 1.1 million ha. Wetland area decreased by 33.5% between 1954 and 1976. The loss rate was largest during the 1976–1986, reaching 44.3%. Since 1986, the loss rate gradually decreased as a result of the publication of wetland conversion policies.

TABLE 15.1 Ecosystem Service Values Per Unit Area of Different Land-Use Types (Dollars (USD)/ha, 2010)

Ecosystem Services	Paddy	Dry Farmland	Forest	Grassland	Water	Wetland
Gas regulation	513.9	310.2	883.1	219.9	879.6	513.9
Climate regulation	263.9	166.7	2,642.3	655.1	1,666.6	263.9
Hydrological regulation	1,259.2	125.0	1,729.1	25,316.8	11,217.4	1,259.2
Soil conservation	4.6	476.8	1,075.2	430.6	1,069.4	4.6
Biodiversity maintenance	97.22	60.2	979.2	592.6	3,643.5	97.2
Agricultural products	629.62	393.5	116.9	370.4	236.1	629.6
Raw materials	41.7	185.2	268.5	106.5	231.5	41.7
Freshwater supply	−1,217.6	9.7	138.9	24,507.8	1,199.1	−1,217.6
Relaxation	41.7	27.8	427.7	458.3	2,189.8	41.7

Source: Fengqin Yan and Shuwen Zhang, 2019.

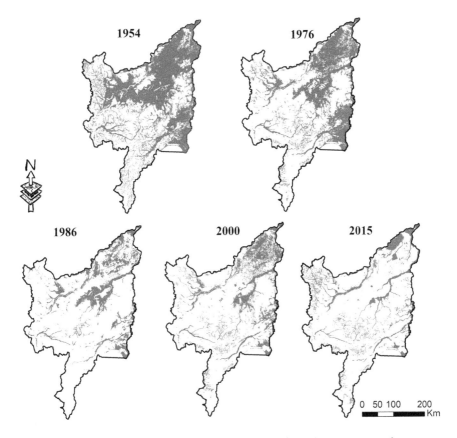

FIGURE 15.1 The distribution of wetlands throughout the Sanjiang Plain between 1954 and 2015.
Source: Fengqin Yan and Shuwen Zhang, 2019.

Wetlands mostly transformed to dry farmland, paddy fields, and grassland. From 1954 to 1976, wetland loss is mostly due to reclamation for dry farmland, drainage for grassland, and reclamation for paddy fields. Wetlands mostly transformed to dry farmland, grassland, and forest during 1976–1986. During 1986–2000, wetlands typically transformed to dry farmland, following by grassland and dry farmland. Between 2000 and 2015, wetlands mostly converted to paddy field, followed by dry farmland and grassland. The area growth of wetlands is primarily due to grassland, dry farmland, and forest conversions.

Ecosystem Service Value Changes

During the past six decades, wetland loss resulted in $57.5 billion (USD) in ESVs (Table 15.2). The ESVs showed a declining trend between 1954 and 2015, especially after 1986. ESVs decreased by $28.0 billion (USD) and $20.3 billion (USD) during 1954–1976 and 1976–1986, respectively. Then, ESV loss dropped to $5.9 billion (USD) and $3.3 billion (USD) during 1986–2000 and 2000–2015, respectively. The ratio of ESV changes decreased from 48.7% in the period 1954–1976 to 5.8% in the period 2000–2015 (Table 15.2). ESVs of agricultural product function increased by $458.6 million (USD) during the past six decades, while ESVs of hydrological regulation, biodiversity maintenance, and relaxation functions showed decreases of $30.5 billion (USD), $9.7 billion (USD), and $5.9 billion (USD), respectively (Table 15.3). The smallest decline was in the raw material function. Since dry farmland provides ESV of raw material, relatively large-scale wetland reclamation for dry farmland guaranteed smaller ESV fluctuations.

TABLE 15.2   ESV Changes Caused by Wetland Loss in Different Periods Million (Billion Dollars (USD))

Time Periods	1954–1976	1976–1986	1986–2000	2000–2015	Total
ESV changes	−28.0	−20.3	−5.9	−3.3	−57.5
Ratio (%)	48.7	35.3	10.2	5.8	100

Source: Fengqin Yan and Shuwen Zhang, 2019.

TABLE 15.3   ESV Variations Caused by Wetland Loss from 1954 to 2015 (Million Dollars (USD))

Ecosystem Services	Paddy	Dry Farmland	Forest	Grassland	Water	Settlement	Other[a]
Gas regulation	−238.5	−973.0	−107.5	25.3	−24.3	0.0	−238.5
Climate regulation	−914.7	−2,563.1	−63.6	38.8	−46.0	0.0	−914.7
Hydrological regulation	−6,493.2	−18,954.1	−3,395.2	−540.9	−309.7	−0.1	−6,493.2
Soil conservation	−694.3	−1,012.6	−130.3	24.5	−29.5	0.0	−694.3
Biodiversity maintenance	−2,312.3	−6,122.9	−1,013.2	117.1	−100.6	0.0	−2,312.3
Agricultural products	256.6	269.0	−45.	−5.2	−6.5	0.0	256.6
Raw materials	−123.8	−79.1	−24.3	4.8	−6.4	0.0	−123.8
Freshwater supply	−1,575.8	−2,033.1	−372.2	−894.3	−33.1	0.0	−1,575.8
Relaxation	−1,400.7	−3,694.3	−642.0	66.4	−60.5	0.0	−1,400.7
Total	−13,496.5	−35,163.1	−5,793.3	−1,163.4	−616.6	−0.0	−13,496.5

Source: Fengqin Yan and Shuwen Zhang, 2019.
[a] Other unused land.

The largest changes in ESVs due to wetland loss were caused by the transformation of wetlands to dry farmland, paddy fields, and grassland. While the conversion of wetlands to dry farmland caused the largest reduction in ESV between 1954 and 2000, the conversion of wetlands to paddy fields caused the largest ESV reduction ($4.5 billion (USD)) during 2000–2015. The largest ESV variations caused by dry farmland conversions occurred during 2000–2015 because the area of dry farmland converted into paddy fields was largest during this period. Conversion between forest and paddy reduced the total ESV in all three periods and reduced the total ESV by $1.2 billion (USD). The conversion from wetlands to forest from 1954 to 1986 reduced the total ESV, while the conversion from forest to wetlands increased the total ESV. Changes between wetlands and grassland increased the total ESVs during 2000–2015 but reduced the total ESV in other periods. The ESV decline caused by the transformation between other ecosystems and wetlands (such as settlements) was relatively small from 1954 to 2015. Wetland loss increased the ESV of agricultural product functions by $458.6 million (USD) although it reduced the ESVs of all other functions during the study period.

Suggestions for Supporting Ecosystem Services

The ES provided by wetlands are not proportional to their area. Although wetlands occupy no larger than 9% of the Earth's land surface, they provide more renewable ES than their small area implies. This study revealed the dramatic decline of ES according to wetland loss over the past six decades in the Sanjiang Plain. Human disturbance and climate changes are important factors that influence wetland changes [6,15]. Our study agrees with previous work, highlighting agricultural expansion as a key reason for wetland loss in the Sanjiang Plain. Therefore, wetland restoration such as returning cultivated land to wetlands should be a priority in this area. Additionally, the succession of wetland restoration should be evaluated timely to improve the benefits of wetland restoration. Wetland damages are sometimes irreversible, and it is unclear to what level ES functions can maintain the restored wetland. For example, studies found that it took decades, even up to a century, for the carbon sequestration ability of restored wetlands to reach that of natural wetlands [16], because restored wetlands do not seem to accumulate carbon rapidly. Therefore, it is more important to protect existing wetlands to the greatest extent possible to maximize ES and prevent further destruction. To protect wetlands in the Sanjiang Plain, the following actions could be taken. First, more wetland nature reserves should be established to strengthen wetland protection. Second, the government could set up an environmental impact assessment system for wetland development projects to reduce the damage of projects on wetlands. Third, public awareness of protecting wetlands should be strengthened in the future by public education to encourage more people to participate in wetland protection activities. In addition to protecting current wetlands, actions should also be taken to restore wetlands. The land with the most potential for restoring wetlands includes croplands converted from wetlands in recent years and croplands with good hydrology and soil conditions. Some actions such as raising fish and crabs in wetlands and species-rich plantings can be taken in protected wetland areas as well as restored wetland areas to increase biodiversity function as well as ES. It should be noted that species-rich plantings should avoid the introduction of invasive species. Wetland restoration results should be assessed timely, and researchers should be included in the decision-making process.

Conclusion

We quantitatively assessed varied ES and the sectors they provide in response to wetland loss. In the last 60 years, wetlands in the Sanjiang Plain have undergone dramatic loss due to agricultural reclamation and drainage for grassland. Our study showed that wetland area has declined by 73.3% (about 2.8 million ha) since 1954, mainly due to the transformation of wetlands to farmland, especially dry farmland. Wetland loss reduced the ESV by $57.5 billion over the past six decades. More actions should be taken to protect wetlands in this area. Additionally, ecosystem service tradeoffs should be integrated into land-use decisions to carry out sustainable land planning in the future. By quantifying ESV variation in response to wetland loss, this study provides support for ecosystem-based land management and wetland protection.

References

1. Zapfack, L., Noiha, N.V. and Tabue, M.R.B. (2016). Economic Estimation of Carbon Storage and Sequestration as Ecosystem Services of Protected Areas: A Case Study of Lobeke National Park. *Journal of Tropical Forest Science* 28(4), 406–415.
2. Zhu, Z.Y., Zhao, Y.J., Rao, L., Qin, B.S., Li, B.B. and Chen, L.H. (2017). Estimation and Spatial Visualization of Ecosystem Services Value in Beijing, Tianjin and Hebei Provinces. *Fresenius Environmental Bulletin* 26(11), 6791–6803.
3. Ringler, C. (2008). The Millennium Ecosystem Assessment: Tradeoffs between Food Security and the Environment. *Turkish Journal of Agriculture and Forestry* 32(3), 147–157.
4. Decleer, K., Wouters, J., Jacobs, S., Staes, J., Spanhove, T., Meire, P. and van Diggelen, R. (2016). Mapping Wetland Loss and Restoration Potential in Flanders (Belgium): An Ecosystem Service Perspective. *Ecology and Society* 21(4), 46.
5. Pattison-Williams, J.K., Pomeroy, J.W., Badiou, P. and Gabor, S. (2018). Wetlands, Flood Control and Ecosystem Services in the Smith Creek Drainage Basin: A Case Study in Saskatchewan, Canada. *Ecological Economics* 147, 36–47.
6. Erwin, K.L. (2009). Wetlands and Global Climate Change: The Role of Wetland Restoration in a Changing World. *Wetlands Ecology and Management* 17(1), 71–84.
7. McNicol, G., Sturtevant, C.S., Knox, S.H., Dronova, I., Baldocchi, D.D. and Silver, W.L. (2017). Effects of Seasonality, Transport Pathway, and Spatial Structure on Greenhouse Gas Fluxes in a Restored Wetland. *Global Change Biology* 23(7), 2768–2782.
8. Fluet-Chouinard, E., Lehner, B., Rebelo, L.-M., Papa, F. and Hamilton, S.K. (2015). Development of a Global Inundation Map at High Spatial Resolution from Topographic Downscaling of Coarse-Scale Remote Sensing Data. *Remote Sensing of Environment* 158, 348–361.

9. Sun, X., Li, Y.F., Zhu, X.D., Cao, K. and Feng, L. (2017). Integrative Assessment and Management Implications on Ecosystem Services Loss of Coastal Wetlands Due to Reclamation. *Journal of Cleaner Production* 163, S101–S112.

10. Liu, J., Liu, M., Tian, H., Zhuang, D., Zhang, Z., Zhang, W., Tang, X. and Deng, X. (2005a). Spatial and Temporal Patterns of China's Cropland During 1990–2000: An Analysis Based on Landsat TM Data. *Remote Sensing of Environment* 98(4), 442–456.

11. Yan, F.Q., Zhang, S.W., Liu, X.T., Chen, D., Chen, J., Bu, K., Yang, J.C. and Chang, L.P. (2016). The Effects of Spatiotemporal Changes in Land Degradation on Ecosystem Services Values in Sanjiang Plain, China. *Remote Sensing* 8(11), 917.

12. Liu, J.Y., Liu, M.L., Tian, H.Q., Zhuang, D.F., Zhang, Z.X., Zhang, W., Tang, X.M. and Deng, X.Z. (2005b). Spatial and Temporal Patterns of China's Cropland During 1990–2000: An Analysis Based on Landsat TM Data. *Remote Sensing of Environment* 98(4), 442–456.

13. Wang, Z.M., Zhang, B., Zhang, S.Q., Li, X.Y., Liu, D.W., Song, K.S., Li, J.P., Li, F. and Duan, H.T. (2006) Changes of Land Use and of Ecosystem Service Values in Sanjiang Plain, Northeast China. *Environmental Monitoring and Assessment* 112(1–3), 69–91.

14. Zhang, L., Yu, X., Jiang, M., Xue, Z., Lu, X. and Zou, Y. (2017). A Consistent Ecosystem Services Valuation Method Based on Total Economic Value and Equivalent Value Factors: A Case Study in the Sanjiang Plain, Northeast China. *Ecological Complexity* 29, 40–48.

15. Sica, Y.V., Quintana, R.D., Radeloff, V.C. and Gavier-Pizarro, G.I. (2016) Wetland Loss Due to Land Use Change in the Lower Parana River Delta, Argentina. *Science of the Total Environment* 568, 967–978.

16. Knox, S.H., Dronova, I., Sturtevant, C., Oikawa, P.Y., Matthes, J.H., Verfaillie, J. and Baldocchi, D. (2017) Using Digital Camera and Landsat Imagery with Eddy Covariance Data to Model Gross Primary Production in Restored Wetlands. *Agricultural and Forest Meteorology* 237, 233–245.

FIGURE 4.1 Right side: A poor fen, dominated by sedges and Sphagnum in the foreground and treed bog in the background. Left side: Continental bog of western Canada characterized by a dense tree layer of black spruce (*Picea mariana*). (Courtesy of Kimberli Scott.)

FIGURE 4.4 *Vaccinium vitis-idaea* (lingonberry) with red berries, growing on a mat of the reindeer lichen, *Cladonia arbuscula/mitis*. (Courtesy of Kimberli Scott.)

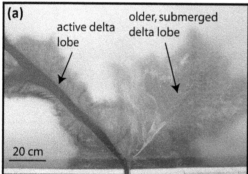
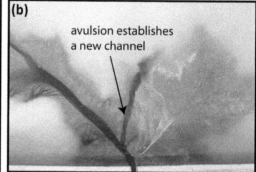

FIGURE 8.3 An example of deltaic avulsion in an experimental delta created in the laboratory. Elapsed time between the images (**a**) and (**b**) is 80 minutes. The water is dyed so that channelized flow is visible.
Source: Figure modified from Edmonds et al.[26]

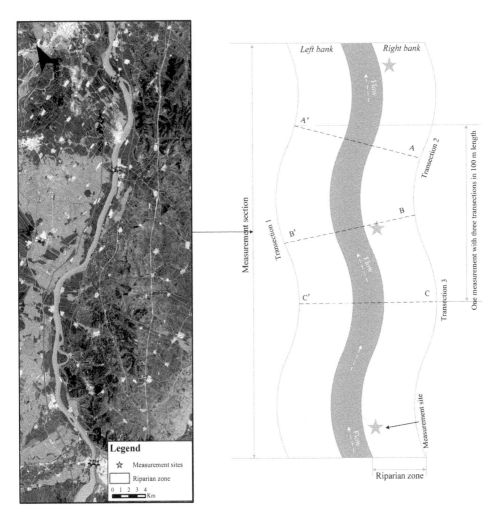

FIGURE 12.1 Landsat eight OLI data (left) acquired on September 06, 2015, illustrated example locations of the measurement sites and the layout of field measurements (right). One measurement section had 3–4 measurement sites, and one measurement site had three transections with 30 m width×50 m length measurement area on both sides of banks.

Source: Fu et al., 2017.

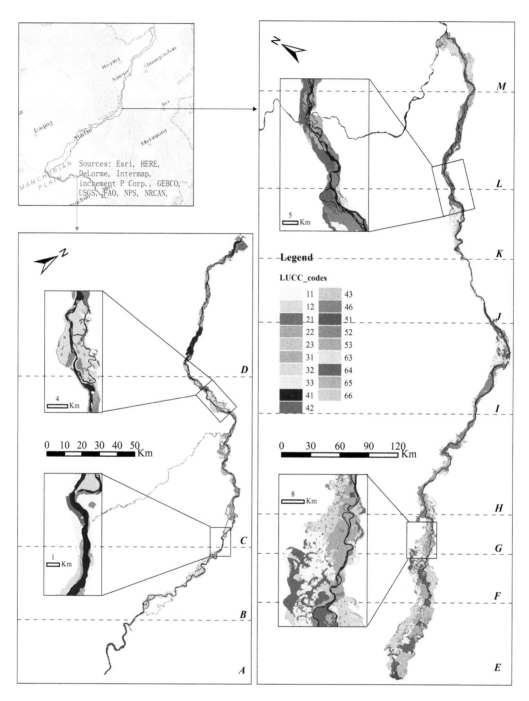

FIGURE 13.2 The land-use data at 1:100,000 scale for riparian zone in 2012. The LUCC_codes represent different land-use types, that is, 11–12: paddy field, glebe field; 21–23: arboreal forest, sparse woodland, and shrub woodland; 31–33: high-, mid-, and low-cover grassland; 41–43: rivers, lakes, and reservoirs; 46: beach; 51–53: residential, rural, and transportation; 63: saline-alkali soil land; 64: wetland; 65–66: barren land.
Source: Fu et al., 2017.

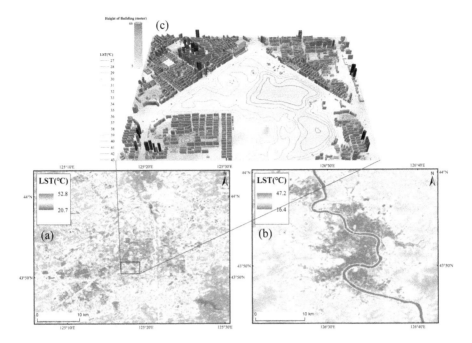

FIGURE 14.4 LST maps of Changchun City and Jilin City.

FIGURE 21.1 Tidal marshes are dominated by herbaceous vegetation comprising a wide range of species. In Chesapeake Bay on the U.S. Atlantic Coast, both tidal freshwater (a) and brackish (b) marshes are abundant. Tidal marshes are less abundant in much of Europe, but include reed-dominated tidal freshwater marshes along the Elbe River on Germany's Atlantic Coast (c), and salt marshes dominated by species, including common cordgrass (*Spartina anglica* C.E. Hubbard) and sea aster (*Aster tripolium* [Jacq.] Dobrocz.) along the North Sea (d). (Courtesy of A.H. Baldwin, University of Maryland.)

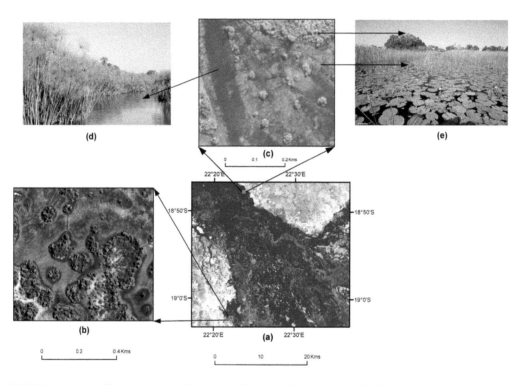

FIGURE 24.1 Spatial heterogeneity and structural diversity of vegetation and land cover in the Okavango Delta, northern Botswana. (**a**) represents part of the delta observed from Landsat TM imagery (RGB:421). (**b**) and (**c**) represent two subset areas as observed from high spatial resolution Quickbird imagery (RGB:421). (**d**) and (**e**) are photographs acquired in the field.

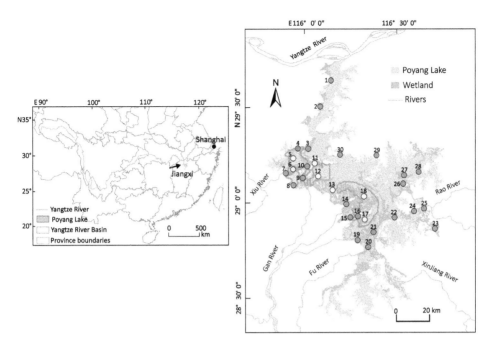

FIGURE 25.1 Locations of 30 sample sites in Poyang Lake wetland. Yellow dots are reference sites, and red dots are impaired sites. Areas enclosed by green colored polygons are national wetland nature reserves. (Modified from Yang, W.; You, Q.; Fang, N.; Xu, L.; Zhou, Y.; Wu, N.; Ni, C.; Liu, Y.; Liu, G.; Yang, T.; Wang, Y. 2018. Assessment of wetland health status of Poyang Lake using vegetation-based indices of biotic integrity. *Ecological Indicators* 90, 79–89.)

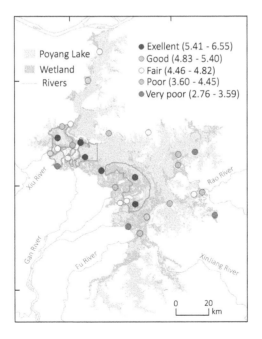

FIGURE 25.4 Health status classifications of 30 sample sites in Poyang Lake wetland based on the V-IBI. (Modified from Yang, W.; You, Q.; Fang, N.; Xu, L.; Zhou, Y.; Wu, N.; Ni, C.; Liu, Y.; Liu, G.; Yang, T.; Wang, Y. 2018. Assessment of wetland health status of Poyang Lake using vegetation-based indices of biotic integrity. *Ecological Indicators* 90, 79–89.)

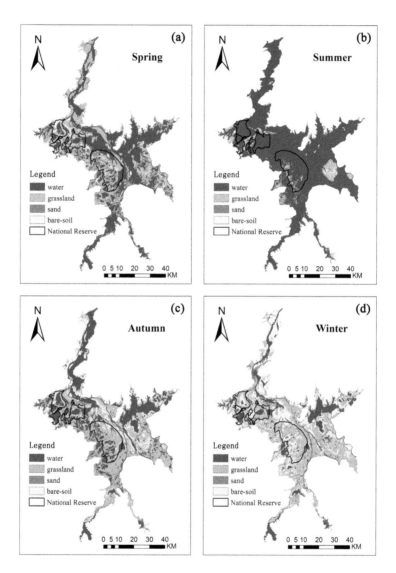

FIGURE 27.2 Land-cover maps of Poyang Lake natural wetland in different seasons.

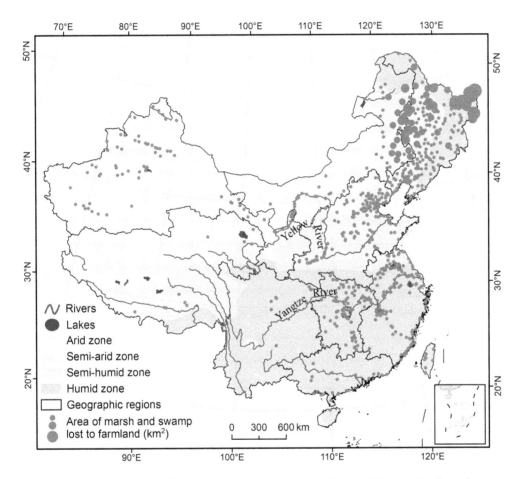

FIGURE 28.1 Spatial variance in China's marshes and swamps lost to farmland (bigger points denote larger areas of marshes and swamps lost to farmland within a county) [8].

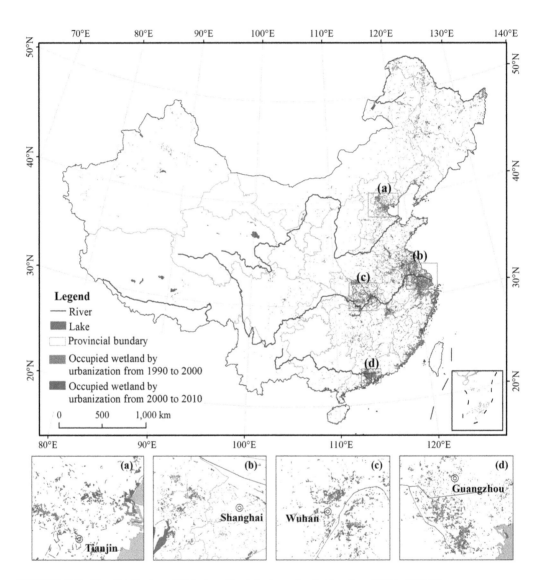

FIGURE 28.2 The distribution of China's wetlands lost to urbanization from 1990 to 2010 [9].

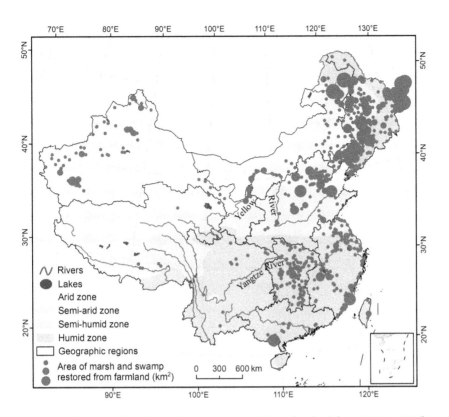

FIGURE 28.5 Spatial pattern of marshes and swamps restored from farmland from 1990 to 2010 (bigger points denote larger areas of marshes and swamps restored from farmland within a county) [8].

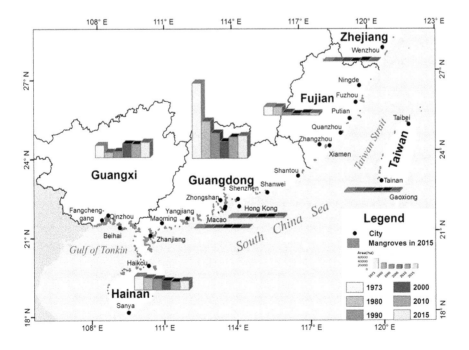

FIGURE 28.6 Spatial pattern of mangrove forests in 2015 and areal extent changes between 1973 and 2015 [14].

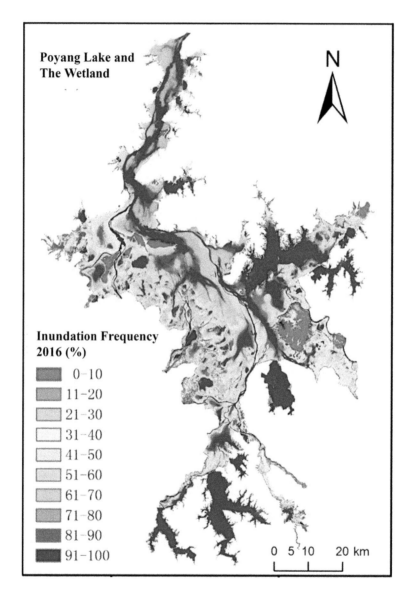

FIGURE 29.1 Variation of 2016 surface water coverage and the inundation frequency of associated wetland in Poyang Lake, China, as derived from GF-1 satellite wide-field view (WFV) sensor.

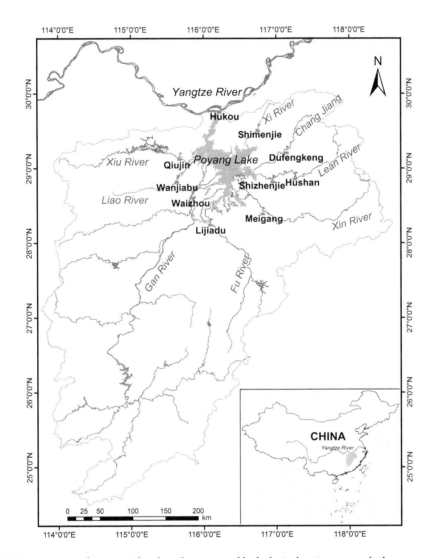

FIGURE 30.1 Location of Poyang Lake, the tributaries, and hydrological stations as marked

FIGURE 30.2 Poyang Lake is one of the most important wintering zones of waterfowl in East Asia, hosting significant proportions of the populations of cranes, geese, and swans, including rare and endangered species such as the Siberian crane (*Grus leucogeranus*). (Photos: Yeqiao Wang.)

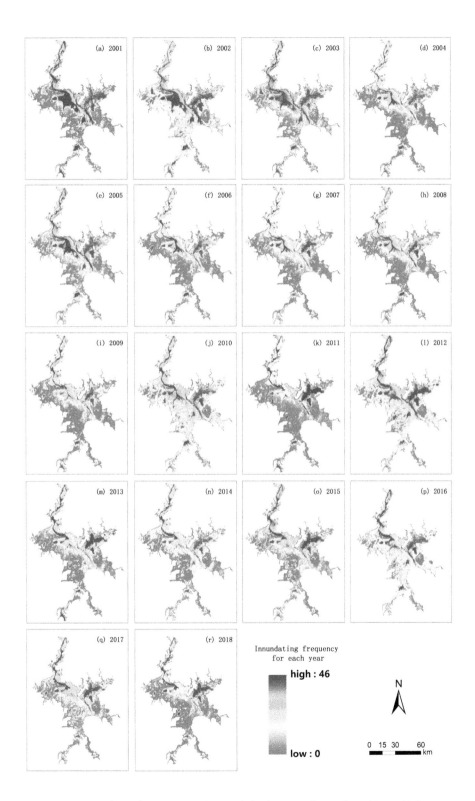

FIGURE 30.3 Variation of inundation areas in Poyang Lake during each year from 2001 to 2018.

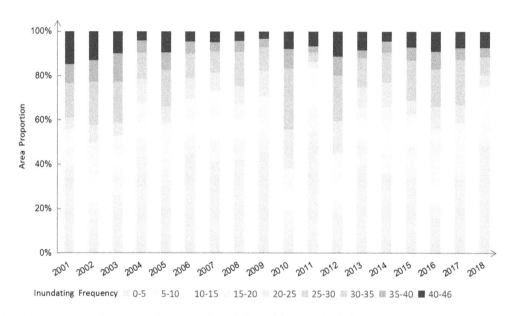

FIGURE 30.4 Inundation area of Poyang Lake at different lake water levels from 2001 to 2018.

16

Wetlands

Ralph W. Tiner
U.S. Fish and
Wildlife Service

Introduction

Wetland is a universal term used to describe the collection of flooded or saturated environments that have been referred to as marshes, swamps, bogs, fens, salinas, pocosins, mangroves, wet meadows, sumplands, salt flats, varzea forests, igapo forests, bottomlands, sedgelands, moors, mires, potholes, sloughs, mangals, palm oases, playas, muskegs, and other regional and local names. It has been defined as a basis for inventorying these natural resources, for conducting scientific studies, and, in some countries, for regulating uses of these areas. Given that wetlands include a diverse assemblage of ecosystems, classification schemes have been developed to separate and describe these different systems and to group similar habitats. Wetlands provide a number of functions that are considered valuable to society (e.g., surface water storage to minimize flood damages, sediment retention and nutrient transformation to improve water quality, shoreline stabilization, streamflow maintenance, and provision of vital habitat for fish, shellfish, wildlife, and plants that yield food and fiber for people). Because of these values and the widespread recognition of wetlands as important natural resources, numerous wetland definitions and classification systems have been developed to inventory these resources around the globe. The purpose of this entry is to provide readers with an understanding of what wetlands are (wetland definition), how they vary globally (wetland types), and their extent as determined by various inventories. This entry should serve as a starting point for learning about wetlands, with the listed references being sources of more detailed information.

Wetland Definitions

Wetlands are aquatic to semiaquatic ecosystems where permanent or periodic inundation or prolonged waterlogging creates conditions favoring the establishment of aquatic life. Wetlands are often located between land and water and have, therefore, been referred to as ecotones (i.e., transitional communities). However, many wetlands are not ecotones between land and water, since they are not associated with a river, lake, estuary, or stream.[1] Wetlands may derive water from many sources, including groundwater, river overflow, surface water runoff, precipitation, snowmelt, tides, melting permafrost, and seepage from impoundments or irrigation projects.

While the term "wetland" has many definitions, all definitions have common elements (see Table 16.1; some definitions even include deepwater habitats.). The presence of water in wetlands may be permanent or temporary. Their water may be salty or fresh. Wetlands may be natural habitats or artificially created. They range from shallow water environments to temporarily wet (i.e., flooded or saturated) areas.

TABLE 16.1 Examples of Wetland Definitions Used for Inventories

Country/Organization	Wetland Definition (Source)
International/Ramsar	"areas of marsh, fen, peatland, or water, whether natural or artificial, permanent or temporary, with water that is static or flowing, fresh, brackish, or salt, including areas of marine water the depth of which at low tide does not exceed 6 m.... ...may incorporate riparian and coastal zone adjacent to wetlands, and islands or bodies of marine water deeper than 6 m at low tide lying within the wetlands."[2]
Australia	"areas of seasonally, intermittently, or permanently waterlogged soils or inundated land, whether natural or artificial, fresh or saline, e.g., waterlogged soils, ponds, billabongs, lakes, swamps, tidal flats, estuaries, rivers, and their tributaries.[12]
Canada	"land that is saturated with water long enough to promote wetland or aquatic processes as indicated by poorly drained soils, hydrophytic vegetation, and various kinds of biological activity which are adapted to a wet environment."[3]
U.S.	"lands transitional between terrestrial and aquatic systems where the water table is usually at or near the surface or the land is covered by shallow water." Wetland attributes include hydrophytic vegetation, undrained hydric soil, or saturated or flooded substrates.[4]

All are wet long enough and often enough to, at least, periodically support hydrophytic vegetation and other aquatic life (including anaerobic microbes), to create hydric soils or substrates, and to activate biogeochemical processes associated with wet environments.

Wetland Types

Differences in climate, soils, vegetation, hydrology, water chemistry, nutrient availability, and other factors have led to the formation of a multitude of wetland types around the globe. In general, wetlands are characterized by their hydrology (e.g., tidal vs. nontidal, inundation vs. soil saturation, frequency and duration of wetness), the presence or absence of vegetation (vegetated vs. nonvegetated), the type of vegetation (forested or treed, shrub, emergent, or aquatic bed), and soil type (e.g., organic vs. inorganic, peatland vs. nonpeatland). Table 16.2 presents brief descriptions of some North American types and Figure 16.1 shows examples of vegetated wetlands.

TABLE 16.2 Brief Nontechnical Descriptions of Some Wetland Types in North America

Wetland Type	General Description
Marsh	Herb-dominated wetland with standing water through all or most of the year, often with organic (muck) soils
Tidal marsh	Herb-dominated wetland subject to periodic tidal flooding
Salt marsh	Herb-dominated wetland occurring on saline soils, typically in estuaries and interior arid regions
Swamp	Wetland dominated by woody vegetation and usually wet for extended periods during the growing season
Mangrove swamp (Mangal)	Tidal swamp dominated by mangrove species
Peatland, mire, moor, or muskeg	Peat-dominated wetland
Bog	Nutrient-poor peatland, typically characterized by ericaceous shrubs, other woody species, and peat mosses
Fen	More or less nutrient-rich peatland, often represented by sedges and/or calciphilous herbs and woody species
Wet meadow	Herb-dominated wetland that may be seasonally flooded or saturated for extended periods, often with mineral hydric soils
Bottomland	Riverside or streamside wetland, usually on floodplain
Flatwood	Forested wetland with poorly drained mineral hydric soils located on broad flat terrain of interstream divides, common on coastal plains and glaciolacustrine plains
Farmed wetland	Wetland cultivated for rice, cranberries, sugar cane, mints, or other crops

These types may be defined differently in other regions.

FIGURE 16.1 Some examples of North American wetlands (top to bottom, left to right): tidal salt marsh, inland marsh, pothole marsh, wet meadow/shrub swamp, northern peatland, bottomland swamp, hardwood swamp, and flatwood wetland.
Source: Photos courtesy of U.S. Fish and Wildlife Service.

Various countries have devised classification systems for describing differences among their wetlands and for categorizing wetlands for natural resources inventories. Scientists have created systems to organize certain wetlands into meaningful groups for analysis and management (e.g., peatland classifications; Tiner[1]). In 1998, the Ramsar Convention Bureau published a multinational classification system to provide consistency for inventorying wetlands and designating wetlands of international importance.[2] This system includes 11 types of marine or coastal wetlands (i.e., shallow water and intertidal habitats; permanent shallow marine waters; marine subtidal aquatic beds; coral reefs; rocky marine shores; sand; shingle or pebble shores; estuarine waters; intertidal mud, sand, or salt flats; intertidal marshes; intertidal forests; coastal brackish/saline lagoons; coastal freshwater lagoons). This system also includes 19 inland wetland types (i.e., permanently flooded aquatic habitats to intermittently flooded sites are represented: permanent inland deltas; permanent rivers/streams/creeks; seasonal, intermittent, or irregular rivers/streams/creeks; permanent freshwater lakes; seasonal or intermittent freshwater lakes; seasonal or intermittent saline/brackish/alkaline lakes and flats; permanent saline/brackish/alkaline marshes and pools; seasonal or intermittent saline/brackish/alkaline marshes and pools; permanent freshwater marshes and pools; seasonal or intermittent freshwater marshes and pools; nonforested peatlands; alpine wetlands including meadows and temporary snowmelt waters; tundra wetlands; shrubby-dominated wetlands; freshwater tree-dominated wetlands on inorganic soils; forested peatlands; freshwater springs and oases; geothermal wetlands; and subterranean karst and cave hydrological systems). Last, nine man-made wetland types (aquaculture ponds; ponds; irrigated land including rice paddies; seasonally flooded agricultural land; salt exploitation sites; water storage areas including impoundments generally more than 8 ha; excavations; wastewater treatment areas; and canals and drainage channels) are also included in the system.

In North America, the Canadian and United States wetland classification systems were developed by government agencies interested in wetland conservation and management. The Canadian system emphasizes wetland origin (class), form, and vegetation in describing the wetland types.[3] Five wetland classes are recognized: bog, fen, marsh, swamp, and shallow water. Within each class, different forms and types are characterized. Eight general vegetation types are defined by the presence or absence of vegetation (treed, shrub, forb, graminoid, moss, lichen, aquatic bed, and nonvegetated). These types may be subdivided into other types (e.g., treed into coniferous or deciduous, shrub into tall, low, and mixed, graminoids into grass, reed, tall rush, low rush, and sedge, aquatic bed into floating and submerged). The U.S. Fish and Wildlife Service's wetland and deepwater habitat classification[4] is the official federal system used for mapping wetlands and for reporting the status and trends of wetlands in the U.S. The features separating wetlands include general ecological and physical factors and specific features such as vegetation, soil/substrate composition, hydrology, water chemistry, and human alterations. Classification follows a hierarchical approach with five main levels designated: ecological system (marine, estuarine, lacustrine, riverine, and palustrine), subsystem, class (vegetated: forested, scrub-shrub, emergent, and aquatic bed; nonvegetated: unconsolidated shore, rocky shore, streambed, and reef), subclass, and modifiers. The modifiers are used to describe a wetland's hydrology (water regime), pH and salinity (water chemistry), soils, and the influence of humans and beaver (special modifiers). Common types include estuarine intertidal emergent wetlands (e.g., salt and brackish marshes), estuarine intertidal unconsolidated shore (e.g., tidal flats and beaches), palustrine emergent wetlands (e.g., marshes, fens, and wet meadows), palustrine forested wetlands, and palustrine scrub-shrub wetlands (e.g., shrub bogs and shrub swamps).

A hydrogeomorphic approach (HGM) to wetland classification has also been developing in the U.S.[5] The HGM system emphasizes abiotic features important for assessing wetland functions. Seven hydromorphic classes are identified: riverine, depressional, slope, mineral soil flats, organic soil flats, lacustrine fringe, and estuarine fringe. The U.S. Fish and Wildlife Service has adapted the HGM approach to provide additional modifiers to its classification system on a pilot basis. These HGM-type descriptors include landscape position (i.e., lotic, lentic, terrene, estuarine, and marine), landform (i.e., slope, basin, interfluve, floodplain, flat, island, and fringe), and water flow path (i.e., inflow, outflow, throughflow, bidirectional flow, isolated, and paludified).[6] These descriptors provide the required information to aid the evaluation of functions of wetlands across watersheds and large geographic areas.

Extent of Wetlands

Comprehensive wetland inventories do not exist in most countries. There are many inconsistencies among the inventories (e.g., different levels of effort, focus on particular types, and artificial wetlands such as rice paddies are often not included in wetland inventories).[7] Consequently, comparative

TABLE 16.3 Estimates of the Current Extent of Wetlands in Different Regions of the World

Region/Country	Wetland Extent (ha)	Source
Africa	121,321,683–124,686,189	[13]
Asia	211,501,790–224,117,790	[14]
Central America		
Mexico	3,318,500 (very incomplete)	[15]
Europe		
Eastern	225,849,930	[16]
Western	28,821,979	[17]
Middle East	7,434,790	[18]
Neotropics	414,996,613	[19]
North America		
Canada	127,199,000–150,000,000	[20]
U.S.	114,544,800	[1]
Oceania	35,748,853	[21]
South America		
Tropical region	200,000,000	[22]

analysis is fraught with problems. Nonetheless, Table 16.3 provides some perspective on the extent of wetlands in many regions. Most of the data came from a series of reports produced for the Bureau of the Ramsar Wetlands Convention.[8] Globally, estimates for wetlands range from about 750 million ha[7] to about 1.5 billion ha. Ten countries have over 2 million ha of peatlands alone, with Canada leading at nearly 130 million ha (represents about 18% of the country) followed by the former U.S.S.R. at 83 million ha.[9,10] About a third of Finland is covered by peatlands (10 million ha). The Pantanal of South America, perhaps the largest wetland in the world, reportedly covers about 200,000 km² (or 2 million ha) during the wet season.[11]

References

1. Tiner, R.W. Wetland Indicators: A Guide to Wetland Identification, Delineation, Classification, and Mapping; Lewis Publishers, CRC Press: Boca Raton, FL, 1999; 392 pp.
2. Ramsar Convention Bureau. Information Sheet on Ramsar Wetlands: Gland, Switzerland, 1998.
3. National Wetlands Working Group. *Wetlands of Canada*; Ecological Land Classification Series No. 21, Land Conservation Branch, Canadian Wildlife Service; Environment Canada: Ottawa, Ont., 1988; 452 pp.
4. Cowardin, L.M.; Carter, V.; Golet, F.C.; LaRoe, E.T. *Classification of Wetlands and Deepwater Habitats of the United States*; FWS/OBS-79/31; U.S. Department of the Interior, Fish and Wildlife Service: Washington, DC, 1979; 131 pp.
5. Brinson, M.M. *A Hydrogeomorphic Classification for Wetlands*; Wetlands Research Program Tech. Rep. WRP-DE-4; U.S. Army Waterways Expt. Station: Vicksburg, MS, 1993; 103 pp.
6. Tiner, R.W. *Keys to Waterbody Type and Hydrogeomorphic-Type Wetland Descriptors for U.S. Waters and Wetlands, Operational Draft*; U.S. Department of the Interior, Fish and Wildlife Service; Northeast Region: Hadley, MA, 2000; 20 pp.
7. Finlayson, C.M.; Davidson, N.C. Global review of wetland resources and priorities for wetland inventory: summary report. In *Global Review of Wetland Resources and Priorities for Wetland Inventory*; Finlayson, C.M., Spiers, A.G., Eds.; Supervising Scientist Report 144; Canberra, Australia, 1999; 9 pp.

8. Finlayson, C.M., Spiers, A.G., Eds.; *Global Review of Wetland Resources and Priorities for Wetland Inventory*; Supervising Scientist Report 144; Canberra, Australia, 1999.

9. Taylor, J.A. Peatlands of the British Isles. In *Mires: Swamp, Bog, Fen, and Moor; Regional Studies*; Gore, A.J.P., Ed.; Elsevier: Amsterdam, the Netherlands; 1–46.

10. Botch, M.S.; Massing, V.V. Mire ecosystems in the U.S.S.R. In *Mires: Swamp, Bog, Fen, and Moor; Regional Studies*; Gore, A.J.P., Ed.; Elsevier: Amsterdam, the Netherlands; 95–152.

11. Swarts, F.A., Ed. *The Pantanal of Brazil, Bolivia, and Paraguay*; Waterland Research Institute; Hudson MacArthur Publishers: Gouldsboro, PA, 2000; 287 pp.

12. Wetland Advisory Committee. *The Status of Wetlands Reserves in System Six*; Report of the Wetland Advisory Committee to the Environmental Protection Authority: Australia, 1977.

13. Stevenson, N.; Frazier, S. Review of wetland inventory information in Africa. In *Global Review of Wetland Resources and Priorities for Wetland Inventory*; Finlayson, C.M., Spiers, A.G., Eds.; Supervising Scientist Report 144; Canberra, Australia, 1999; 94 pp.

14. Watkins, D.; Parish, F. Review of wetland inventory information in Asia. In *Global Review of Wetland Resources and Priorities for Wetland Inventory*; Finlayson, C.M., Spiers, A.G., Eds.; Supervising Scientist Report 144; Canberra, Australia, 1999; 26 pp.

15. Olmstead, I. Wetlands of Mexico. In *Wetlands of the World: Inventory, Ecology, and Management Volume I*; Whigham, D.F., Dykyjova, D., Heiny, S., Eds.; Kluwer Academic Publishers: Dordrecht, the Netherlands, 1993; 637–677.

16. Stevenson, N.; Frazier, S. Review of wetland inventory information in Eastern Europe. In *Global Review of Wetland Resources and Priorities for Wetland Inventory*; Finlayson, C.M., Spiers, A.G., Eds.; Supervising Scientist Report 144; Canberra, Australia, 1999; 53 pp.

17. Stevenson, N.; Frazier, S. Review of wetland inventory information in Western Europe. In *Global Review of Wetland Resources and Priorities for Wetland Inventory*; Finlayson, C.M., Spiers, A.G., Eds.; Supervising Scientist Report 144; Canberra, Australia, 1999; 57 pp.

18. Frazier, S.; Stevenson, N. Review of wetland inventory information in the Middle East. In *Global Review of Wetland Resources and Priorities for Wetland Inventory*; Finlayson, C.M., Spiers, A.G., Eds.; Supervising Scientist Report 144; Canberra, Australia, 1999; 19 pp.

19. Davidson, I.; Vanderkam, R.; Padilla, M. Review of wetland inventory information in the Neotropics. In *Global Review of Wetland Resources and Priorities for Wetland Inventory*; Finlayson, C.M., Spiers, A.G., Eds.; Supervising Scientist Report 144; Canberra, Australia, 1999; 35 pp.

20. Davidson, I.; Vanderkam, R.; Padilla, M. Review of wetland inventory information in the North America. In *Global Review of Wetland Resources and Priorities for Wetland Inventory*; Finlayson, C.M., Spiers, A.G., Eds.; Supervising Scientist Report 144; Canberra, Australia, 1999; 35 pp.

21. Watkins, D. Review of wetland inventory information in Oceania. In *Global Review of Wetland Resources and Priorities for Wetland Inventory*; Finlayson, C.M., Spiers, A.G., Eds.; Supervising Scientist Report 144; Canberra, Australia, 1999; 26 pp.

22. Junk, W.J. Wetlands of tropical South America. In *Wetlands of the World: Inventory, Ecology, and Management Volume* I; Whigham, D.F., Dykyjova, D., Heiny, S., Eds.; Kluwer Academic Publishers: Dordrecht, the Netherlands, 1993; 679–739.

17

Wetlands: Biodiversity

Jean-Claude
Lefeuvre
University of Rennes

Virginie Bouchard
The Ohio State University

Role of Hydrology

Wetlands differ significantly in their water source and seasonal hydrologic regime. Hydrological patterns (i.e., flooding frequency, duration and hydroperiod) influence physical and chemical characteristics (e.g., salinity, oxygen and other gas diffusion rates, reduction–oxidation potential, nutrient solubility) of a wetland. In return, these internal parameters and processes control flora and fauna distribution as well as ecosystem functions. Plants, animals and microbes are often oriented in predictable ways along the hydrological gradient (Figure 17.1). Conversely, the biotic component affects the hydrology by eventually modifying flow or water level in a wetland.[1,2] Species also influence nutrient cycles and other ecosystem functions.[3]

A Landscape Perspective

Although the hydrology is part of the ecological signature of an individual wetland, wetlands are considered as neither aquatic nor terrestrial systems. They have characteristics from both systems and are defined as ecotones placed under this dual influence.[4] Because wetlands are located at the interface of multiple systems, they assure vital functions (e.g., wildlife habitats) beneficial at the landscape level. Reduction of wetland area often reduces biodiversity in the landscape.[2,5] Increases in biodiversity occur when wetlands are created or restored in a disturbed landscape.[6]

Wetland Flora

Wetland plants are adapted to a variety of stressful abiotic conditions (e.g., immersion, wave abrasion, water level fluctuation, low oxygen conditions). Identical adaptations to common environmental features have led taxonomically distinct species to sometimes look similar in terms of morphology, life cycle and life forms.[7] Traditionally, wetland plants have been classified into groups of different life forms, primarily in relation with hydrological conditions. Helophytes are defined as plant species with over-wintering buds in water or in the submerged bottom.[8] They are differentiated from hydrophytes in that their vegetative organs are partially raised above water level.[8] Hejny's classification is based on relatively stable vegetative features that determine the ability of wetland plants to survive two unfavorable conditions, cold and drought.[7] This classification uses the types of photosynthetic

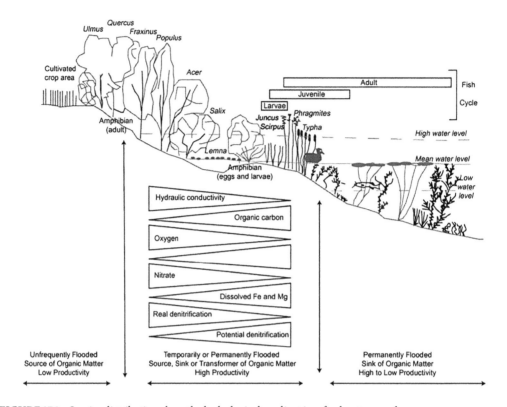

FIGURE 17.1   Species distribution along the hydrological gradient in a freshwater marsh

organs present in both the growth and flowering phases. Other classifications include both life form and growth form. When a species has a range of growth form, it is classified under the form showing the greatest achievement of its potential.[7]

Plant communities in wetlands can be more or less homogeneous, mosaic-like, or distributed along a gradient resulting in a clear zonation of species. A gradient exists if one or several habitat parameters change gradually in space. This phenomenon is common in fresh water marshes that present a gradient of water depth and water saturated soils (Figure 17.1). Such a gradient is often accompanied by differences in peat accumulation that is influenced by waves or currents. General principles of the zonation of aquatic plants have been largely described.[7,9,10] Littoral vegetation can belong to several types of communities, which derived from the general principle that, from deeper water to the shore, we may expect successively submergent, floating, and emergent macrophytes. The most important habitat factor is water depth, depending on slope and peat accumulation.[10] Other factors may be poor irradiance caused by high turbidity or exposure to waves or flow.[7,10]

Riparian ecosystems are found along streams and rivers that occasionally flood beyond confined channels or where riparian sites are created by channel meandering in the stream network. Riparian or bottomland hardwood forests contain unique tree species that are flood tolerant. Species distribution is associated with floodplain topography, flooding frequency, and flooding duration.[11,12] In Southeastern U.S. bottomland, seasonally flooded forests are colonized by *Platanus occidentalis* (sycamore), *Ulmus americana* (American elm), *Populus deltoides* (cottonwood) and are flooded between 2% and 25% of the growing season (Figure 17.1). Other species such as *Fraxinus pennsylvanica* (green ash), *Celtis laevigata* (sugarberry) and *Carya aquatica* (water hickory) colonized bottomlands that are flooded by less than 2% of the growing season.[11] Freshwater marshes are dominated with emergent macrophytes rooted in the bottom with aerial leaves (i.e., helophytes). Species such as *Typha* (cattail), *Phragmites* (reed grass) and *Scirpus* (bulrush) are often clonal. A plant community is usually organized in sequence

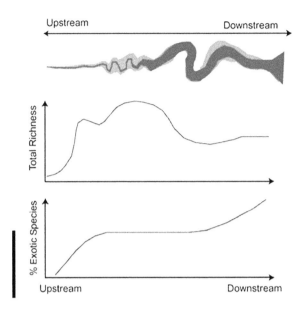

FIGURE 17.2 Total richness and invasive species distribution along a stream longitudinal gradient.
Source: Adapted from Planty-Tabacchi.[14]

of patches that are dominated by one species. The second plant groups are the rooted plants with float-ing leaves *(Nymphaeid).* Lotus *(Nelumbo)* and water lilies *(Nymphaea)* have very identical morphology (i.e., similar leaves and flowers) but a genetic analysis showed that lotus is more closely related with plane-tree than with water lilies.[10] Submerged plants include elodeids (i.e., cauline species whose whole life cycle can be completed below the water surface or where only the flowers are emergent) and isoetids (i.e., species growing on the bottom whose whole life cycle can be completed without contact with the surface). Submerged species include species such as coontail *(Ceratophyllum demersum)* and water mil-foil *(Myriophyllum* spp.). Plant species found in salt marshes are called halophytes (i.e., plants which complete their cycle in saline environments). A saltmarsh can be divided into low, middle, and upper marsh, according to flooding frequency and duration. Each zone is dominated by different plant species according to their tolerance to saline immersion.[10]

A dominant competitive species—often a clonal species—can modify the theoretical zonation. Change in water chemistry (i.e., eutrophication) or hydrology may favor a particular species over the natural plant community. Highly competitive species are often invasive and aggressive in displacing native spe-cies. The expansion of *Phragmites australis* into tidal wetlands of North America causes a reduction in biodiversity as many native species of plants are replaced by a more cosmopolitan species.[13] In riparian ecosystems, biodiversity is usually higher in the intermediate zones, whereas it is lower upstream and downstream (Figure 17.2). The percentage of exotic species is low in upstream areas, but can represent up to 40 in downstream zones (Figure 17.1).[14]

Wetland Fauna

Diversity of vertebrate and invertebrate species is the result of a community composed of resident and transient species, which use the space differently and at various times of day and year. The density and variety of animal populations at a particular wetland site are also explained by climatic events that affect geographic areas on a large scale. For example, the population of waterfowl during winter is largely dependent on climate variations in the northern part of the continents.[15]

Resident species are often dependent on the type of vegetation. Animal communities are gener-ally distributed along a zonation pattern, parallel to the plant communities, which is driven by the hydrological gradient. A few species depend entirely on a single plant species for their survival. One beetle *(Donacia* spp.) may depend on reeds, at least during its larval stage where another beetle *(Galerucella* spp.) uses only water lilies as its habitat and diet. For many other residents, their habitats extend to several plant Source: Adapted from Planty-Tabacchi.[14] communities during their life span. It is generally the case for many vertebrates such as amphibians, rodents, passerines, and waterfowl.

Many amphibian species depend on wetland or riparian zones for reproduction and larval stage. Likewise, wet meadows are necessary as a reproduction zone and nursery for a number of freshwater fishes. About 220 animals and 600 plant species are threatened by a serious reduction of wetlands in California, and the state's high rate of wetland loss (91% since the 1780s) is partly responsible.[15] Many waterfowl species are sensitive to areas of reduction, patch size and distribution, wetland density, and proximity to other wetlands.[16] When the Marais Poitevin (France), one of the principal wintering and passages sites for waterfowl in the Western Europe, underwent agricultural intensification in the 1980s, the population of ducks and waders declined tremendously. This decline was partly due to a 50% reduction of wet meadows between 1970 and 1995, primarily caused by the conversion to arable farmlands[17] Thus, maintenance of biodiversity depends on the existence of inter connections between wetlands, and between aquatic and terrestrial ecosystems. In fact, some authors have pointed out that an increase in biodiversity occurs when wetlands are created or restored in a disturbed landscape.[6,18]

Conclusions

Despite the importance of wetlands, conservation efforts have ignored them for a long time. It is urgent to conserve the existing wetlands, and also to restore and create wetland ecosystems. A wide range of local, state, federal, and private programs are available to support the national policies of wetland "No Net Loss" in the U.S., and around the world. From a biodiversity perspective, on-going wetland protection policies may not be working because restored or created wetlands are often very different from natural wet-lands.[19] Created wetlands often result in the exchange of one type of wetland for another, and result in the loss of biodiversity and functions at the landscape level.[19] We know now that it takes more than water to restore a wet- land,[20] even if an important place should be given to the ability of self-design of wetland ecosystems.[21]

References

1. Naiman, R.J.; Johnston, C.A.; Kelley, J.C. Alteration of North American streams by beaver. Bioscience **1988**, *38*, 753–762.
2. National Research Council. *Wetlands: Characteristics and Boundaries*; National Academy Press: Washington, DC, 1995; 306 pp.
3. Naiman, R.J.; Pinay, G.; Johnston, C.A.; Pastor, J. Beaver influences on the long-term biogeochemical characteristics of boreal forest drainage networks. Ecology **1994**, *75*, 905–921.
4. Hansen, A.J.; Di Castri, F.; Eds.; Landscape boundaries. In *Consequences for Biotic Diversity and Ecological Flows; Ecological Studies*; Springer: Berlin, 1983; 452 pp.
5. Gibbs, J.P. Wetland loss and biodiversity conservation. Conser. Biol. **2000**,*14*, 314–317.
6. Weller, M.W. *Freshwater Marshes*, 3rd Ed.; University of Minnesota Press: Minneapolis, U.S., 1994; 192.
7. Westlake, D.F.; Kvet, J.; Szczepanski, A.; Eds.; *The Primary Ecology of Wetlands*; Cambridge University Press: 1998; 568 pp.
8. Raunkiaer, C. *The Fife Forms of Plants and Statistical Plant Geography*; Oxford University Press: Oxford, England, 1934; 632 pp.
9. Grevilliot, F.; Krebs, L.; Muller, S. Comparative importance and interference of hydrological conditions and soil nutrient gradients in floristic biodiversity in flood meadows. Biodivers. Conserv. **1998**, *7*, 1495–1520.

10. Mitsch, W.J.; Gosselink, J.G. *Wetlands*, 3rd Ed.; Wiley: New York, 2000; 920 pp.

11. Clark, J.R.; Benforado, J., Eds.; *Wetlands of Bottomland Hardwood* Forest; Elsevier: Amsterdam, the Netherlands, 1981; 401 pp.

12. Keogh, T.M.; Keddy, P.A.; Fraser, L.H. Patterns of tree species richness in forested wetlands. Wetlands **1999**, *19*, 639–647.

13. Chambers, R.M.; Meyerson, L.A.; Saltonstall, K. Expansion of *Phragmites australis* into tidal wetlands of North America. Aquat. Bot. **1999**, *64*, 261–273.

14. Planty-Tabacchi, A.M. Invasions Des Corridors Riverains Fluviaux Par Des Esépces Végétales D'origine Ètrangére. These University Paul Sabatier Toulouse III, France, 1993; 177 pp.

15. Hudson, W.E., Ed.; *Landscape Linkages and Biodiversity*; Island Press: Washington, DC, 1991.

16. Leibowitz, S.C.; Abbruzzese, B.; Adams, P.R.; Hughes, L.E.; Frish, J. *A Synoptic Approach to Cumulative Impact Assessment: A Proposed Methodology*, 1992; 138 pp.

17. Duncan, P.; Hewison, A.J.M.; Houte, S.; Rosoux, R.; Tournebize, T.; Dubs, F.; Burel, F.; Bretagnolle, V. Long-term changes in agricultural practices and wildfowling in an internationally important wetland, and their effects on the guild of wintering ducks. J. Appl. Ecol. **1999**, *36*, 11–23.

18. Hickman, S. Improvement of habitat quality for resting and migrating birds at the Des Plaines River wetlands demonstration project. Ecol. Engng. **1994**, *3*, 485–494.

19. Whigham, D.F. Ecological issues related to wetland preservation, restoration, creation and assessment. Sci. Total Environ. 1999, *240*, 31–40.

20. Zedler, J.B. Progress in wetland restoration ecology. Trends Ecol. Evol. **2000**, *15*(10), 402–407.

21. Mitsch, W.J.; Wu, X.; Nairn, R.W.; Weihe, P.E.; Wang, N.; Deal, R.; Boucher, C.E. Creating and restoring wetlands: a whole ecosystem experiment in self-design. BioScience **1998**, *48*, 1019–1030.

18

Wetlands: Classification

Arnold G. van
der Valk
Iowa State University

Introduction

There are many kinds of wetlands around the world that differ because of their geographic location (tropical, temperate, arctic, etc.), geomorphological setting (channel, coastal or lake fringe, basin, slope, etc.), water chemistry (freshwater, brackish water, salt water, etc.), soils (mineral, organic), and vegetation (herbaceous, shrub, forested, mosses, etc.). In many parts of the world, wetlands are utilized by local people, for example, for grazing cattle and other domestic animals in sub-Saharan Africa and crop cultivation in Asia, and this can result in significant impacts on their size, hydrology, flora, and fauna. Although most wetlands are natural features of the landscape, there are also wetlands that were purposefully constructed or that developed in association with dams and irrigation systems. Nevertheless, all these wetlands have three things in common: (1) They are shallowly flooded or have saturated soils; (2) their soils are developed to some extent under anaerobic conditions, and consequently have characteristics that differ from those of adjacent terrestrial soils; and (3) their flora and fauna are different from those of adjacent terrestrial ecosystems. Wetlands, however, are often highly dynamic systems: their hydrology and vegetation can change from year to year; and because they can occur in multiple states, this complicates their classification.

The need to classify wetlands has grown over the years, as the perceived societal and scientific value of wetlands has increased. Wetlands historically were often viewed as nuisances or even dangerous places that harbored a variety of human diseases. Wetland classification and inventorying were initially driven by the need to identify wetlands to drain. As waterfowl began to decline in the 20th century, the importance of wetlands as habitat for waterfowl stimulated early efforts to classify them. Waterfowl interests wanted to preserve those wetlands that were important as breeding or overwintering habitat to ensure the survival of recreationally important waterfowl species. By the late 20th century, wetlands came to be valued not only as wildlife habitat but for their other services to society, such as pollutant removal, flood water storage, groundwater recharge, etc. In some places, this resulted in legal protection being given to wetlands in order to preserve these wetland services. Legal protection required that all wetlands in a jurisdiction be identified and classified.

In order to classify wetlands, we first have to define what a wetland is. In 1995, a Committee on Characterization of Wetlands[1] in the United States provided a formal definition of a wetland: "A wetland is an ecosystem that depends on constant or recurrent, shallow inundation or saturation at or near the surface of the substrate. The minimum essential characteristics of a wetland are recurrent, sustained inundation or saturation at or near the surface and the presence of physical, chemical, and biological features reflective of recurrent, sustained inundation or saturation. Common diagnostic features of wetlands are hydric soils and hydrophytic vegetation. These features will be present except where specific physicochemical, biotic, or anthropogenic factors have removed them or prevented their development." This definition or its minor variants is now universally accepted by wetland ecologists. In short, wetlands are what they are because of their hydrology, and their hydrology creates physical and chemical conditions that produce distinct soils and habitats that can be exploited only by subsets of the local flora and fauna that are adapted to these conditions. Among the important fauna found in wetlands are waterfowl, capybaras, water buffalo, alligators, and many kinds of fish. Among the most important plants are cattails (*Typha* spp.), water lilies (*Nymphaea* spp.), pondweeds (*Potamogeton* spp.), mangroves (e.g., *Avicennia* spp.), papyrus (*Cyperus papyrus*), and common reed (*Phragmites* spp.)

There are two ways in which the term wetland is commonly used: (1) an entire landscape feature (basin, fringe, flat, channel) that qualifies as a wetland because of its hydrology, plants, and soils; for example, the Everglades or Okavango Delta; (2) a distinguishable portion or area of such a landscape feature, usually because of its dominant vegetation, for example, a sawgrass flat or wet prairie within the Everglades. In a wetland classification, it is important to distinguish between these two uses of the term wetland. They represent different scales at which a classification can be done. Landscape features are often not homogeneous units with regard to hydrology, soils, or vegetation, but they can be delimited, mapped, measured, and counted. Distinguishable portions or areas of landscape features are homogeneous units with regard to hydrology, water chemistry, vegetation, soils, or some other feature. This distinction was sometimes ignored in some older classification systems. Most wetlands when classified as a landscape feature contain a complex of plant communities that reflect local topographic gradients (i.e., water depths, soils, water chemistry), related plant distribution patterns, and secondary patterns caused by alterations of the original landscape by plants and animals, for example, tree islands in the Everglades. Consequently, classifications based solely on landscape features collect very different information than those based on homogeneous distinguishable areas. Modern classification systems often combine both approaches with higher units in the system based on landscape features (rivers, lakes, depressions, flats, etc.) and subunits based on homogenous distinguishable areas.

Ecologists in Europe and North America began describing wetlands in some detail in the 19th century.[2] They distinguished between coastal/saline wetlands, interior/freshwater wetlands, and peat bogs; and they recognized many subtypes of these on the basis of their vegetation, for example, reed swamp, carr, forest swamp, fen, etc. Early ecologists distinguished between wetlands (marshes, swamps, tidal marshes) and peatlands (mires, bogs, fens) and treated them as different kinds of entities. Today, this is no longer the case, and peatlands are considered to be wetlands with organic soils.

Although plant ecologists and geographers in Europe in the 19th century had described a variety of different types of wetland types, one of the first attempts to develop a national wetland classification occurred in the United States in the 1890s, and the first national wetland inventories were conducted in 1906 and 1922. These inventories were carried out by federal government agencies and were done using questionnaires, soil survey and other maps, drainage information, and various kinds of state reports. These early inventories used very crude classification systems (e.g., the 1922 survey recognized only five wetland types). Their main purpose was to identify wetlands for "reclamation," that is, drainage. Limited and crude as these classification systems and inventories were, they did provide the first information about the state of wetlands in the United States.[2] The next generation of wetland classification systems developed in the United States were more positively disposed to wetlands and initially focused on evaluating and conserving wetlands as habitat for waterfowl. But as wetlands became more visible in public policy and research arenas in the 1970s, the need arose for more general wetland classification systems that could be used for multiple purposes (conservation, restoration, legal, management, research) and especially for implementing national wetland inventories and wetland protection policies. Wetland classification systems in the United States since the 1950s will be reviewed first and then selected contemporary classifications that were developed outside the United States. Only those classification systems that were adopted or proposed by a national agency or international organization will be considered.

American Classification Systems

The designers of any wetland classification to be used in inventorying wetlands have to deal with a number of major issues. What will be the "unit" classified? How will data on wetlands be collected? How accurate do the data have to be? Who will use the data generated? In what form will it be stored? How will inventory data be made available/distributed/retrieved? In the United States, the first two national classification systems developed in the last half of the 20th century were intended to be used as a basis for a national wetland inventory (NWI). Consequently, the practical realities (people, time, money) of collecting data at a large geographic scale were major factors in their design. The practical realities of collecting data and distributing them, however, changed dramatically from the 1950s to the late 1970s. The advent of easily available, high-resolution aerial photography and satellite imagery; inexpensive computers; geographic information system software for collecting, storing, and analyzing spatial data; high speed graphics printers; the Internet; etc. has had an impact on the design of classification systems and on the distribution of inventory data collected using them. This can be seen through the comparison of the *Classification of Wetlands of the United States*[3] commonly referred to as the Martin et al. (1953) system and its replacement *Classification of Wetlands and Deepwater Habitats of the United States*[4] commonly referred to as the Cowardin et al. (1979) system.

Martin et al. (1953) developed a national wetland classification system to inventory waterfowl habitat. This classification system was horizontal in that its 20 wetland types, which covered all inland and coastal wetlands in the conterminous United States, had no subdivisions or subtypes. One of the main limitations faced by its developers was the small number of staff available to collect data in the field and their limited expertise in wetland ecology. Wetland types (e.g., inland fresh water meadows, inland shallow fresh marshes, inland deep marshes, and inland open fresh water) were vegetation types largely defined by certain characteristic species and water depth: features easy to observe in the field. This classification system was not consistent in the kinds of categories used, and it mixed vegetation types like inland shallow fresh marshes (distinguishable units) with landscape unit like bogs and sounds and bays. It also put wetlands from different parts of the country with little in common (Cypress-Tupelo Swamps from the southeastern and Black Spruce forests from northeastern United States) into the same category, namely, wooded swamps. The use of water depth to distinguish some freshwater wetland types also caused problems because this was highly dependent on the time of year a wetland was visited. The uneven nature of the categories reflected the main purpose of this classification system to find out how much waterfowl habitat of various types still existed in the United States. It emphasized types of wetlands used by waterfowl (six types of inland freshwater wetlands) and plays down those that were not considered important to waterfowl (e.g., bogs). This classification was used to do a detailed inventory of American wetlands, and the results of this inventory were published in 1956 in *Wetlands of the United States. Their Extent and Their Value to Waterfowl and Other Wildlife*.[5] The results of this inventory were only available in this Fish and Wildlife report, commonly referred to as Circular 39. A number of regional wetland classifications systems were developed in the United States in the 1960s and 1970s that dealt with many of the shortcomings of the Martin et al. (1953) system.[2]

By the time that the next American NWI was developed 20 years later, advances in technology, especially the availability of high quality aerial photography and computers to store and retrieve spatial information, greatly influenced the development of the *Classification of Wetlands and Deepwater Habitats of the United States*.[4] The more widespread interest in wetlands in the 1970s is reflected in the makeup of the committee that developed this new classification system. All the authors of the Martin et al. system worked for the U.S. Fish and Wildlife Service. Although the U.S. Fish and Wildlife Service was still heavily involved, the authors of the new classification also came from other federal agencies (U.S. Geological Survey, National Oceanographic and Atmospheric Administration) as well as academia. One of the main differences between the new classification and its predecessor is that it is a hierarchical system. Much like the hierarchical system used to classify plant and animal species, it is possible to select and examine several levels of aggregation. Overall, the developers of the new classification system had many major objectives for it: (1) to build the system using consistent and uniform terminology that could be applied to any wetland in the United States; (2) to be able to group ecologically similar wetlands together at all levels of the classification hierarchy; (3) to make it readily applicable to mapping all the wetlands of the United States from aerial photographs as part of a NWI. The system has three levels in its hierarchy: system, subsystem, and class. Systems and subsystems are based on hydrology (Figure 18.1). There are five systems: marine, estuarine, riverine, lacustrine, and palustrine. Except for the palustrine, each system has two or more subsystems. In turn, each system or subsystem has multiple classes. Classes are mostly types of plant communities (aquatic bed, moss-lichen wetland, forested wetland, etc.) or various kinds of non-vegetated substrates (bedrock, rubble, sand, etc.) that can be identified on aerial photographs. Classes can further be subdivided into subclasses and dominance types. While classes and subclasses are based primarily on dominant life form or easily recognized substrate features for unvegetated areas, dominance types are described by their most abundant plant species or sessile macro invertebrate species (e.g., oysters). In addition, water regime, soil, water chemistry, and special modifiers can also be used to characterize wetlands. Special modifiers are used to describe human-made wetlands (excavated, impounded, diked, farmed, etc.). In short, the Cowardin et al. system is a classification in which the basic units are distinguishable areas on aerial photographs. This classification system has been used successfully by the NWI in the United States for over 30 years, and the NWI, which is part of the Fish and Wildlife Service, periodically issues status and trend reports for the country that document changes in different kinds of wetlands. NWI maps are now distributed on the World Wide Web (http://www.fws.gov/wetlands/).

Although it has served its primary purpose well, the Cowardin et al. (1979) wetland classification system has proven to be inadequate for other purposes, especially the evaluation of wetland functions and values. In 1993, another wetland classification system was published by Mark Brinson[6] for this purpose that he called A *Hydrogeomorphic Classification for Wetlands*. It is most commonly known as the HGM system. Unlike the Cowardin et al. system, the basic unit in the HGM system is a landscape unit. The HGM system identifies wetland types believed to have similar functions. The position of the wetland in the landscape and the source(s) of water and direction of flow into and out of the wetland are the major classificatory variables. In the HGM system, seven basic HGM classes of wetlands are recognized: depressional, riverine, slope, lacustrine, mineral soil flats, organic soil flats, and estuarine (tidal). The HGM system deals only with wetlands as defined by the Committee on Characterization of Wetlands. It excludes the deepwater systems that are also covered in the Cowardin et al. classification system. Various characteristics of a wetland and its watershed can then be used to assess its functions (flood peak attenuation, sediment retention, water storage, nutrient retention, habitat value, carbon sequestration, etc.). A series of regional manuals have been developed to do functional assessments using the HGM system. These can be downloaded from the U.S. Army Corps of Engineers website (http://el.erdc. usace.army.mil/wetlands/wlpubs.html).

Other Classification Systems

Besides those in the United States, wetland classifications have been developed in many other countries.[2,7] For example, Canada developed a national classification system that was built on a variety of provincial and regional systems proposed in the 1970s and 1980s. *Wetlands of Canada*,[8] which was published in 1988, describes the Canadian system in detail. Like the Cowardin et al. system in the USA, the Canadian system is hierarchical, but its levels (class, form, and type) are completely different. Its classes are landscape units. These are widespread wetlands that are distinguished primarily by their dominant vegetation and soils (bog, fen, swamp, marsh, and shallow open water). Forms are based on landscape position, surface morphology of the wetland, its sources of water, presence of permafrost, etc. It is an open-ended category and more forms can be added as needed. Types are based on the physiognomy of the dominant plant species (trees, shrubs, graminoid, moss, lichen, etc.) of the wetland's vegetation. Subforms and phases are also recognized based on species composition, depth of water, water chemistry, etc. The lower level units are homogeneous, distinguishable units. Thus, the Canadian system includes both landscape and distinguishable area units.

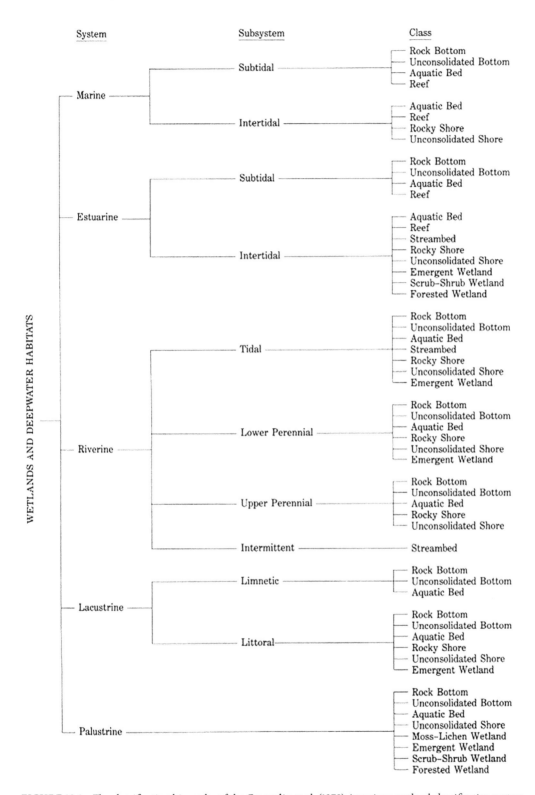

FIGURE 18.1 The classification hierarchy of the Cowardin et al. (1979) American wetland classification system.
Source: Cowardin et al.[4]

In Australia, Semeniuk and Semeniuk (1995) developed a geomorphic classification system for non-coastal wetlands that has much in common with the HGM system.[9] Like the HGM system, it is built around various landforms (basins, channels, flats, slopes, and highlands) and hydrology. They, however, emphasized water longevity (permanent inundation, seasonal inundation, intermittent inundation, and seasonal waterlogged) rather than source and direction of water flow. When landform and length of inundation/saturated are combined, they recognized 13 different wetland types. Likewise, Australia has developed its own classification system for inventory purposes.[10] It recognizes six general wetland types: lakes, swamps, lands subject to inundation, river and creek channels, tidal flats, and coastal water bodies (estuaries, lagoons). Each major type is subdivided into multiple subtypes, mostly based on length of inundation or stream flow for inland wetlands and patterns of tidal inundation for tidal flats.

South Africa has spent a great deal of time and effort developing a classification system that is suitable for the diverse conditions found in southern Africa. As in the United States, a number of regional wetland classifications had developed, but they were inadequate for a NWI. In 2006, the *National Wetland Inventory: Development of a Wetland Classification System for South Africa* was published[11] and an update, *Further Development of a Proposed National Wetland Classification System for South Africa*, was published in 2009.[12] There is also a user's manual, *Classification System for Wetlands and Other Aquatic Ecosystems in South Africa*.[13] Like the American Cowardin et al. system, the South African system is hierarchical, but, unlike it, the South African system incorporates an HGM approach at some levels. It uses primary discriminators to classify wetland (HGM) units in the first three levels of the hierarchy (Figure 18.2). At level 1, the highest level, three basic types of wetland systems are recognized: marine, estuarine, and inland. At level 2, bioregional differences (e.g., Namaqua Inshore Bioregion off

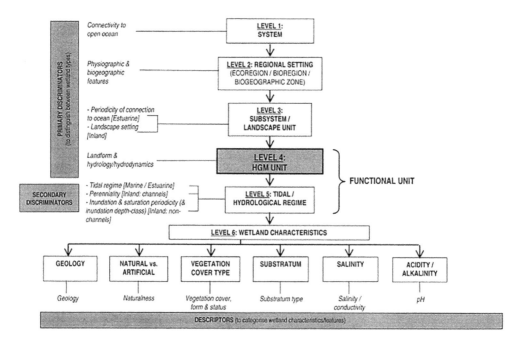

FIGURE 18.2   The structure of the proposed South African wetland classification system. Primary discriminators are used in levels 1–3 to classify the hydrogeomorphic (HGM) units in level 4. Secondary discriminators in level 5 are used to subdivide the HGM units in level 4. In level 6, there are various descriptors that can be used to further characterize the HGM units in level 4.
Source: SANBI.[12]

the western coast) are the main discriminators rather than wetland types. Level 3 is based on subsystems of each of the level 1 systems. For estuarine systems, connectivity (permanently open, temporarily open) to the ocean is used to delimit subsystems and for the inland system, four landscape units (slope, valley floor, plain, and bench) are recognized. There are no marine subsystems. At level 4, HGM units are used to classify wetlands. These HGM units are based on land-form; sources and movement of water into, through, and out of the wetland; and hydrodynamics (direction and strength of water flow), as per Brinson's (1993) HGM system in the United States. Secondary discriminators are used at level 5 to subdivide the HGM units at level 4. These include tidal regimes for estuarine and marine wetlands and water regimes for inland wetlands. Level 6 includes various descriptors that can be used to further characterize the wetland. There are six classes of descriptors: geology, natural vs. artificial, vegetation type, substratum, salinity, and pH. The South African wetland classification system is, in effect, a fusion of the two contemporary American classification systems, Cowardin et al.[4] and Brinson's HGM[6] with HGM units nested within broader (system) units. Like the Cowardin et al. system, the South African system includes not only inland and coastal wetlands but also deepwater habitats.

During the 1980s and 1990s, wetland inventories were done in many countries or regions around the world.[7,14] Some of these consisted primarily of a list of nationally important wetlands with a brief description of their major features. In spite of being an important source of information about wetlands globally, these various efforts did not use a standardized method or classification system, and thus the results are uneven and sometimes not comparable. The Ramsar Convention Bureau has spearheaded efforts to develop a truly international classification system.[15] The main goal of the Ramsar Convention was to protect wetland habitats important to migratory water birds. Countries that are signatories to the Ramsar Convention agree to designate one wetland as a wetland of international importance. These are widely known as Ramsar Sites. Each signatory government (160 as of 2011) is also expected to implement a plan to promote the conservation and wise use, as far as possible, of its wetlands. This includes an inventory of its wetlands.

The Ramsar Convention's approach to wetlands is both practical and flexible. Almost anything that is shallowly flooded is considered to be a wetland, regardless of its origin or current use. It has its own definition of wetlands:[15]."… areas of marsh, fen, peatland or water, whether natural or artificial, permanent or temporary, with water that is static or flowing, fresh, brackish or salt, including areas of marine water the depth of which at low tide does not exceed six metres." Although the Ramsar definition of a wetland recognizes that hydrology (flooding/saturation) is the defining feature of a wetland, it differs from the Committee on Characterization of Wetlands' definition in that it does not explicitly mention hydric soils or plants. Like the Cowardin et al. and South African systems, it also includes deepwater marine habitats. The Ramsar classification system (Table 18.1) recognizes six major wetland types: marine, estuarine, lacustrine, riverine, palustrine, and human made.[15] The Ramsar Wetland Classification system is designed to enable all countries to carry out a rapid inventory of the wetlands, especially of wetlands that are nominated as Ramsar sites. The current version (2011) of the system recognizes 42 wetland types grouped into three broad classes (marine and coastal wetlands, inland wetlands, and human-made wetlands). Although at first sight, it appears to be a horizontal classification much like the Martin et al. (1953) American classification, it can also be viewed as a hierarchical system as demonstrated by Scott and Jones (1995).[7] In fact, the Ramsar system closely parallels the Cowardin et al. (1979) American classification system, except that in the Ramsar classification human-made wetlands are at the same level in the hierarchy as hydrologic systems while they are only descriptors of classes in the Cowardin et al. system This reflects the difference between extant wetlands in much of the United States, which are generally less impacted and exploited by people than their counterparts in other parts of the word, particularly in Europe, northern Africa, and most of southern Asia. Like the Martin et al. system, however, it is a mixed classification system with both landscape units (coastal brackish/saline lagoons) and homogeneous distinguishable units (intertidal forested wetlands).

TABLE 18.1 Ramsar Classification System and Codes for Wetland Types

Marine/Coastal Wetlands

A – Permanent shallow marine waters in most cases less than 6 m deep at low tide; includes sea bays and straits.

B – Marine subtidal aquatic beds; includes kelp beds, sea-grass beds, tropical marine meadows.

C – Coral reefs.

D – Rocky marine shores; includes rocky offshore islands, sea cliffs.

E – Sand, shingle or pebble shores; includes sand bars, spits and sandy islets; includes dune systems and humid dune slacks.

F – Estuarine waters; permanent water of estuaries and estuarine systems of deltas.

G – Intertidal mud, sand or salt flats.

H – Intertidal marshes; includes salt marshes, salt meadows, saltings, raised salt marshes; includes tidal brackish and freshwater marshes.

I – Intertidal forested wetlands; includes mangrove swamps, nipah swamps and tidal freshwater swamp forests.

J – Coastal brackish/saline lagoons; brackish to saline lagoons with at least one relatively narrow connection to the sea.

K – Coastal freshwater lagoons; includes freshwater delta lagoons.

Zk(a) – Karst and other subterranean hydrological systems, marine/coastal.

Inland Wetlands

L – Permanent inland deltas.

M – Permanent rivers/streams/creeks; includes waterfalls.

N – Seasonal/intermittent/irregular rivers/streams/creeks.

O – Permanent freshwater lakes (over 8 ha); includes large oxbow lakes.

P – Seasonal/intermittent freshwater lakes (over 8 ha); includes floodplain lakes.

Q – Permanent saline/brackish/alkaline lakes.

R – Seasonal/intermittent saline/brackish/alkaline lakes and flats.

Sp – Permanent saline/brackish/alkaline marshes/pools.

Ss – Seasonal/intermittent saline/brackish/alkaline marshes/pools.

Tp – Permanent freshwater marshes/pools; ponds (below 8 ha), marshes and swamps on inorganic soils; with emergent vegetation water-logged for at least most of the growing season.

Ts – Seasonal/intermittent freshwater marshes/pools on inorganic soils; includes sloughs, potholes, seasonally flooded meadows, sedge marshes.

U – Non-forested peatlands; includes shrub or open bogs, swamps, fens.

Va – Alpine wetlands; includes alpine meadows, temporary waters from snowmelt.

Vt – Tundra wetlands; includes tundra pools, temporary waters from snowmelt.

W – Shrub-dominated wetlands; shrub swamps, shrub-dominated freshwater marshes, shrub carr, alder thicket on inorganic soils.

Xf – Freshwater, tree-dominated wetlands; includes freshwater swamp forests, seasonally flooded forests, wooded swamps on inorganic soils.

Xp – Forested peatlands; peat swamp forests.

Y – Freshwater springs; oases.

Zg – Geothermal wetlands.

Zk(b) – Karst and other subterranean hydrological systems, inland.

Human-Made Wetlands

1 – Aquaculture (e.g., fish/shrimp) ponds.

2 – Ponds; includes farm ponds, stock ponds, small tanks; (generally below 8 ha).

3 – Irrigated land; includes irrigation channels and rice fields.

4 – Seasonally flooded agricultural land (including intensively managed or grazed wet meadow or pasture).

5 – Salt exploitation sites; salt pans, salines, etc.

6 – Water storage areas; reservoirs/barrages/dams/impoundments (generally over 8 ha).

(Continued)

TABLE 18.1 (**Continued**)   Ramsar Classification System and Codes for Wetland Types

7 – Excavations; gravel/brick/clay pits; borrow pits, mining pools.

8 – Wastewater treatment areas; sewage farms, settling ponds, oxidation basins, etc.

9 – Canals and drainage channels, ditches.

Zk(c) – Karst and other subterranean hydrological systems, human-made.

Note: "Floodplain" is a broad term used to refer to one or more wetland types, which may include examples from the R, Ss, Ts, W, Xf, Xp, or other wetland types. Some examples of floodplain wetlands are seasonally inundated grassland (including natural wet meadows), shrub lands, woodlands, and forests. Floodplain wetlands are not listed as a specific wetland type herein.

The categories are intended to provide only a very broad framework to aid in the rapid identification of main wetland habitats represented at each site.

Source: Ramsar Convention Secretariat.[15]

Conclusion

Although a number of successful national and international wetland classifications have been developed, we do not have a universally accepted international wetland classification system. Current national and international wetland classifications often use different definitions of wetlands, terminology to describe them, and methods for delineating them. Consequently, some wetland classification systems do not include aquatic systems that are recognized as wetlands in others.[13,14] Inconsistencies in the treatment of man-made wetlands are a particular problem. To date, the Ramsar classification system is the only international classification that is being used by multiple countries, mostly those that do not have a national system. Although very useful for classifying wetlands at a coarse scale, the Ramsar classification, in its current form, is too crude to be used for detailed national inventories of wetlands like that of the U.S. NWI.

Advances in aerial photography and satellite-born sensors as well as in the acquisition, processing, and storage of spatial data have greatly reduced the time and cost of collecting wetland inventory data. Nevertheless, the need to develop and implement a global inventory of wetlands is hampered by the lack of a universally accepted classification system. The United States historically has led the way in the development of wetland classification systems and in carrying out a detailed NWI, but new developments in other parts of the world like South Africa suggest that more sophisticated wetland classification systems can be developed that could become the basis of a new international wetland classification system. Before such a new international system can be developed, however, a universally accepted definition of what constitutes a wetland needs to be adopted.

References

1. Commission on Characteristics of Wetlands. *Characteristics and Boundaries*; National Academy Press: Washington, DC, USA, 1995.
2. Tiner, R.W. Wetland Indicators: A Guide to Wetland Identification, Delineation, Classification, and Mapping; Lewis Publishers: Boca Raton, FL, USA, 1999.
3. Martin, A.C.; Hotchkiss, N.; Uhler, F.M.; Bourn, W.S. Classification of Wetlands of the United States. Fish and Wildlife Service Spec. Sci. Rep., Wildlife No. 20; U.S. Department of the Interior Fish and Wildlife Service: Washington, DC, USA, 1953; 1–14.
4. Cowardin, L.W.; Carter, Virginia; Golet, F.C.; LaRoe, E.T. Classification of wetlands and deepwater habitats of the United States. Fish and Wildlife Service FWS/OBS-79/31; U.S. Department of the Interior Fish and Wildlife Service: Washington, DC, USA, 1979; 1–131.

5. Shaw, S.P.; Fredine, C.G. Wetlands of the United States. Their Extent and Their Value to Waterfowl and Other Wildlife. U.S. Fish and Wildlife Service Circular 39; U.S. Department of the Interior Fish and Wildlife Service: Washington, DC, USA, 1956; 1–67.

6. Brinson, M.M. A Hydrogeomorphic Classification of Wetlands. U.S. Army Corps of Engineers Wetlands Research Program Tech. Rept. WRP-DE-4; U.S. Army Corps of Engineers Waterways Experiment Station: Vicksburg, MS, USA, 1993; 1–92.

7. Finlayson, C.M.; van der Valk, A.G., Eds. Classification and Inventory of the World's Wetlands; Kluwer Academic Press; Dordrecht, 1995.

8. National Wetlands Working Group. Wetlands of Canada. Ecological Land Classification No. 24. Sustainable Development Branch Canadian Wildlife Service, Environment Canada; Ottawa, Canada.

9. Semeniuk, C.A.; Semeniuk, V.A. Geomorphic approach to global classification for inland wetlands. In Classification and Inventory of the World's Wetlands; Finlayson, C.M.; van der Valk, A.G., Eds.; Kluwer Academic Press; Dordrecht, 1995; 103–124. 1988.

10. Paijamans, K.; Galloway, R.W.; Faith, D.P.; Fleming, P.M.; Haantjens, H.A.; Heyligers, P.C.; Kalma, J.D.; Loffler, E. Aspects of Australian Wetlands. Division of Water and land Resources Tech. Paper No. 44, Commonwealth Scientific and Industrial Research Organization: Melbourne, Australia, 1985, 1–71.

11. Ewart-Smith, J.L.; Ollis, D.J.; Malan, H.L. National Wetland Inventory: Development of a Wetland Classification System. Water Research Commission WRC Report No. KV174/06: Water Research Commission: Pretoria, South Africa, 2006.

12. SANBI. Further Development of a Proposed National Wetland Classification System for South Africa. Primary Project Report. South African National Biodiversity Institute: Pretoria, South Africa, 2009.

13. Ollis D.J.; Snaddon, C.D.; Job, N.M.; Mbona, N. Classification System for wetlands and other aquatic ecosystems in South Africa. SANBI Biodiversity Series 22. South African National Biodiversity Institute: Pretoria, South Africa, 2013.

14. Finlayson, C.M.; Davidson, N.C.; Spiers, A.G.; Stevenson, N.J. Global wetland inventory-current status and future priorities. Mar. Freshwater Res. **1999**, 50, 717–727.

15. Ramsar Convention Secretariat, The Ramsar Convention Manual: A Guide to the Convention on Wetlands (Ramsar, Iran, 1971), 5th ed. Ramsar Convention Secretariat; Gland, Switzerland, 2011.

19

Wetlands: Ecosystems

Sherri DeFauw
University of Maine

Introduction

Wetlands perform key roles in the global hydrologic cycle. These transitional ecosystems vary considerably in their capacity to store and subsequently redistribute water to adjacent surface water systems, groundwater, the atmosphere, or some combination of these. Saturation in the root zone or water standing at or above the soil surface is key to defining a wetland. When oxygen levels in water-logged soils decline below 1%, anaerobic (or reducing) conditions prevail. Most, but not all, wetland soils exhibit redoximorphic features formed by the reduction, translocation, and oxidation of iron (Fe) and manganese (Mn) compounds; the three basic kinds of redoximorphic features include redox concentrations, redox depletions, and reduced matrix.[1] Microbial transformations in flooded soils also impact other biogeochemical cycles (C, N, P, S) at various spatial and temporal scales. Several of the most rapidly disappearing wetland ecosystems in North America are profiled here, in terms of properties and processes.

Hydrologic Considerations

Wetland water volume and source of water are heavily influenced by landscape position, climate, soil properties, and geology. Wetlands may be surface flow dominated, precipitation dominated, or groundwater discharge dominated systems (Figure 19.1). Surface flow dominated wetlands include riparian swamps and fringe marshes. In unregulated settings (i.e., no dams or diversions) these ecosystems are subject to large hydrologic fluxes, and vary the most in terms of soil development, sediment loads, and nutrient exchanges. Precipitation-dominated wetlands (e.g., prairie potholes and bogs) reside in landscape depressions and typically have a relatively impermeable complex of clay and/or peat layers that retard infiltration (or recharge) and also impede groundwater discharge (or inflow). Groundwater dominated wetlands (e.g., fens and seeps) may form in riverine settings, at slope breaks, or in areas where abrupt to rather subtle changes in substrate porosity occur. Groundwater contributions to wetlands are complex, dynamic, and rather poorly understood.[2]

Frequency and duration of flooding, and the long-term amplitude of water level fluctuations in a landscape are the three most important hydrologic parameters that "shape" the aerial extent of a

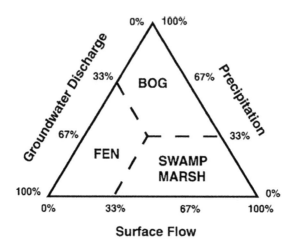

FIGURE 19.1 The relative contributions of groundwater discharge, precipitation, and surface flow determine main wetland types. Swamps and marshes are distinguished by the frequency and duration of surface flows. **Source:** Modified from Keddy.[3]

wetland complex as well as determine the relative abundance of four intergrading wetland settings (i.e., swamps, wet meadows, marshes, and aquatic ecosystems).[3] Wetland hydrodynamics control soil redox conditions. The hydroperiod–redox linkage, in turn, controls plant macronutrient concentrations (N, P, K, Ca, Mg, S), micronutrient availability (B, Cu, Fe, Mn, Mo, Zn, Cl, Co), pH, organic matter accumulation, decomposition, and influences plant zonation. Oftentimes, the zonation of plants (as determined by competition and/or physiological tolerances) provides key insights into the hydrodynamics and biogeochemistry of an area.[3]

Ecosystem Profiles

Four of the most rapidly disappearing wetland ecosystems in North America are summarized here in terms of key properties and processes. These profiles include comments on geographical extent, geomorphology, soils, hydrodynamics, biogeochemistry, vegetation structure, and/or indicator species, as well as recent estimates of ecosystem losses and ecological significance.

Riparian Swamps

Bottomland hardwood forests once dominated the river floodplains of the eastern, southern, and central United States. In the Mississippi Alluvial Plain, an estimated 8.6 million hectare area of bottomland hardwoods has been reduced to 2 million forested hectares remaining.[4] Although their true extents are not well-documented (due to difficulties in determining upland edges), these hydrologically open, linear landscape features have been logged, drained, and converted to other uses (predominantly agriculture) at alarming rates. Between 1940 and 1980, bottomland hardwood forests were cleared at a rate of 67,000 ha yr^{-1}.[5]

Riparian wetlands are unique, vegetative zonal expressions of both short- and long-term fluvial processes. A widely used classification scheme relates flooding conditions (frequency and flood duration during the growing season) with zonal associations of hardwood species.[6] Zone II (intermittently exposed "swamp"—bald cypress and water tupelo usually dominate the canopy) and Zone III (semipermanently flooded "lower hardwood wetlands"—overcup oak and water hickory are common) are accepted as wetlands by most; however, Zones IV (seasonally flooded "medium hardwood wetlands" that include laurel oak, green ash, and sweetgum) and V (temporarily flooded "higher hardwood wetlands"—typically an oak–hickory association with loblolly pine) have been controversial when dealing with wetland management issues.[7] Typically the complexity of floodplain microtopography abates smooth transitions from one zone to the next, with unaltered floodplain levees oftentimes exhibiting the highest plant diversity.[8]

Riparian wetlands process large influxes of energy and materials from upstream watersheds and lateral runoff from agroecosystems. As a result of these inputs, combined with decomposition of resident biomass, the organic matter content of these alluvial soils usually ranges from 2% to 5%.[8] Clay-rich bottomland tracts have higher concentrations of N and P as well as higher base saturations (i.e., Ca, Mg, K, Na) compared to upland areas. Watersheds dominated by riparian ecosystems export large amounts of organic C in dissolved and particulate forms.[9]

Prairie Potholes

Prairie potholes comprise a regional wetland mosaic that includes parts of the glaciated terrains of the Dakotas, Iowa, Minnesota, Montana, and Canada. Originally, this region encompassed 8 million hectares of wetland prior to drainage for agriculture; an estimated 4 million hectares remained at the close of the 1980s.[10] These landscape depressions are the most important production habitat in North America for most waterfowl.[11]

Most prairie potholes are seasonally flooded wetlands dependent on snowmelt, rainfall, and groundwater. Four main hydrological groupings are recognized: ephemeral, intermittent, semi-permanent, and permanent.[11] Water level fluctuations are as high as 2–3 m in some settings. Measures of soil hydraulic conductivity demonstrate that groundwater flow is relatively slow (0.025–2.5 m yr^{-1}); therefore, the potholes are hydrologically isolated from each other in the short-term.[12]

Long-term ecological studies at the Cottonwood Lake Study Area, NPWRC-USGS[11] have revealed the intricacies of several prairie pothole phases. During periods of drought, marsh soils, sediments, and seed banks are exposed (i.e., dry marsh phase). Seed banks in natural sites have 3000–7000 seeds/m^2. A mixture of annuals (usually the dominant group) and emergent macrophyte species germinate on the exposed mudflats and a wet meadow develops. As water levels increase, the annuals decline and emergent macrophytes rapidly recolonize (i.e., regenerating marsh). If the flooding is consistently shallow, emergent macrophytes will eventually dominate the entire pothole. Sustained deep water flooding results in extensive declines in emergent macrophytes (i.e., degenerating marsh), and intensive grazing by muskrats may culminate in a lake marsh phase as submersed macrophytes become established. When water levels recede, emergent macrophytes re-establish. The rich plant communities that develop in these dynamic marshland complexes are also controlled by two additional environmental gradients, namely, salinity and anthropogenic disturbances (involving conversion to agriculture and extensive irrigation well pumping).[11]

Northern Peatlands

Deep peat deposits, in the United States, occur primarily in Alaska, Michigan, Minnesota, and Wisconsin, and are scattered throughout the glaciated northeast and northwest, as well as mountaintops of the Appalachians. The most extensive peatland system, in North America, is the Hudson Bay lowlands of Canada that occupies an estimated 32 million hectares.[13] The Alaskan and Canadian peatlands are relatively undisturbed, and the least threatened by developmental pressures. Elsewhere, peatlands have either been converted for agricultural use (including forestry) or mined for fuel and horticultural materials.

Bogs and fens are the two major types of peatlands that occupy old lake basins or cloak the landscape. The most influential, interdependent physical factors shaping these ecosystems include: 1) water level stability; 2) fertility; 3) frequency of fire; and 4) grazing intensity.[3] Northern bogs are dominated by oligotrophic *Sphagnum* moss species, and may be open, shrubby, or forested tracts. These predominantly rainfed (ombrogenous) systems have low water flow (with a water table typically 40–60 cm below the peat surface), are extremely low in nutrients (especially poor in basic cations), and accumulate acidic peats (pH 4.0–4.5).[14] Fens are affected by mineral-bearing soil waters (groundwater and/or surface water flows), and possess water levels at or near the peat surface. Fens may be subdivided into three hydrologic types: soligenous (heavily influenced by flowing surface water); topogenous (largely influenced by stagnant groundwater); or limnogenous (adjacent to lakes and ponds).[14] These minerogenous ecosystems range from acidic (pH 4.5) to basic (pH 8.0); vegetation varies from open, sedge-dominated settings to shrubby, birch-willow dominated associations to forested, black spruce-tamarack tracts. Nutrient availability gradients do not necessarily coincide with the ombrogenous–minerogenous gradient; recent investigations indicate higher P availability in more ombrogenous peatlands, and greater N availability in more minerogenous peatlands.[15]

Northern peatlands represent an important, long-term carbon sink, with an estimated 455 Pg (1 Petagram = 10^{15} g) stored worldwide.[16] An estimated 220 Pg of C is currently stored in North American peatlands, compared to about 20 Pg in storage during the last glacial maximum.[17] High latitude peatlands also release about 60% of the methane generated by natural wetlands.[18] In addition, sponge-like living *Sphagnum* carpets facilitate permanently wet conditions, and the high cation exchange capacity of cells retains nutrients and serves to acidify the local environment[19]

Pocosins

The Pocosins region of the Atlantic coastal plain extends from Virginia to the Georgia–Florida border.[20] These nonalluvial, evergreen shrub wetlands are especially prevalent in North Carolina; in fact, pocosins once covered close to 1 million hectare in this state.[21] Derived from an Algonquin Indian word for "swamp-on-a-hill," pocosins are located on broad, flat plateaus and sustained by waterlogged, acidic, nutrient-poor sandy, or peaty soils usually far removed from large streams. Wetland losses are high, with 300,000 hectare drained for agriculture and forestry uses between 1962 and 1979.[22]

Pocosins are characterized by a dense, ericaceous shrub layer; an open canopy of pond pine may be present or absent.[21] A typical low pocosin ecosystem [less than 1.5 m (5 ft) tall] includes swamp cyrilla (or titi), fetterbush, bayberry, inkberry, sweetbay, laurel-leaf greenbrier, and sparsely distributed, stunted pond pine.[22] Pocosin soils may be either organic (with a deep peat layer—e.g., Typic Medisaprist) or mineral (usually including a water restrictive spodic horizon—e.g., Typic Endoaquod). As peat depth decreases, the stature of the vegetation increases. High pocosin [with shrubby vegetation 1.5–3.0 m (5–10 ft) tall and canopy trees approximately 5 m (16 ft) in height] usually occurs on peat deposits of 1.5 m (5 ft) or less in thickness or on wet sands.[20] The major natural disturbance to these wetlands is periodic burning (with a fire frequency of about 15–50 years).

Pocosin surface and subsurface waters are similar to northern ombrogenous bogs, but are more acidic with higher concentrations of sodium, sulfate, and chloride ions.[23] Carbon : Phosphorus (C : P) ratios increase sharply during the growing season; phosphorus availability limits plant growth and probably plays a crucial role in controlling nutrient export. Undisturbed pocosins export organic N and inorganic phosphate in soil water.[23]

Conclusions

Despite existing preservation policies, U.S. wetland conversions are anticipated to continue at a rate of 290,000–450,000 acres (117,408–182,186 ha) annually.[24] It is widely known that wetlands are the product of many environmental factors acting simultaneously; perturbations in one realm (e.g., hydrology) not only impact local wetland properties and processes, but also have consequences in linked ecosystems as well. Wetlands are major reducing systems of the biosphere, transforming nutrients and metals, and regulating key exchanges between terrestrial and aquatic environments.

Acknowledgment

William J. Rogers, B. A. Howell, and Van Brahana commented on an earlier draft; their reviews are genuinely appreciated.

References

1. Soil survey staff. *Keys to Soil Taxonomy,* 8th Ed.; USDA-NRCS: Washington, DC, 1998.
2. http://water.usgs.gov/nwsum/WSP2425/hydrology.html (accessed July 2002).
3. Keddy, P.A. *Wetland Ecology: Principles and Conservation*; Cambridge University Press: Cambridge, 2000.
4. Llewellyn, D.W.; Shaffer, G.P.; Craig, N.J.; Creasman, L.; Pashley, D.; Swan, M.; Brown, C. A decision support system for prioritizing restoration sites on the Mississippi River alluvial plain. Conserv. Biol. **1996**, *10*(5), 1446–1455.
5. Kent, D.M. *Applied Wetlands Science and Technology*, 2nd Ed.; Lewis Publishers: Boca Raton, FL, 2001.
6. Clark, J.R., Benforado, J., Eds.; *Wetlands of Bottomland Hardwood Forests*; Elsevier: Amsterdam, 1981.
7. Mitsch, W.J.; Gosselink, J.G. *Wetlands*; Van Nostrand Reinhold Company: New York, 1986.
8. Wharton, C.H.; Kitchens, W.M.; Pendleton, E.C.; Sipe, T.W. *The Ecology of Bottomland Hardwood Swamps of the Southeast: A Community Profile*; U.S. Fish and Wildlife Service, Biological Services Program FWS/OBS-81/37: Washington, DC, 1982; 1–133.
9. Brinson, M.M.; Swift, B.L.; Plantico, R.C.; Barclay, J.S. *Riparian Ecosystems: Their Ecology and Status*; U.S. Fish and Wildlife Service, Biological Services Program FWS/ OBS-81/17: Washington, DC, 1981; 1–151.
10. Leitch, J.A. Politicoeconomic overview of prairie potholes. In *Northern Prairie Wetlands*; van der Valk, A.G., Ed.; Iowa State University Press: Ames, 1989; 2–14.
11. http://www.npwrc.usgs.gov/clsa (accessed July 2002).
12. Winter, T.C.; Rosenberry, D.O. The interaction of ground water with prairie pothole wetlands in the cottonwood lake area, East-Central North Dakota, 1979–1990. Wetlands **1995**, *15*(3), 193–211.
13. Wickware, G.M.; Rubec, C.D.A. *Ecoregions of Ontario*, Ecological Land Classification Series No. 26; Environment Canada, Sustainable Development Branch: Ottawa, Ontario, 1989.
14. http://www.devonian.ualberta.ca/peatland/peatinfo.htm (accessed July 2002).
15. Bridgham, S.D.; Updegraff, K.; Pastor, J. Carbon, nitrogen, and phosphorus mineralization in Northern Wetlands. Ecology **1998**, *79*, 1545–1561.
16. Gorham, E. Northern peatlands: role in the carbon cycle and probable responses to climatic warming. Ecol. Appl. **1991**, *1*(2), 182–195.
17. Halsey, L.A.; Vitt, D.H.; Gignac, L.D. *Sphagnum* dominated peatlands in North America since the last glacial maximum: their occurrence and extent. The Bryologist **2000**, *103*, 334–352.
18. Cicerone, R.J.; Ormland, R.S. Biogeochemical aspects of atmospheric methane. Global Biogeochem. Cycles **1988**, *2*(2), 299–327.
19. van Breeman, N. How *Sphagnum* bogs down other plants. Trends Ecol. Evol. **1995**, *10*(7), 270–275.
20. Sharitz, R.R.; Gresham, C.A. Pocosins and Carolina bays. In *Southern Forested Wetlands: Ecology and Management*; Messina, M.G., Conner, W.H., Eds.; Lewis Publishers: Boca Raton, FL, 1998; 343–389.

21. Richardson, C.J. Pocosins: an ecological perspective. Wetlands **1991**, *11*(S), 335–354.
22. Richardson, C.J.; Evans, R.; Carr, D. Pocosins: an ecosystem in transition. In *Pocosin Wetlands: An Integrated Analysis of Coastal Plain Freshwater Bogs in North Carolina*; Richardson, C.J., Ed.; Hutchinson Ross Publishing Company: Stroudsburg, PA, 1981; 3–19.
23. Wallbridge, M.R.; Richardson, C.J. Water quality of pocosins and associated wetlands of the Carolina coastal plain. Wetlands **1991**, *11*(S), 417–439.
24. Schultink, G.; van Vliet, R. *Wetland Identification and Protection: North American and European Policy Perspectives*; Agricultural Experiment Station Project 1536; Michigan State University, Department of Resource Development: East Lansing, MI, 1997.

20

Wetlands: Freshwater

Brij Gopal
Centre for Inland
Waters in South Asia

Introduction

Wetland is the term used since the late 1960s for a very large diversity of habitats that were previously known by common terms such as bog, fen, mire, marsh, swamp, and mangrove and hundreds of local names in different countries. Wetlands occur in all climates, across a wide range of latitudes, and from sea level to more than 5000 m altitude (as in the Himalaya). The unifying characteristic of these areas on the Earth's surface is that the land remains either water-saturated or under shallow water for a period ranging from several weeks to the whole year and is generally covered with vegetation that differs from that in the adjacent areas (see also[1]). Wetlands are sometimes transitional in character between deep-water and terrestrial habitats, and are often located between them. These habitats may depend for their water entirely upon the precipitation (bogs and prairie potholes) or on the surface water from the rivers (riparian/floodplain wetlands), lakes (littoral zones), or the oceans (coastal wetlands). Wetlands also develop where the ground water is discharged naturally onto the surface (fens). Wetlands are usually grouped into freshwater and marine wetlands or into inland and coastal wetlands. The inland wetlands, occurring above the mean sea level, also include those developing in saline waters (salt lakes) besides those experiencing estuarine/brackish water conditions (lagoons and backwaters).

Kinds of Wetlands

There are many ways of categorizing wetlands on the basis of diverse characteristics such as hydrology, geomorphology, nutrient status, production, salinity, and so on. Freshwater wetlands can be readily grouped into bogs, fens, marshes, swamps, and shallow waters. The bogs and fens occur in temperate climates and accumulate peat, whereas marshes are dominated by herbaceous plants (mostly grasses and sedges), swamps are dominated by trees and shrubs, and shallow waters may have more of submerged and floating plants. *Taxodium* swamps in southeast United States, *Melaleuca* swamps of Malaysia and Lower Mekong basin, *Myristica* swamps of southwest India, peat swamp forests (usually dominated by dipterocarps) of Indonesia and Maputaland (South Africa), and the floodplain forests of the Amazonia are examples of important swamps. Well-known marshes include the reed and papyrus marshes of Okavango, Mesopotamian marshes of Iraq, reed marshes of the Danube River delta, papyrus marshes of Egypt, Camargue marshes of southern France, and the extensive Everglades of Florida (U.S.A.). Wetlands also develop in depressions that are flooded seasonally for only a few months.

Humans have also created wetlands around the world, especially in arid and semi-arid regions, by constructing small and large reservoirs and dugout ponds to store water for various human uses. Another kind of wetlands includes the extensive paddy fields and aquaculture ponds (or fish ponds), which are modified by humans from the naturally occurring wetlands for specific production systems. Among the human-made wetlands, mention may also be made of the 200 ha Putrajaya wetlands (Malaysia) that were constructed in late 1990s to treat storm water and wastewater from the new capital township, Putrajaya.[2]

Wetland Characteristics

Wetlands can be distinguished from lakes with deep water by the dominant community of primary producers. The planktonic organisms dominate the deep water lakes, whereas macrophytes (including macroalgae, bryophytes, and vascular plants) are usually the dominant producers in wetlands. Wetlands generally support high densities and diversity of fauna, particularly birds, fish, and macro-invertebrates. Another characteristic of most freshwater wetlands is the hydric soils, which develop under prolonged waterlogging or submergence that eliminates oxygen from the soil. The hypoxic (reduced supply of oxygen) to anoxic (total absence of O_2) conditions in the soil affect the root respiration and consequently the plant growth. Wetland plants have a variety of morphological, anatomical, and physiological adaptations against hypoxia. These adaptations include the presence of air spaces within their roots, stems, and leaves that allow the storage and transport of gases from the atmosphere to the roots. Wetland plants are particularly tolerant to many toxic substances, such as heavy metals, which are either precipitated on the surface of their roots or stored in isolated areas of their roots and stems. In the absence of oxygen, anaerobic micro-organisms mediate several chemical transformations that often result in the reduction of C, S, N, Fe, Mn, and other substances to their toxic forms. The reduction of carbon results in the production of methane (commonly known as marsh gas), whereas the reduction of nitrates (denitrification) results in harmful ammonia or loss of nitrogen to the atmosphere. Nitrogen deficiency in wetlands is often explained by the occurrence of insectivorous plants such as *Drosera* and *Utricularia* in wetlands. The reduction, translocation, and oxidation of iron (Fe) and manganese (Mn) compounds in the soil impart characteristic redoximorphic features to many wetland soils.

Besides climate and geomorphic features of the area, the hydrological regimes (i.e., the duration, depth, and annual amplitude of changes as well as the frequency and timing of flooding) primarily determine all wetland characteristics. (For more on wetland hydrology, see [1].) Flooding regimes control soil chemistry, pH and nutrient availability, microbial decomposition and organic matter accumulation, and distribution and zonation of biota. Nutrients, salinity, and disturbance are other important modifiers that affect wetland structure and function.[3] Peat (partly decomposed organic matter) accumulation occurs in both temperate and tropical climates. In northern temperate regions, peat is formed largely by moss *(Sphagnum* species), whereas in the Amazon basin and tropical Southeast Asia, peat develops from woody material in swamp forests.

Wetland Biodiversity

Wetlands are usually very rich in biodiversity.[4] Often, their biodiversity is underestimated because one or two species of plants may dominate large areas (e.g., Lake Neusiedler See in Austria with almost exclusive occurrence of *Phragmites australis)* and similarly large populations of one animal species may appear dominant. However, many wetlands are extraordinarily rich in species of higher plants alone: 1150 species in the Everglades (U.S.A.), 672 species in Teici (Latvia), 650 species in Kushiro (Japan), 873 species in Dongdong Tinghu (China), 1256 species in Okavango (Botswana), and approximately 900 species of only trees in the whitewater floodplains of Amazon basin. The faunal diversity is exceptionally large considering the fact that wetland animals not only include the residents in the wetland proper and regular migrants from deepwater habitats, terrestrial uplands, and other wetlands but also occasional visitors and those indirectly dependent on wetland biota.[5] On a regional scale, the beta-diversity of wetlands is quite large as 15–20% of all biota of the region occurs in wetlands.

Wetland Productivity

Wetlands are usually considered to be highly productive systems. However, this is true only for warmer climates because primary production is greatly constrained by low temperatures in northern temperate regions. Peat bogs are usually very low in production (0.5 $g/m^2/yr$), whereas marshes and swamps (such as Amazonian swamp forests and Indonesian peat swamp forests) have very high production – as high as or even higher than that of rainforests. Nutrient availability is also a major constraint to wetland production. Wetlands are accordingly classified as oligo-trophic or eutrophic depending on their nutrient status.

Wetland Use and Human Impacts

Humans have extensively used, impacted, and transformed freshwater wetlands in all parts of the world. Wetlands have provided, and are still used as a valuable source of food, fuel, fodder, and fiber. Many wetland plants have been a source of medicine and various chemicals. In the valleys of Tigris– Euphrates and Indus, wetlands formed the cornerstone of human civilizations that grew there. In most of South and Southeast Asia, wetlands were closely integrated into the social and cultural life of the people, and were even considered sacred. In temperate regions, although peat bogs were greatly despised, peat formed a major source of fuel and reeds (*Phragmites*) were extensively used for thatching the cottages. However, with growing agricultural and urban development, wetlands were drained and reclaimed on a large scale. Forested wetlands (swamps) were also cleared. Marshes and shallow water bodies were converted to paddy fields and fish ponds. Within the United States, more than half of the wetlands were lost to drainage during 1780–1980, and more than 90% of wetlands in New Zealand were lost within a few decades after colonization started in the mid-1800s. Wetland losses were relatively small in Asia where they were converted rapidly and until recently, still lower in Africa and South America where developmental and human pressure had remained rather low.

Values and Ecosystem Services

After large areas of wetlands were lost worldwide, people started realizing their importance as critical habitats for waterfowl in the early 1950s. By the 1970s, wetlands were recognized as ecosystems and researchers discovered their important functions and values to human beings. Various studies demonstrated their major role in regulating water quality, controlling erosion, supporting fisheries and wildlife, and above all in the hydrological cycle. The Millennium Ecosystem Assessment[6] brought into focus the term "ecosystem services" for the benefits derived by the humans from the functioning of ecosystems. Wetlands also offer a wide range of ecosystem services. They provide freshwater and a variety of plant and animal resources, regulate floods and droughts, climate and soil erosion, support high biodiversity and formation of soils, and have many cultural, recreational, and aesthetic benefits to different communities. However, it is important to note that wetlands differ greatly in their ability to provide various ecosystem services and no wetland can provide all of them.

Threats to Wetlands

Wetlands in general and freshwater wetlands in particular, are among the most threatened ecosystems as humans continue to alter, directly and indirectly, the hydrological regimes. Surface water flows are extensively diverted and impounded for irrigation, hydropower, and other human uses and thereby affect the downstream wetlands. The abstraction of both surface and ground water, drainage, and the wastewater discharge also alter the hydrological regimes directly. Indirect alterations are caused by land use changes throughout the wetland catchments. Discharge of domestic wastewater and industrial effluents into wetlands directly or through lakes and rivers, seriously affect the biota, particularly the benthic and microbial communities, and a number of ecosystem processes.

Another common threat to freshwater wetlands throughout the world has been the introduction of exotic invasive species, mostly plants. Water hyacinth tops the list in all tropical countries and is becoming a problem in its native Brazil as well. *Lythrum salicaria, Melaleuca quinquenervia, Typha angustifolia, Phragmites australis, Mimosa pigra*, and *Arundo donax* are among the invasives that are causing large changes in wetland plant and animal communities and their functioning in different parts of the world.

Both natural and human-induced fires are common threats to wetlands, although fire has also been used as a management tool in some areas. For example, the Southeast Asian peat swamp forests are now affected by increased frequency of fire—largely caused by drainage, land use changes, and long dry spells. These fires not only cause the loss of wetlands with high biodiversity but also have an impact on wildlife, human health, and economy.

Wetlands and Climate Change

Wetlands play an important role in regulating climate change. On one hand, they contribute significantly to carbon sequestration through their high primary production and low decomposition rates, and on the other, they are an important source of the greenhouse gas, methane. It is projected that the draining of peatlands and fire in peat swamps will add significantly to the climate change processes. At the same time, wetlands themselves are directly threatened by climate change as the melting of glaciers and the increased variability in precipitation regimes will impinge upon the hydrological regimes of wetlands in different parts of the world. Climate change is also projected to favor the spread of invasive species.

Conservation and Restoration

During the past few decades, there have been many efforts at national and international level to conserve wetlands and not only protect them against degradation but also to restore them. The Ramsar Convention on Wetlands of International Importance, which was agreed upon in 1971 and came into force in 1980, is the only global convention dealing with an ecosystem type. It has served as the single most important instrument for wetland conservation. It requires the contracting parties to not only designate important wetlands within their country but also to include wetland conservation in their national land-use planning so as to promote the wise use of wetlands, to establish wetland nature reserves, and to promote training in the fields of wetland research and management. As of October 12, 2013, 168 nations had designated 2165 wetlands (including marine and saline wetlands) covering a total area of 205,830,125 ha under the Ramsar Convention.

Wetland restoration requires the restoration of hydrological regimes, reduction in nutrient loadings and wastewater discharges, and protection against other threats to the specific wetland. Ramsar Convention also promotes wetland restoration through an analysis of change in their ecological character, and along with its approach for wise use of wetlands.

References

1. Mitsch, W.J.; Gosselink, J.G. *Wetlands*; 4th Ed.; John Wiley: New York, 2007; 582 pp.
2. Shutes, R.B.E. Artificial wetlands and water quality improvement. Environ. Int. **2001, 26,** 441–447.
3. Keddy, P.A. *Wetland Ecology*, 2nd Ed.; Cambridge University Press: Cambridge, 2010; 497 pp.
4. Gopal, B. Biodiversity in wetlands. In *The Wetlands Handbook*; Maltby, E., Barker, T., Eds.; Blackwell Science: Oxford, U.K, 2009; 65–95.
5. Gopal, B.; Junk, W.J. Biodiversity in wetlands: An introduction. In *Biodiversity in Wetlands: Assessment, Function and Conservation*, Vol. 1.; Gopal, B., Junk, W.J., Davis, J.A., Eds.; Backhuys Publishers: Leiden, 2000; 1–10.
6. Millennium Ecosystem Assessment. *Ecosystems and Human Well-Being: Wetlands and Water Synthesis*. World Resources Institute: Washington, DC, 2005; 68 pp.

Bibliography

Brinson, M.M.; Swift, B.L.; Plantico, R.C.; Barclay, J.S. Riparian Ecosystems: Their Ecology and Status; U.S. Fish and Wildlife Service, Biological Services Program FWS/ OBS-81/17: Washington, DC, 1981; 1–151.

Fraser, L.H.; Keddy, P.A. *The World's Largest Wetlands*: Ecology and Conservation. Cambridge University Press: New York, 2005; 488.

Grobler, R.; Moning, C.; Sliva, J.; Bredenkamp, G.; Grundling, P.-L. Subsistence farming and conservation constraints in coastal peat swamp forests of the Kosi Bay Lake system, Maputaland, South Africa. *Géocarrefour* 2004, 79/4, [online] URL: http://geocarrefour.revues.org/842.

Lähteenoja, O.; Ruokolainen, K.; Schulman, L.; Alvarez, J. Amazonian floodplains harbour minerotrophic and ombrotrophic peatlands. Catena **2009, 79,** 140–145.

Lähteenoja, O.; Rojas Reátegui, Y.; Räsänen, M.; Del Castillo Torres, D.; Oinonen, M.; Page, S. The large Amazonian peatland carbon sink in the subsiding Pastaza-Marañón foreland basin, Peru. Global Change Biol. **2012, 18,** 164–178.

LePage, B.A. *Wetlands: Integrating Multidisciplinary Concepts*; Springer: New York, 2011; 261 p.

Lewis, W.M. *Wetlands: Characteristics and Boundaries*; National Academies Press: Washington, DC, 1995; 307 p.

Schöngart, J.; Wittmann, F.; Worbes, M. Biomass and net primary production of Central Amazonian floodplain forests. In *Amazonian Floodplain Forests: Ecophysiology, Biodiversity, and Sustainable Management*; Junk, W.J.; Piedade, M.T.F.; Wittmann, F.; Schöngart, J., Parolin, P., Eds.; Springer: New York, 2010; 347–388.

Wittmann, F.; Schöngart, J.; Montero, J.C.; Motzer, T.; Junk, W.J.; Piedade, M.T.F.; Queiroz, H.L.; Worbes, M. Tree species composition and diversity gradients in whitewater forest across the Amazon basin. J. Biogeogr. **2006, 33,** 1334–1347.

Website

http://www.ramsar.org/.

21

Wetlands: Tidal

William H.
Conner and Jamie
A. Duberstein
Clemson University

Andrew H. Baldwin
University of Maryland

Introduction

Tidal wetlands occur in some of the most dynamic areas of the Earth and are positioned at the interface between land and sea where two of the most powerful forces acting on the planet's waters collide. Light and heat from solar energy evaporate water and move it to the atmosphere, while gravity from the moon and sun propels the ebbing and flowing of tides, resulting in low and high water events on diurnal timescales. Vegetation, also fueled by solar energy, plays a crucial role in stabilizing coastal boundaries. At the mouths of the rivers, edges of embayments, and lagoons, these two great forces interact in complex ways to form the hydrodynamic framework for the development of tidal wetlands.

Tidal wetlands reduce the energy of tides, storms, and floods through complex interactions with biotic and abiotic constituents. At low tide, most energy dissipation is a result of linear force being converted to helical motion and undulating bottom structure, and friction with the channel surface. As the banks are topped at high tide, vegetation plays an increasingly important role in energy dissipation. Sedimentation and deposition of nutrient-laden materials are important in maintaining this same vegetation, as is the salinity profile determined by the relative contribution of riverine and oceanic water. Animals can also play an important role through such means as oyster bar development, benthic burrows, beaver dams, and excavations. The complexity of the interactions makes characterizing tidal wetlands a difficult task. In this entry, we briefly review the three primary tidal wetland types that occur globally, including tidal marshes, mangroves, and freshwater forested wetlands.

Tidal Marshes

Tidal marshes (Figure 21.1) are coastal wetlands dominated by herbaceous plants, unlike tidal freshwater forested wetlands and tropical tidal forests (mangroves). Marshes occur along coastlines at middle to high latitudes of both the Northern and Southern Hemispheres and are less common at tropical latitudes, where mangrove trees are the dominant form of intertidal wetland vegetation [1]. Contrary

FIGURE 21.1 (See color insert.) Tidal marshes are dominated by herbaceous vegetation comprising a wide range of species. In Chesapeake Bay on the U.S. Atlantic Coast, both tidal freshwater (a) and brackish (b) marshes are abundant. Tidal marshes are less abundant in much of Europe, but include reed-dominated tidal freshwater marshes along the Elbe River on Germany's Atlantic Coast (c), and salt marshes dominated by species, including common cordgrass (*Spartina anglica* C.E. Hubbard) and sea aster (*Aster tripolium* [Jacq.] Dobrocz.) along the North Sea (d). (Courtesy of A.H. Baldwin, University of Maryland.)

to popular belief, not all tidal marshes are saline. Rather, this ecosystem spans salinity regimes ranging from fresh to concentrations above that of seawater (Figure 21.2) and includes tidal freshwater (<0.5 ppt salinity), oligohaline (0.5–5.0 ppt salinity), brackish (5.0–18 ppt salinity), and salt (>18 ppt salinity) marshes [2–4]. There is a great deal of uncertainty in estimates of the area of these wetlands depending upon where one draws the line between tidal and nontidal areas. Worldwide, there are an estimated 12×10^6 ha of tidal marshes with nearly 25% of them in the United States [5].

Soil and Biogeochemistry

Tidal marshes in sheltered estuaries and deltas receive inputs of mineral riverine sediment, while in marine settings, they may receive considerable inputs of marine sediment [6]. Due to high primary productivity, most tidal marshes accumulate organic matter from roots, stems, and leaves. Organic matter content of tidal marsh soils may be low in areas receiving abundant mineral sediment, such as salt marshes of the Yangtze River in China [7], or high (>80% organic matter) where there is little mineral sediment input such as the floating marshes of the Louisiana Delta in the United States [8]. Organic matter deposition is an important contribution to vertical accretion of the marsh surface in many tidal marshes, helping them to maintain elevation under rising sea-level conditions. For example, across 76 tidal freshwater marshes in North America and Europe, organic matter was responsible for 62% of marsh vertical accretion, with the remainder coming from mineral matter [9].

As in other wetlands, anaerobic conditions predominate in soils, driving a series of sequential reduction reactions as soil oxidation-reduction (redox) potential decreases [10]. In tidal freshwater marshes,

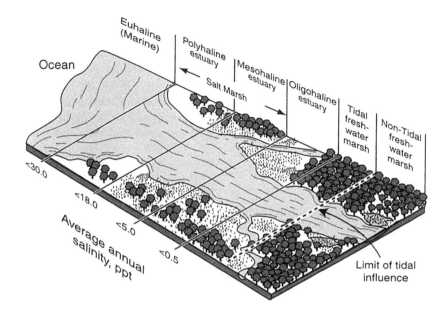

FIGURE 21.2 The salinity of tidal marshes varies across estuaries. Most large estuaries contain a tidal freshwater zone of varying width that experiences tidal fluctuation but little or no intrusion of saline water. The salinity of estuarine water has an overriding influence on the species composition of plants and animals present and on biogeochemical processes.

Source: © John Wiley & Sons, Inc., 2009 as included in Wetland Ecosystems. Reprinted with permission.

methanogenesis is generally the terminal redox reaction, but when sulfate is present due to ocean water intrusion, for example, in saline marshes, sulfate reduction is likely to predominate because it is energetically more favorable than methanogenesis [11,12]. One implication of these redox processes is that increases in sea level may result in sulfate intrusion into tidal freshwater marshes, resulting in a shift from methanogenesis to sulfate reduction, accelerating decomposition of organic matter, and potential loss of wetlands due to reduced vertical accretion [14].

Vegetation and Fauna

Because they span a range of inundation frequencies and salinity regimes, tidal marshes support spatially diverse assemblages of plant and animal communities. In general, the number of plant species in a given wetland increases as elevation increases (i.e., inundation decreases) and salinity decreases [14]. However, due in part to mixing of freshwater and brackish species, the transition between the tidal freshwater and brackish zones of estuaries may contain the most species-rich tidal marshes [15]. Among the most globally widespread taxa of tidal marsh plants are common reed (*Phragmites australis* [Cav.] Trin. ex Steud.), bullrushes (*Schoenoplectus* spp.), and cordgrasses (*Spartina* spp.) [16–18]. Across inundation gradients, vegetation often forms distinct bands of differing plant communities (horizontal zonation), with more distinct zonation occurring in salt marshes than in tidal freshwater marshes [3]. The distribution of plant species across estuarine salinity gradients is driven by a tradeoff between competitive ability and salinity stress tolerance. Salt-tolerant species can grow in tidal freshwater marshes, but only in the absence of competition from neighboring freshwater species, while freshwater species die or exhibit reductions in growth due to salt stress when transplanted to saline marshes [18,19]. While many of the dominant species of tidal marshes are clonal perennials, seed dispersal and seed banks are important sources of propagules for the maintenance or regeneration of vegetation in both tidal freshwater and salt marshes and may play a role in developing resilient native plant communities in wetland restoration [20–23].

In some tidal freshwater marshes, annual species comprise half or more of the species richness and biomass production [24,25]. In addition to vascular plants, benthic and epiphytic microalgae are important but often overlooked primary producers in marshes across estuarine salinity gradients [3,26–28].

The high primary production of tidal marshes supports abundant and diverse faunal communities, including aquatic and terrestrial invertebrates, fish, birds, mammals, reptiles, and (in tidal freshwater zones) amphibians [3]. Fish use marsh edges and, to a lesser extent, marsh interiors as nursery habitats [29], and fish and invertebrate species composition differs between salinity regimes [3,30]. Terrestrial invertebrate communities (spiders, beetles, isopods, and others) also vary along salinity gradients in tidal marshes [31].

Ecosystem Processes

Primary production from vascular plants, benthic algae, and phytoplankton in tidal marshes and associated creeks can be high [5]. Aboveground primary production in tidal freshwater marsh emergent plants varies considerably between vegetation types and species, but can be above 2,000 g/m²/yr, and total production including belowground may reach 8,500 g/m²/yr [32,33]. Primary production in emergent salt marsh plants is also common in the range of 1,000–2,000 g/m²/yr aboveground and may exceed 8,000 g/m²/yr, including belowground [5,28].

The high primary production of tidal marshes results in abundant litter and detritus that are important in internal carbon (C) and nutrient cycling and export of materials supporting food webs in adjacent waters. In general, decomposition rates of plant material are faster in tidal freshwater wetlands than in salt and brackish tidal wetlands because the freshwater plant species have relatively lower C:N ratios [3]. However, decomposition rates may differ considerably between species within a wetland of a particular salinity regime, and even within the same individual plant due to differences in lignin content, for example, between stems and leaves [34]. Much of the organic matter and nutrients flow out of tidal marshes into adjacent marshes, with estimates of the amount of primary production exported varying considerably and depending on factors such as geomorphology, hydrology, and weather [5].

Trends, Restoration, and Management

Tidal marshes are subject to alteration and loss from both natural and anthropogenic causes. Among the most important threats to tidal marshes are global climate change, coastal eutrophication, and invasive species, as well as continued direct physical impacts on marshes via draining, dredging, or filling. Acceleration in the rate of sea-level rise may exceed the ability of marshes to accrete vertically, resulting in their conversion to open water systems [35], although vegetation allows some marshes to maintain elevation under conditions of rising sea level [36,37]. Higher atmospheric concentration of CO_2 favors C_3 over C_4 species [38], but nitrogen (N) addition in combination with added CO_2 promotes encroachment of C_4 plants, negating increases in community biomass that would otherwise occur due to CO_2 alone [39]. Excess nutrients alter plant community composition in both tidal freshwater and saline marshes [40,41]. On the Atlantic and Gulf of Mexico coasts of North America, Eurasian genotypes of common reed are displacing native marsh plant communities in both tidal freshwater and brackish marshes [42], although some faunal communities are not strongly impacted [43,44]. Similarly, the dominant species at lower elevation in U.S. Atlantic and Gulf coast salt marshes, smooth cordgrass (*Sporobolus alterniflorus* (Loisel.) P.M. Peterson & Saarela), has been introduced in estuaries around the globe, including the U.S. west coast and the Yangtze River Delta in China [45,46]. In addition to these impacts, salt marsh dieback is a recent phenomenon that is not fully understood but may be linked to drought and changes in soil biogeochemistry [47], and that can alter soil structure and sedimentation dynamics [48,49].

Restoration and management of tidal marshes have offset some of their losses and changes. Provided suitable tidal hydrology, vegetation appears to regenerate fairly quickly (5–10 years) in restored salt marshes, although development of soil carbon and nitrogen pools may require much longer time frames

[50,51]. In tidal freshwater marshes, seed dispersal may rapidly establish a diverse plant community and seed banks in restored sites, but in an urban environment constraints on vegetation development (altered hydrology, nutrients, invasive species) may prevent establishment of vegetation similar to that seen in marshes within rural environments [22,52–54]. Control of invasive plants may help restore native marsh vegetation, at least temporarily, but may have little impact on invertebrate or fish communities that are not altered by the presence of the invader [55]. Efforts to divert water from the Mississippi River that introduce sediments and fresh water have the potential to prevent marsh loss, but excess nitrogen may reduce resilience in sediment-starved areas by reducing root growth [56,57]. Over small areas, sediment can be sprayed onto the surface to maintain surface elevation in rapidly subsiding wetlands [58,59].

Tropical Tidal Forests

This tidal habitat type is dominated by a group of plant species known as mangroves [60]. With some exceptions, mangroves are mostly woody, facultative halophytes sensu [61] that dominate tropical coastlines throughout the world due to a number of strong, and unique, life history characteristics (Figure 21.3) [62]. While mangroves are also found in subtropical and some warm temperate environments (e.g., New Zealand's North Island, northern Gulf of Mexico), they readily outcompete tidal salt marshes in the tropics [63], differentiating them from marsh distribution more in terms of climate than in their similar hydrodynamic requirements [64]. Mangrove distribution includes 123 countries comprising between 13,776,000 and 15,236,100 ha worldwide [65,66], and they provide critical goods and services to many inhabitants of low-lying, underdeveloped nations [67]. Mangrove forests are threatened by development on nearly every coastline they exist [68,69].

Soil and Biogeochemistry

While the marine intertidal environment can be quite harsh to vegetation, process transformations of nutrients and energy from anaerobic soils—despite growing in stressful conditions—define mangrove ecosystems and enable high degrees of productivity. In fact, Alongi [70; p. 108] states that "highly evolved and energetically efficient plant-soil-microbe relations are a major factor in explaining why mangroves are highly productive in harsh tropical environments." Paradoxically, there are no "normal" mangrove soils; pedogenic properties range from coralline to highly organic, acidic to alkaline (pH 5.8–8.5), and anoxic to suboxic (redox, −200 to +300 mv) [70]. The type of vegetation can also influence the biogeochemistry of mangrove soils [71–73]; the tight connections between plant and soil in mangroves provide a model for the entire field of ecology [70].

FIGURE 21.3 Mangrove forest at low tide along the Shark River in Everglades National Park, Florida, the United States (Courtesy of K.W. Krauss, U.S. Geological Survey.)

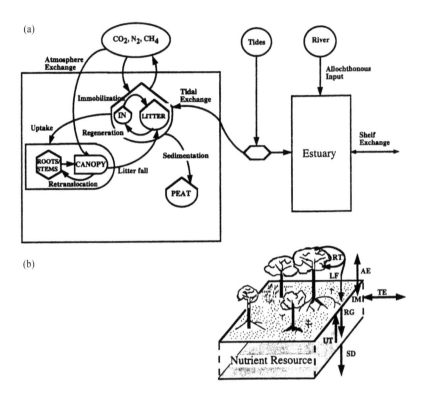

FIGURE 21.4 (a) The suite of organic matter and nutrient fluxes between mangrove soils and the remainder of the estuary. (b) Conceptual overview of spatial linkages among ecological processes in mangroves, with application to all tidal wetlands being reviewed herein [66]. Abbreviations: AE, atmosphere exchange; IM, immobilization; LF, litterfall; RT, retranslocation; RG, regeneration; SD, sedimentation; TE, tidal exchange; UT, uptake.
Source: © Lewis Publishers 1998 as included in Southern Forested Wetland Ecology and Management. Reprinted with permission.

Twilley [74] described the fluxes of organic matter and nutrients in South Florida mangrove ecosystems as a balance among all sources and sinks (Figure 21.4); whether a particular mangrove wetland is considered a source, sink, or conveyor of specific nutrients relates strongly to the process of material exchange between wetland and estuary and in situ transformations in lieu of geophysical processes in the estuary [74]. For example, N in the form of NO_3 and NH_4 is taken up by fine roots within the soil during material flux through the wetland, by one account, leaving only about 4%–12% as the fraction of N buried within the mangrove soil [75]. Denitrification (efflux of N_2O and N_2 to the atmosphere) and dissolved fluxes of organic and inorganic N can represent an important conduit for N loss from mangroves. However, where growth limitations have been observed, studies have often implicated phosphorus (P) availability [76]. Feller [77] discovered that fertilizing stunted mangroves in Belize with N and P produced a P-only response, both in growth and in altering the relative sclerophylly of leaves.

Carbon fluxes from mangroves are large when modeled on a global perspective (Table 21.1); the net export of dissolved inorganic C from all mangrove wetlands ranges from 112 to 160 Tg C/year [70,78]. Mangrove soils can be incredibly proficient at sequestering and storing atmospheric C [79], rating high among tropical and coastal ecosystems in C burial rates (226 g C/m^2 yr) [80]. The export of dissolved organic carbon and particulate organic carbon to the coastal zone is also significant and may account for an atmospheric C sink similar to burial [81]. The volume of litterfall from mangrove ecosystems that can serve as a source of organic C to the soil ranges globally from 1.3 to 18.7 t/ha/yr and is directly related to standing forest structural attributes (Figure 21.5). While productivity is important, the true potential

TABLE 21.1   Summary of Material Carbon Fluxes (T_g C/yr) from Mangrove Wetlands Globally, Assuming an Area of 160,000 ha [62]

Vegetation		Tidal Fluxes	
GPP	+735	POC	−29
R_c	−423	DOC	−14
NPP	+309	Rwater	−35
Wood	+67		
Litter	+68	**Total Respiration**	−500
Root	+174		
		Export	−160
Soil Fluxes			
Burial	+29		
R_s	−42		

Abbreviations: GPP, gross primary productivity; NPP, net primary productivity; R, respiration; total respiration, $R_{canopy} + R_{soil} + R_{water}$; "+," C import/sequestration; "−," C export/loss.

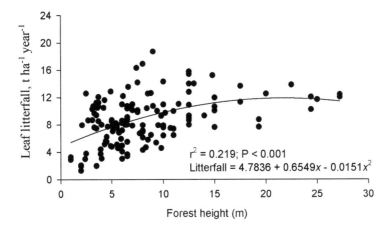

FIGURE 21.5   Mangrove litterfall (mt/ha/yr) as a function of mean forest height (*m*) for both Atlantic East Pacific (AEP) and Indo West Pacific (IWP) biogeographic regions (sensu [76]). (Adapted from [143–158].)

for C sequestration within mangrove soils lies in the high relative rates of root growth balanced against low rates of decomposition in low-oxygen soils [82]. Root growth has been documented to be as high as 525 g/m²/yr¹ from biogenic Caribbean atolls, contributing directly to a layer of peat 10 m deep and 7000–8000 years old [83].

Vegetation and Fauna

Globally, there are nearly 73 species and/or hybrids of mangroves [84], and forests range in height from submeter, stunted mangroves in the Caribbean [77] to tall forests in the 63 m range in Gabon, equatorial Africa [85]. From an ecophysiological perspective, mangroves stand out among tropical forest vegetation and tolerate at least five environmental regimes that would lead to rapid mortality in many plants: (i) Mangroves are facultative halophytes that often flourish in high salinity, in fact, optimal growth rates vary by species but range from 5% to 75% full-strength seawater [62]; (ii) mangrove leaves absorb high, persistent light levels inherent to the tropics without undergoing photoinhibition [86]; (iii) mangroves often occur in low-nutrient environments [87] and are especially efficient at nutrient resorption prior to leaf senescence [88,89]; (iv) mangroves have high levels of water-use efficiency at both the individual leaf and tree levels [80] to ensure less water demand against osmotic gradients; and (v) mangroves tolerate tidal flooding quite well [90,91].

Mangrove faunal assemblages are much more diverse than vegetation assemblages, and include crustaceans, fish, insects, spiders, reptiles, and fur-bearing mammals. Mangroves provide critical habitat for economically important species, such as the mud crab (*Scylla serrata*), which can affect local economies considerably [92]. However, of the faunal guilds present, most research has focused on the effects that insects and crustaceans have on litter and propagule decomposition, recycling of organic matter and nutrients, and structuring of forests. Insects can affect trees in several ways; Robertson and Duke [93] found that among 14 tree species surveyed in an Australian mangrove wetland, 8%–100% of the leaves were damaged, accounting for 0.3%–35% of the total leaf area lost. Individual tree photosynthetic capacity and ecosystem nutrient cycling are impacted by such patterns of herbivory [94], which itself can be influenced by different available nutrient regimes. Likewise, insect herbivory transcends life form in mangroves to include stems [77] and propagules [95] along with leaves.

Research has also focused on the effects that crabs have on influencing which mangrove tree species regenerate, and potentially which tree species are allowed to dominate the overstory [96]. Globally, over 54% of all reproductive propagules or seeds that fall to the floor of a mangrove forest are consumed by crabs [97]. This predation is highly species specific, with mangroves producing palatable propagules (e.g., *Avicennia* spp.), sometimes harboring greater predation pressure from crabs than those maintaining less palatable propagules or seeds (e.g., *Xylocarpus* spp., *Rhizophora* spp.). Crabs often have primacy over many environmental variables in determining the regeneration potential of a mangrove forest, especially considering how tolerant mangrove seedlings are to a diverse array of other stressors, which often prove more detrimental to seedlings in other types of tropical forests [98].

Ecosystem Processes

Because of the high rates of productivity and potential for organic matter export (see Table 21.1), a central paradigm was proposed ("Outwelling Hypothesis") [99], implicating tidal wetlands (including mangroves) in enabling high productivity in adjacent coastal environments. Enhanced material export is thought to promote aquatic primary productivity, trophic exchange between autotrophic and heterotrophic communities, and support, for example, local fisheries. A rate of C export of 64–333 g C/m²/yr from mangroves (*low* [100]; *high* [101]) echoes scaled rates of 112–160 Tg C/yr previously referenced to suggest strong potential for supporting energetic transformations in adjacent nearshore environments. However, Alongi [70] provides the caveat that "the number of factors (influencing export from mangroves) and their nature is such that each system is unique; some mangroves export nutrients and some do not" (p. 130).

Mangroves also provide important protection for coastal communities: In 2004, mangroves and other coastal vegetation protected human communities in India during the Asian tsunami [102] and suppressed storm surge at a rate of 9.4 cm per linear km inland during a hurricane in South Florida [103]. The capacity for storm surge suppression through mangroves may be even higher [104]. Further, mangroves respond to sea-level rise actively along many coastlines through sediment trapping, enhanced root growth, or both [105]. The potential for mangrove migration inland as sea level rises is also strong, but many mangrove communities occur in prime real estate areas and thus become "squeezed" by development. This considerably diminishes the capacity of mangroves to protect coastlines, filter nutrients, and support high levels of coastal productivity.

Trends, Restoration, and Management

Impacts to mangroves include land clearing, hydrological alteration, overharvesting both of trees and fauna, coastal eutrophication, pruning of canopies, sea-level rise, road construction, and maricultural operations [106]. Overall, mangrove loss has historically been large globally at a rate of 1%–2% per

FIGURE 21.6 Mangrove nursery in Bali, Indonesia, established to make an effective use of natural tides for mass propagation of mangroves to support forestry operations and ecological restoration projects. (Courtesy of K.W. Krauss, U.S. Geological Survey.)

year [69], but because of coastal protection and restoration declined to around 0.26%–0.66% per year between 2000 and 2012 [107,108]. The potential ecological services that mangroves provide, including strong recent interest in their potential for C sequestration [80], have prompted efforts to restore mangroves. Indeed, treatises on the silvics of mangroves have been around for over a century [109], culminating from a classic overview for Malaysia [90]. However, such attention to tree regeneration has taken the focus off of hydrological restoration, which must precede any efforts to restore mangrove communities through planting in areas where they have been destroyed [110]. Timber production, fisheries and wildlife enhancement, mitigation and legislative compliance, social enrichment, and restoration of ecosystem services provide the necessary impetus for mangrove restoration [111], and nurseries have been established to support wide-scale plantings in some areas (Figure 21.6). Restoration (or rehabilitation, sensu) [111] of mangroves is certainly more than tree planting; where successful, appropriate attention is made to re-establishing tidal and dispersal connections within degraded former wetlands [110].

Management of most mangroves involves passive approaches, from allowing natural forest dynamics to providing larger protected areas that include mangroves (e.g., Everglades National Park [the United States], Sungei Buloh Wetland Reserve [Singapore]). However, only about 6.9% of mangrove forests are in protected areas globally [66]. The aim of management of these unique systems "should be to maintain the use of mangroves as a renewable resource, providing fisheries and forestry products and possessing an inherent amenity value based on their geomorphological, recreational and scientific characteristics" [109; p.275]. Often, management of ecological services can be melded with forestry operations to provide needed revenue for land purveyors. This type of management is perhaps most advanced in Malaysia and Thailand, and benefits greatly by encouraging local community involvement. Yet, mangrove timber management per se dates to the late 1700s in the Sundarbans of India and Bangladesh [112]. While scientific discoveries have been largely ignored when establishing management strategies [113], some managers have made use of discoveries, specifically related to zonation of forests along specific inundation regimes to establish harvest and developmental zones [114]. Where mangroves are undergoing reduced growth or succumbing to sea-level rise, harvest and human activity is sometimes discouraged [115]. However, the ultimate responsibility for establishing official public use, harvest, or exploitation guidelines lies with a diverse array of community, regional, and/or national governments overlaying the global distribution of mangroves, complicating their sustainable management.

Temperate Tidal Freshwater Forests

While global in distribution, nearly all studies on tidal freshwater forests originate from the southeastern United States. These forests are generally found at the outlets of coastal rivers with low gradient and low topographic relief at or near sea level [116]. The actual extent of tidal freshwater forests in other parts of the world is not well documented. These forests are unique in hydrologic and salinity characteristics. They are flooded and drained regularly by freshwater overflow attributed to local high tides, and they are also prone to saltwater influx during low river flows (as occurs in drought years) or high storm tides (usually during hurricanes). Tide patterns vary across the range of tidal forests from diurnal, semidiurnal, or mixed and are of different amplitudes and range depending upon the location [116]. These forests are found in the upper estuaries of U.S. Atlantic and Gulf coast river systems where there is sufficient freshwater flow to maintain annual average salinities of less than 0.5 ppt [116], but insufficient flow to dampen upstream tidal movement [117]. Hydrology within these forests can be difficult to interpret because of river discharge, tidal stage, local precipitation, evapotranspiration, groundwater, and prevailing winds [118]. River discharge and tidal stage vary on a seasonal basis and have potential implications for wetland saltwater intrusion. These forests are subject to regular flooding from <1 m in the western Gulf of Mexico (Apalachicola Bay in Florida to the southern tip of Texas) to between 2 and 3 m on the Atlantic Coast (southern North Carolina to northeast Florida) [116]. Studies show that these tidal wetlands store more carbon than many other coastal wetland types and warrant inclusion along with tidal marshes and mangroves in "blue carbon" initiatives [119], and they support high rates of annual carbon sequestration (uptake of CO_2 from the atmosphere) and lateral carbon export into aquatic environments that can influence critical near-shore and marine energy transformations [120].

Soil and Biogeochemistry

There is very little published research on tidal freshwater forest soils, but those that have included the entire tidal forest range have confirmed a wide range of soil characteristics [121,122]. For example, an analysis of soil profiles from the Suwannee River found soils representing 7 orders and 18 taxonomic subgroups, ranging from upland in appearance to deep mucks [122]. Factors such as microtopography, local climate, elevation, proximity to river mouth, vegetative cover, and physiographic origin lead to highly variable soil conditions [123]. Generally, soils of tidal freshwater forests tend to be highly organic, a result of suppressed decomposition under anaerobic conditions and moderate to high plant production [124].

Biogeochemistry of tidal freshwater forests is highly complex because of being influenced by coastal tides [123]. Those forests also receive water at different frequencies from river floods, saltwater surges, and groundwater sources, with the quantity and timing of each source varying seasonally and annually. Because of anaerobic conditions and high organic matter content, tidal freshwater forest soils have the potential to retain nutrients and other material through sedimentation, plant uptake/detritus storage, sorption, and microbial immobilization, and can promote N loss through denitrification [123]. However, elevated microsites ("hummocks") that exist in healthy tidal freshwater forests have relatively aerobic conditions, resulting in nitrification rates twice as high as the surrounding floodplain [125]; hummock-hollow topography is not prevalent in swamps with long-term elevated soil salinity. The accumulation of silts and clays contributes to the retention of P and other nutrients [5], while high organic matter content contributes to high cation-exchange capacity and the retention of heavy metals and other potential pollutants [117]. Similar to tidal marshes undergoing salinization, mineralization of soil organic matter may be promoted by greater sulfate availability (associated with saltwater) [13], resulting in potential losses of soil surface elevation as freshwater forests become more degraded (but see [126]).

Vegetation and Fauna

Canopy tree richness tends to be low in tidal freshwater forests. Dominant tree species are baldcypress (*Taxodium distichum* (L.) Rich.), water tupelo (*Nyssa aquatica* L.), and swamp tupelo (*Nyssa biflora* Walter) in lower-elevation areas and ash (*Fraxinus* spp.), red maple (*Acer rubrum* L.), sweetgum (*Liquidambar styraciflua* L.), American hornbeam (*Carpinus caroliniana* Walter), and sweetbay (*Magnolia virginiana* L.) in higher-elevation areas. These trees are sensitive to saltwater intrusion, however, with bald cypress generally being the most salt tolerant, although its growth is reduced considerably at mean annual salinity concentrations above 2 ppt [127,128]. Understory trees, shrub, and herb layers are generally sparse and low in diversity because of dense canopy and frequent flooding in the upper reaches of the river. However, as salinity levels increase, tree canopy decreases, resulting in more extensive and species-rich subcanopies and herbaceous layers [128]. Common species vary from river system to river system [e.g., 122,129,130].

Low-salinity tidal freshwater forests likely serve as reservoirs for abundant infaunal populations [131], with oligochaetes, especially from the taxonomic families Tubificidae and Lumbriculidae, being dominant [127]. Faunal communities in tidal freshwater forests are likely affected by both hydrological patterns and salinity levels, but little is known about faunal populations of these areas. Surveys of reptiles and amphibians in tidal freshwater forests produced few results despite apparent habitat suitability provided by a perpetual water source [132]. While structural differences related to woody versus herbaceous vegetation were expected to have a significant impact on infauna and epifauna, researchers in the Cape Fear River, North Carolina, found salinity variations to have a stronger direct impact on fauna [127].

Ecosystem Processes

There are few estimates of primary productivity for tidal freshwater forests, but these forests are expected to benefit from tidal subsidies of nutrients and energy just as marshes do [5]. There is a broad range in productivity values since forests near their downstream limit are likely to be stunted by higher salinities and flooding frequency. Upstream areas may be more productive due to nutrient subsidies from flood waters and reduced saltwater intrusion. In South Carolina, litterfall in downstream areas was 88–118 g/m^2/yr compared to 563–686 g/m^2/yr in upstream forests [133]. Similar trends have been found in Louisiana where litterfall productivity of tidal forests was 170 g/m^2/yr versus 700 g/m^2/yr for freshwater forests without any salinity impact [134].

Nutrient dynamics in tidal freshwater forests can be variable and change along the estuarine gradient [5]. Tidal export of detritus matter is highly significant, but little data are available for these systems. The export of C, N, and P from forested wetlands contributes to the productivity of the lower estuary [135]. The major pulse of materials to the lower estuary coincides with the time of high detrital formation and the arrival of migrant species entering the estuary for growth and spawning purposes.

Trends, Restoration, and Management

When settlers arrived in the United States, rivers were the main means of travel and trade in coastal areas [136]. The adjacent tidal freshwater forests provided material for construction activities, trade, and fuel as well as land for agricultural practices. The result was a much reduced expanse of forested area. In South Carolina, 40,000 ha of tidal forests were cleared for rice production, most of which today still remains as marsh or pond [137]. Extensive logging between 1890 and 1925 resulted in nearly every virgin stand of bald cypress disappearing. In Louisiana, oil and gas field canals, shipping channels, and pipelines crisscross the coastal zone, creating conduits for saltwater intrusion into freshwater swamps. Signs of saltwater stress and forest dieback are so severe along these channels that there has been die-off of trees resulting in "ghost forests" [116]. Unfortunately, we have no records of how much tidal freshwater forest existed in the past. The actual extent of tidal freshwater forests in the southeastern United States today is conservatively estimated to be over 200,000 ha [138].

Eustatic sea-level rise and land subsidence have resulted in widespread hydrological changes in many freshwater forested wetlands [139–141]. The most widespread change is increased flooding depth and duration, followed by more prevalent and pervasive events of saltwater intrusion [142]. The frequency and severity of droughts and hurricanes are major natural factors that influence the extent and concentration of saltwater distribution that contributes to forest dieback in the coastal zone. Projected sea-level rise and changing climate are expected to accelerate the process and extent of saltwater intrusion into coastal freshwater forested wetlands, further impacting these habitats and restoration efforts [116].

There is an overall paucity of large-scale restoration efforts in tidal freshwater forests. Most efforts have been of small scale and represent a hodgepodge of funding sources over the past 25 years. Restoration efforts may benefit from being tied to large engineering designs such as freshwater or sediment diversions, which could provide continuous freshwater for mitigating persistent salinity incursion. While tidal bald cypress plantings may avoid stress in any given year without such measures, site elevation relative to mean sea level makes plantings susceptible to salt-induced mortality during droughts or hurricanes. The physical difficulties associated with propagating good trees, planting seedlings on a large scale, and monitoring require a dedicated multi-institutional approach.

Acknowledgments

We thank Drs. Michael J. Osland, Alex T. Chow, and Ken W. Krauss for helpful comments on this manuscript. This paper is technical contribution, No. 6779, of the Clemson University Experiment Station. This chapter is based upon work supported by the USGS Land Change Science Research and Development Program and by NIFA/USDA, under project number SC-1700531.

References

1. Sharitz, R.R.; Batzer, D.P.; Pennings, S.C. Ecology of freshwater and estuarine wetlands: an introduction. In Ecology of Freshwater and Estuarine Wetlands, 2nd Ed.; Batzer, D.P.; Sharitz, R.R., Eds.; University of California Press: Berkeley, 2014; 1–22.
2. Gosselink, J.G. *The Ecology of Delta Marshes of Coastal Louisiana: A Community Profile*; U.S. Fish and Wildlife Service, Department of the Interior: Washington, DC, 1984.
3. Odum, W.E. Comparative ecology of tidal fresh-water and salt marshes. *Annu. Rev. Ecol. Syst.* 1988, *19*, 147–176.
4. Barendregt, A.; Whigham, D.F.; Baldwin, A.H., Eds. *Tidal Freshwater Wetlands*; Backhuys: Leiden, the Netherlands, 2009.
5. Mitsch, W.J.; Gosselink, J.G. *Wetlands*, 5th Ed.; John Wiley & Sons: New York, 2015.
6. Kolka, R.K.; Thompson, J.A. Wetland geomorphology, soils, and formative processes. In *Ecology of Freshwater and Estuarine Wetlands*; Batzer, D.P., Sharitz, R.R., Eds.; University of California Press: Berkeley, 2006; 7–42.
7. Zhou, J.L.; Wu, Y.; Kang, Q.S.; Zhang, J. Spatial variations of carbon, nitrogen, phosphorous and sulfur in the salt marsh sediments of the Yangtze Estuary in China. *Estuar. Coast. Shelf Sci.* 2007, *71*, 47–59.
8. Sasser, C.E.; Gosselink, J.G.; Holm, G.O.; Visser, J.M. Tidal freshwater wetlands of the Mississippi River deltas. In *Tidal Freshwater Wetlands*; Barendregt, A., Whigham, D.F.; Baldwin, A.H., Eds.; Backhuys: Leiden: the Netherlands, 2009; 167–178.
9. Neubauer, S.C. Contributions of mineral and organic components to tidal freshwater marsh accretion. *Estuar. Coast. Shelf Sci.* 2008, *78*, 78–88.

10. Megonigal, J.P.; Hines, M.E.; Visscher, P.T. Anaerobic metabolism: Linkages to trace gases and aerobic processes. In Biogeochemistry; Schlesinger, W.H., Ed.; Elsevier-Pergamon: Oxford, UK, 2004; 317–424.

11. Howarth, R.W.; Teal, J.M. Sulfate reduction in a New England salt marsh. *Limnol. Oceanogr.* 1979, *24*, 999–1013.

12. Neubauer, S.C.; Givler, K.; Valentine, S.K.; Megonigal, J.P. Seasonal patterns and plant-mediated controls of subsurface wetland biogeochemistry. *Ecology* 2005, *86*, 3334–3344.

13. Weston, N.B.; Vile, M.A.; Neubauer, S.C.; Velinsky, D.J. Accelerated microbial organic matter mineralization following salt-water intrusion into tidal freshwater marsh soils. *Biogeochemistry* 2011, *102*, 135–151.

14. Engels, J.G.; Jensen, K. Patterns of wetland plant diversity along estuarine stress gradients of the Elbe (Germany) and Connecticut (USA) Rivers. *Estuar. Coast.* 2009, *2*, 301–311.

15. Sharpe, P.J.; Baldwin, A.H. Patterns of wetland plant species richness across estuarine gradients of Chesapeake Bay. *Wetlands* 2009, *29*, 225–235.

16. Yang, S.L. Trapping effect of tidal marsh vegetation on suspended sediment, Yangtze Delta. *Estuar. Coast. Shelf Sci.* 1998, *47*, 227–233.

17. Byrd, K.B.; Kelly, M. Salt marsh vegetation response to edaphic and topographic changes from upland sedimentation in a Pacific estuary. *Wetlands* 2006, *26*, 813–829.

18. Engels, J.G.; Jensen, K. Role of biotic interactions and physical factors in determining the distribution of marsh species along an estuarine salinity gradient. *Oikos* 2010, *119*, 679–685.

19. Crain, C.M.; Silliman, B.R.; Bertness, S.L. Physical and biotic drivers of plant distribution across estuarine salinity gradients. *Ecology* 2004, *85*, 2539–2549.

20. Huiskes, A.H.L.; Koutstaal, B.P.; Herman, P.M.J.; Beeftink, W.G.; Markusse, M.M.; De Munck, W. Seed dispersal of halophytes in tidal salt marshes. *J. Ecol.* 1995, *83*, 559–567.

21. Leck, M.A.; Simpson, R.L. Ten-year seed bank and vegetation dynamics of a tidal freshwater marsh. *Am. J. Bot.* 1995, *82*, 1547–1557.

22. Neff, K.P.; Baldwin, A.H. Seed dispersal into wetlands: Techniques and results for a restored tidal freshwater marsh. *Wetlands* 2005, *25*, 392–404.

23. Hazelton, E.L.G.; Downard, R.; Kettenring, K.M.; McCormick M.K.; Whigham, D.F. Spatial and temporal variation in brackish wetland seedbanks: Implications for wetland restoration following *Phragmites* control. *Estuaries Coasts* 2018, *41* (Suppl 1), S68–S84.

24. Whigham, D.F.; Simpson, R.L. Annual variation in biomass and production of a tidal freshwater wetland and comparison with other wetland systems. *Va. J. Sci.* 1992, *43*, 5–14.

25. Baldwin, A.H.; Egnotovich, M.S.; Clarke, E. Hydrologic change and vegetation of tidal freshwater marshes: Field, greenhouse, and seed-bank experiments. *Wetlands* 2001, *21*, 519–531.

26. Rizzo, W.M.; Wetzel, R.L. Intertidal and shoal benthic community metabolism in a temperate estuary: Studies of spatial and temporal scales of variability. *Estuaries* 1985, *8*, 342–351.

27. Visser, J.M.; Midway, S.; Baltz D.M.; Sasser, C.E. Ecosystem structure of tidal saline marshes. In *Coastal Wetlands: An Integrated Ecosystem Approach*, 2nd Ed.; Perillo, G.M.E.; Wolanski, E., Cahoon, D.R.; Hopkinson, C.S., Eds.; Elsevier: the Netherlands, 2019; 519–538.

28. Tobias, C.; Neubauer, S.C. Salt marsh biogeochemistry – An overview. In *Coastal Wetlands: An Integrated Ecosystem Approach*, 2nd Ed.; Perillo, G.M.E.; Wolanski, E., Cahoon, D.R.; Hopkinson, C.S., Eds.; Elsevier: the Netherlands, 2019; 539–596.

29. Minello, T.J.; Able, K.W.; Weinstein, M.P.; Hays, C.G. Salt marshes as nurseries for nekton: Testing hypotheses on density, growth and survival through meta-analysis. *Mar. Ecol.-Prog. Ser.* 2003, *246*, 39–59.

30. Mathieson, S.; Cattrijsse, A.; Costa, M.J.; Drake, P.; Elliot, M.; Gardner, J.; Marchand, J. Fish assemblages of European tidal marshes: a comparison based on species, families and functional guilds. *Mar. Ecol.-Prog. Ser.* 2000, *204*, 225–242.

31. Desender, K.; Maelfait, J.P. Diversity and conservation of terrestrial arthropods in tidal marshes along the River Schelde: A gradient analysis. *Biol. Conserv.* 1999, *8*, 221–229.

32. Whigham, D.F. Primary production in tidal freshwater wetlands. In *Tidal Freshwater Wetlands*; Barendregt, A.; Whigham, D.F.; Baldwin, A.H., Eds.; Backhuys: Leiden, the Netherlands, 2009; 115–122.

33. Whigham, D.F.; Baldwin, A.H.; Barendregt, A. Tidal freshwater wetlands. In *Coastal Wetlands: An Integrated Ecosystem Approach*, 2nd Ed.; Perillo, G.M.E.; Wolanski, E., Cahoon, D.R.; Hopkinson, C.S., Eds.; Elsevier: the Netherlands, 2019; 619–640.

34. Findlay, S.E.G.; Nieder, W.C.; Ciparis, S. Carbon flows, nutrient cycling, and food webs in tidal freshwater wetlands. In *Tidal Freshwater Wetlands*; Barendregt, A., Whigham, D.F.; Baldwin, A.H., Eds.; Backhuys: Leiden, the Netherlands, 2009; 137–144.

35. Craft, C.; Clough, J.; Ehman, J.; Joye, S.; Park, R.; Pennings, S.; Guo, H.; Machmuller, M. Forecasting the effects of accelerated sea-level rise on tidal marsh ecosystem services. *Front. Ecol. Environ.* 2009, *7*, 73–78.

36. Morris, J.T.; Sundareshwar, P.V.; Nietch, C.T.; Kjerfve, B.; Cahoon, D.R. Responses of coastal wetlands to rising sea level. *Ecology* 2002, *83*, 2869–2877.

37. Wiberg, P.L.; Fagherazzi, S.; Kirwan, M.L. Improving predictions of salt marsh evolution through better integration of data and models. *Annu. Rev. Mar. Sci.* 2020, *12*, 6.1–6.25.

38. Erickson, J.E.; Megonigal, J.P.; Peresta, G.; Drake, B.G. Salinity and sea level mediate elevated CO_2 effects on C_3-C_4 plant interactions and tissue nitrogen in a Chesapeake Bay tidal wetland. *Global Change Biol.* 2007, *13*, 202–215.

39. Langley, J.A.; Megonigal, J.P. Ecosystem response to elevated CO_2 levels limited by nitrogen-induced plant species shift. *Nature* 2010, *466*, 96–99.

40. Levine, J.M.; Brewer, J.S.; Bertness, M.D. Nutrients, competition, and plant zonation in a New England salt marsh. *J. Ecol.* 1998, *86*, 285–292.

41. Baldwin, A.H. Nitrogen and phosphorus differentially affect annual and perennial plants in tidal freshwater and oligohaline wetlands. *Estuar. Coast.* 2013, *36*, 547–558.

42. Chambers, R.M.; Meyerson, L.A.; Saltonstall, K. Expansion of *Phragmites australis* into tidal wetlands of North America. *Aquat. Bot.* 1999, *64*, 261–273.

43. Fell, P.E.; Weissbach, S.P.; Jones, D.A.; Fallon, M.A.; Zeppieri, J.A.; Faison, E.K.; Lennon, K.A.; Newberry, K.T.; Reddington, L.K. Does invasion of oligohaline tidal marshes by reed grass, Phragmites australis (Cav.) Trin. ex Steud., affect the availability of prey resources for the mummichog, Fundulus heteroclitus L.? *J. Exp. Mar. Biol. Ecol.* 1998, *222*, 59–77.

44. Posey, M.H.; Alphin, T.D.; Meyer, D.L.; Johnson, J.M. Benthic communities of common reed *Phragmites australis* and marsh cordgrass *Spartina alterniflora* marshes in Chesapeake Bay. *Mar. Ecol.-Prog. Ser.* 2003, *261*, 51–61.

45. Ayres, D.R.; Smith, D.L.; Zaremba, K.; Klohr, S.; Strong, D.R. Spread of exotic cordgrasses and hybrids (*Spartina* sp.) in the tidal marshes of San Francisco Bay, California, USA. *Biol. Invasions* 2004, *6*, 221–231.

46. Xiao, D.R.; Zhang, L.Q.; Zhu, Z.C. The range expansion patterns of *Spartina alterniflora* on salt marshes in the Yangtze Estuary, China. *Estuar. Coast. Shelf Sci.* 2010, *88*, 99–104.

47. Alber, M.; Swenson, E.M.; Adamowicz, S.C.; Mendelssohn, I.A. Salt marsh dieback: An overview of recent events in the U.S. Estuarine Coast. *Shelf Sci.* 2008, *80*, 1–11.

48. Crawford, J.T.; Stone, A.G. Relationships between soil composition and *Spartina alterniflora* dieback in an Atlantic salt marsh. *Wetlands* 2015, *35*, 13–20.

49. Colman, D.J.; Kirwan, M.L. The effect of a small vegetation dieback event on salt marsh sediment transport. *Earth Surf. Processes Landforms* 2018, *44*, 944–952.

50. Eertman, R.H.M.; Kornman, B.A.; Stikvoort, E.; Verbeek, H. Restoration of the Sieperda tidal marsh in the Scheldt estuary, the Netherlands. *Restor. Ecol.* 2002, *1*, 438–449.

51. Noll, A.; Mobilian, C.; Craft, C. Five decades of wetland soil development of a constructed tidal salt marsh, North Carolina, USA. *Ecol. Restor.* 2019, *37*, 163–170.

52. Leck, M.A. Seed-bank and vegetation development in a created tidal freshwater wetland on the Delaware River, Trenton, New Jersey, USA. *Wetlands* 2003, *23*, 310–343.

53. Baldwin, A.H. Restoring complex vegetation in urban settings: The case of tidal freshwater marshes. *Urban Ecosyst.* 2004, *7*, 137.

54. Neff, K.P.; Rusello, K.; Baldwin, A.H. Rapid seed bank development in restored tidal freshwater wetlands. *Restor. Ecol.* 2009, *17*, 539–548.

55. Warren, R.S.; Fell, P.E.; Grimsby, J.L.; Buck, E.L.; Rilling, G.C.; Fertik, R.A. Rates, patterns, and impacts of *Phragmites australis* expansion and effects of experimental *Phragmites* control on vegetation, macroinvertebrates, and fish within tidelands of the lower Connecticut River. *Estuaries* 2001, *24*, 90–107.

56. Kearney, M.S.; Riter, J.C.A.; Turner, R.E. Freshwater river diversions for marsh restoration in Louisiana: twenty-six years of changing vegetative cover and marsh area. *Geophy. Res. Lett.* 2011, *38*, doi:10.1029/2011GL047847.

57. Elsey-Quirk, T.; Graham, S.A.; Mendelssohn, I.A.; Snedden, G.; Day, J.W.; Twilley, R.R.; Shaffer, G.; Sharp, L.A.; Pahl, J.; Lane, R.R. Mississippi river sediment diversions and coastal wetland sustainability: Synthesis of responses to freshwater, sediment, and nutrient inputs. *Estuar. Coast. Shelf Sci.* 2019, *221*, 170–183.

58. Ford, M.A.; Cahoon, D.R.; Lynch, J.C. Restoring marsh elevation in a rapidly-subsiding salt marsh by thin-layer deposition of dredge material. *Ecol. Eng.* 1999, *12*, 189–205.

59. Thorne, K.M.: Freeman, C.M.: Rosencranz, J.A.: Ganju, N.K., Guntenspergen, G.R. Thin-layer sediment addition to an existing salt marsh to combat sea-level rise and improve endangered species habitat in California, USA. *Ecol. Eng.* 2019, *136*, 197–208.

60. Tomlinson, P.B. *The Botany of Mangroves*; Cambridge University Press: Cambridge, UK, 1986.

61. Ball, M.C. Ecophysiology of mangroves. *Trees* 1988, *2*, 129–142.

62. Krauss, K.W.; Lovelock, C.E.; McKee, K.L.; López-Hoffman, L.; Ewe, S.M.L.; Sousa, W.P. Environmental drivers in mangrove establishment and early development: A review. *Aquat. Bot.* 2008, *89*, 105–127.

63. Saintilan, N.; Rogers, K.; McKee, K.L. Salt marsh-mangrove interactions in Australasia and the Americas. In *Coastal Wetlands: An Integrated Ecosystem Approach*; Gerardo, M.E.P., Wolanski, E., Cahoon, D.R., Brinson, M.M., Eds.; Elsevier: Amsterdam, Netherlands, 2009; 855–883.

64. Friess, D.A.; Krauss, K.W.; Horstman, E.M.; Balke, T.; Bouma, T.J.; Galli, D.; Webb, E.L. Are all intertidal wetlands naturally created equal? Bottlenecks, thresholds and knowledge gaps to mangrove and saltmarsh ecosystems. *Biol. Rev.* 2012, *87*, 346–366.

65. Spalding, M; Kainuma, M.; Collins, L. *World Atlas of Mangroves*; Earthscan: London, 2010.

66. Giri, C.; Ochieng, E.; Tieszen, L.L.; Zhu, Z.; Singh, A.; Loveland, T.; Masek, J.; Duke, N. Status and distribution of mangrove forests of the world using earth observation satellite data. *Global Ecol. Biogeogr.* 2011, *20*, 154–159.

67. Ewel, K.C.; Twilley, R.R.; Ong, J.E. Different kinds of mangrove forests provide different goods and services. *Global Ecol. Biogeogr. Lett.* 1998, *7*, 83–94.

68. Valiela, I.; Bowen, J.L.; York, J.K. Mangrove forests: One of the world's threatened major tropical environments. *BioScience* 2001, *51*, 807–815.

69. Duke, N.C. Mangrove floristics and biogeography. In *Tropical Mangrove Ecosystems*; Robertson, A.I.; Alongi, D.M., Eds.; American Geophysical Union: Washington, 1992; 63–100.

70. Alongi, D.M. *The Energetics of Mangrove Forests*; Springer: New York, 2009.

71. McKee, K.L. Soil physicochemical patterns and mangrove species distribution – reciprocal effects? *J. Ecol.* 1993, *81*, 477–487.

72. Alongi, D.M.; Tirendi, F.; Dixon, P.; Trott, L.A.; Brunskill, G.J. Mineralization of organic matter in intertidal sediments of a tropical semi-enclosed delta. *Estuar. Coast. Shelf Sci.* 1999, *48*, 451–467.

73. Gleason, S.M.; Ewel, K.C.; Hue, N. Soil redox conditions and plant-soil relationships in a Micronesian mangrove forest. *Estuar. Coast. Shelf Sci.* 2003, *56*, 1065–1074.

74. Twilley, R.R. Mangrove wetlands. In *Southern Forested Wetlands Ecology and Management*; Messina, M.G.; Conner, W.H., Eds.; Lewis Publishers: Boca Raton, FL, 1998; 445–473.

75. Alongi, D.M.; Trott, L.A.; Wattayakorn, G.; Clough, B. Below-ground nitrogen cycling in relation to net canopy production in mangrove forests of southern Thailand. *Mar. Biol.* 2002, *140*, 855–864.

76. Sherman, R.E.; Fahey, T.J.; Howarth, R.W. Soil-plant interactions in a neotropical mangrove forest: iron, phosphorus and sulfur dynamics. *Oecologia* 1998, *115*, 553–563.

77. Feller, I.C. Effects of nutrient enrichment on growth and herbivory of dwarf red mangrove (*Rhizophora mangle*). *Ecol. Monogr.* 1995, *65*, 477–505.

78. Bouillon, S.; Borges, A.V.; Castañeda-Moya, E.; Diele, K.; Dittmar, T.; Duke, N.C.; Kristensen, E.; Lee, S.Y.; Marchland, C.; Middelburg, J.J.; Rivera-Monroy, V.H.; Smith III, T.J.; Twilley, R.R. Mangrove production and carbon sinks: a revision of global budget estimates. *Global Biogeochem. Cycle* 2007, *22*, GB2013.

79. Donato, D.C.; Kauffman, J.B.; Murdiyarso, D.; Kurnianto, S.; Stidham, M.; Kanninen, M. Mangroves among the most carbon-rich forests in the tropics. *Nat. Geosci.* 2011, *4*, 293–297.

80. McLeod, E.; Chmura, G.L.; Bouillon, S.; Salm, R.; Björk, M.; Duarte, C.M.; Lovelock, C.E.; Schlesinger, W.H.; Silliman, B.R. A blueprint for blue carbon: Toward an improved understanding of the role of vegetated coastal habitats in sequestering CO_2. *Front. Ecolo Environ.* 2011, *9*, 552–560.

81. Maher, D.T.; Call, M.; Santos, I.R.; Sanders, C.J. Beyond burial: Lateral exchange is a significant atmospheric carbon sink in mangrove forests. *Biol. Lett.* 2018, *14*, 20180200.

82. Middleton, B.A.; McKee, K.L. Degradation of mangrove tissues and implications for peat formation in Belizean island forests. *J. Ecol.* 2001, *89*, 818–828.

83. McKee, K.L.; Cahoon, D.R.; Feller, I.C. Caribbean mangroves adjust to rising sea level through biotic controls on change in soil elevation. *Global Ecol. Biogeogr.* 2007, *16*, 545–556.

84. Duke, N.C.; Ball, M.C.; Ellison, J.C. Factors influencing biodiversity and distributional gradients in mangroves. *Global Ecol. Biogeogr. Lett.* 1998, *7*, 27–47.

85. Simard, M.; Fatoyinbo, L.; Smetanka, C.; Rivera-Monroy, V.H.; Castaneda-Moya, E.; Thomas, N.; Van der stocken, T. Mangrove canopy height globally related to precipitation, temperature and cyclone frequency. *Nat. Geosci.* 2019, *12*, 40–45.

86. Cheeseman, J.M. The analysis of photosynthetic performance in leaves under field conditions: a case study using *Bruguiera* mangroves. *Photosynth. Res.* 1991, *29*, 11–22.

87. Reef, R.; Feller, I.C.; Lovelock, C.E. Nutrition of mangroves. *Tree Physiol.* 2010, *30*, 1148–1160.

88. Feller, I.C.; Whigham, D.F.; O'Neill, J.P.; McKee, K.L. Effects of nutrient enrichment on within-stand cycling in a mangrove forest. *Ecology* 1999, *80*, 2193–2205.

89. Lovelock, C.E.; Feller, I.C.; Ball, M.C. Testing the growth rate vs. geochemical hypothesis for latitudinal variation in plant nutrients. *Ecol. Lett.* 2007, *10*, 1154–1163.

90. Watson, J.G. Mangrove forests of the Malay Peninsula. *Malay. For. Rec.* 1928, *6*, 1–275.

91. Chapman, V.J. *Mangrove Vegetation*; J. Cramer: Vaduz, 1976.

92. Alberts-Hubatsch, H.; Lee, S.Y.; Meynecke, J.; Diele, K.; Nordhaus, I.; Wolff, M. Life-history, movement, and habitat use of *Scylla serrata* (Decapoda, Portunidae): Current knowledge and future challenges. *Hydrobiologia* 2016, *763*, 5–21.

93. Robertson, A.I.; Duke, N.C. Insect herbivory on mangrove leaves in North Queensland. *Aust. J. Ecol.* 1987, *12*, 1–7.

94. Hogarth, P.J. *The Biology of Mangroves*; Oxford University Press: Oxford, 1999.

95. Farnsworth, E.J.; Ellison, A.M. Global patterns of predispersal propagule predation in mangrove forests. *Biotropica* 1997, *29*, 318–330.

96. Smith, T.J., III. Seed predation in relation to dominance and distribution in mangrove forests. *Ecology* 1987, *68*, 266–273.

97. Allen, J.A.; Krauss, K.W.; Hauff, R.D. Factors limiting the intertidal distribution of the mangrove species *Xylocarpus granatum*. *Oecologia* 2003, *135*, 110–121.

98. Lindquist, E.S.; Krauss, K.W.; Green, P.T.; O'Dowd, D.J.; Sherman, P.M.; Smith, III, T.J. Land crabs as key drivers in tropical coastal forest recruitment. *Biol. Rev.* 2009, *84*, 203–223.

99. Odum, E.P. The status of three ecosystem-level hypotheses regarding salt marsh estuaries: tidal subsidy, outwelling, and detritus-based food chains. In *Estuarine Perspectives*; Kennedy, V.S., Ed.; Academic Press: New York, 1980; 485–495.

100. Twilley, R.R. The exchange of organic carbon in basin mangrove forests in a southwest Florida estuary. *Estuar. Coast. Shelf Sci.* 1985, *20*, 543–557.

101. Alongi, D.M. *Coastal Ecosystem Processes*; CRC Press: Boca Raton, 1998.

102. Danielsen, F.; Sorensen, M.K.; Olwig, M.F.; Selvam, V.; Parish, F.; Burgess, N.D.; Hiraishi, T.; Karunagaran, V.M.; Rasmussen, M.S.; Hansen, L.B.; Quarto, A.; Suryadiputra, N. The Asian tsunami: A protective role for coastal vegetation. *Science* 2005, *310*, 643.

103. Krauss, K.W.; Doyle, T.W.; Doyle, T.J.; Swarzenski, C.M.; From, A.S.; Day, R.H.; Conner, W.H. Water level observations in mangrove swamps during two hurricanes in Florida. *Wetlands* 2009, *29*, 142–149.

104. Zhang, K.; Liu, H.; Li, Y.; Xu, H.; Shen, J.; Rhome, J.; Smith, T.J., III. The role of mangroves in attenuating storm surges. *Estuar. Coast. Shelf Sci.* 2012, *102–103*, 11–23.

105. McKee, K.L. Biophysical controls on accretion and elevation change in Caribbean mangrove ecosystems. *Estuar. Coast. Shelf Sci.* 2011, *91*, 475–483.

106. Lewis, R.R., III; Milbrandt, E.C.; Brown, B.; Krauss, K.W.; Rovai, A.S.; Beever, J.W., III; Flynn, L.L. Stress in mangrove forests: early detection and preemptive rehabilitation are essential for future successful worldwide mangrove forest management. *Mar. Pollut. Bull.* 2016, *109*, 764–771.

107. Feller, I.C.; Friess, D.A.; Krauss, K.W.; Lewis, R.R., III. The state of the world's mangroves in the 21st century under climate change. *Hydrobiologia* 2017, *803*, 1–12.

108. Friess, D.A.; Rogers, K.; Lovelock, C.E.; Krauss, K.W.; Hamilton, S.E.; Lee, S.Y.; Lucas, R.; Primavera, J.; Rajkaran, A.; Shi, S. The state of the world's mangrove forests: Past, present, and future. *Ann. Rev. Environ. Resour.* 2019, *44*, 16.1–16.27.

109. Saenger, P. *Mangrove Ecology, Silviculture and Conservation*; Kluwer Academic Publishers: Dordrecht, 2002.

110. Lewis, R.R. Ecological engineering for successful management and restoration of mangrove forests. *Ecol. Eng.* 2005, *24*, 403–418.

111. Field, C.D. Rehabilitation of mangrove ecosystems: An overview. *Mar. Poll. Bull.* 1998, *37*, 383–392.

112. Chowdhury, R.A.; Ahmed, I. History of forest management. In *Mangroves of the Sundarbans*, Volume 2; Hussain, Z., Acharya, G., Eds.; IUCN Wetlands Program: Switzerland, 1994; 155–180.

113. Kairo, J.G.; Dahdouh-Guebas, F.; Bosire, J.; Koedam, N. Restoration and management of mangrove ecosystems–a lesson for and from the East African region. *S. Afr. J. Bot.* 2001, *67*, 383–389.

114. Aksornkoae, S. Scientific mangrove management in Thailand. In *Tropical Forestry in the 21st Century*; Aksornkoae, S., Puangchit, L., Thaiutsa, B., Eds.; Kasetsart University: Bangkok, 1997; 118–126.

115. Krauss, K.W.; Cahoon, D.R.; Allen, J.A.; Ewel, K.C.; Lynch, J.C.; Cormier, N. Surface elevation change and susceptibility of different mangrove zones to sea-level rise on Pacific high islands of Micronesia. *Ecosystems* 2010, *13*, 129–143.

116. Doyle, T.W.; O'Neil, C.P.; Melder, P.V.; From, A.S.; Palta, M.M. Tidal freshwater swamps of the Southeastern United States: Effects of land use, sea-level rise and climate change. In Ecology of Tidal Freshwater Forested Wetlands of the Southeastern United States; Conner, W.H., Doyle, T.W., Krauss, K.W., Eds.; Springer: Dordrecht, 2007; 1–28.

117. Simpson, R.L.; Good, R.E.; Leck, M.A.; Whigham, D.F. The ecology of freshwater tidal wetlands. *BioScience* 1983, *33* (4), 255–259.

118. Anderson, C.J.; Lockaby, B.G. Soils and biogeochemistry of tidal freshwater forested wetlands. In *Ecology of Tidal Freshwater Forested Wetlands of the Southeastern United States*; Conner, W.H., Doyle, T.W., Krauss, K.W., Eds.; Springer: Dordrecht, 2007; 65–88.

119. Lovelock, C.E.; Duarte, C.M. Dimensions of blue carbon and emerging perspectives. *Biol. Lett.* 2019, *15*, 20180781. doi:10.1098/rsbl.2018.0781.

120. Krauss, K.W.; Noe, G.B.; Duberstein, J.A.; Conner, W.H.; Stagg, C.L.; Cormier, N.; Jones, M.C.; Bernhardt, C.J.; Lockaby, B.G.; From, A.S.; Doyle, T.W.; Day, R.; Ensign, S.H.; Pierfelice, K.N.; Hupp, C.R.; Chow, A.T.; Whitbeck, J.L. The role of the upper tidal estuary in wetland blue carbon storage and flux. *Global Biogeochemical Cycles* 2018, *32*, 817–839.

121. Doumlele, D.G.; Fowler, K.; Silberhorn, G.M. Vegetative community structure of tidal freshwater swamp in Virginia. *Wetlands* 1984, *4*, 129–145.

122. Light, H.M.; Darst, M.R.; Lewis, L.J.; Howell, D.A. Hydrology, vegetation, and soils of riverine and tidal forests of the Lower Suwannee River, Florida, and potential impacts of flow reductions. Professional Paper 1656A. U.S. Geological Survey, Tallahassee, FL, 2002.

123. Anderson, C.J.; Lockaby, B.G. Seasonal patterns of river connectivity and saltwater intrusion in tidal forested freshwater wetlands. *River Res. Appli.* 2012, *28*, 814–826.

124. Wharton, C.H.; Kitchens, W.M.; Pendleton, E.C.; Sipe, T.W. *The Ecology of Bottomland Hardwood Swamps of the Southeast: A Community Profile, FWS/OBS-81/37*; U.S. Fish and Wildlife Service, Biological Services Program: Washington, DC, 1982.

125. Noe, G.B.; Krauss, K.W.; Lockaby, B.G.; Conner, W.H.; Hupp, C.R. The effect of increasing salinity and forest mortality on soil nitrogen and phosphorus mineralization in tidal freshwater forested wetlands. *Biogeochemistry* 2013, *114*, 225–244.

126. Hackney, C.T.; Posey, M.; Leonard, L.L.; Alphin, T.; Avery, G.B. *Monitoring the Effects of a Potential Increased Tidal Range in the Cape Fear River Ecosystem Due to Deepening Wilmington Harbor, North Carolina.* Year 1: June1, 2003-May 31, 2004; U.S. Army Corps of Engineers, Wilmington District (Contract No. DACW 54-00-R-0008), Wilmington, NC, 2005.

127. Hackney, C.T.; Avery, G.B.; Leonard, L.A.; Posey, M.; Alphin, T. Biological, chemical, and physical characteristics of tidal freshwater swamp forests of the Lower Cape Fear River/Estuary, North Carolina. In Ecology of Tidal Freshwater Forested Wetlands of the Southeastern United States; Conner, W.H., Doyle, T.W., Krauss, K.W., Eds.; Springer: Dordrecht, 2007; 183–221. http://onlinelibrary.wiley.com/doi/10.1002/rra.1489/abstract (accessed August 2011).

128. Krauss, K.W.; Duberstein, J.A.; Doyle, T.W.; Conner, W.H.; Day, R.H.; Inabinette, L.W.; Whitbeck, J.L. Site condition, structure, and growth of bald cypress along tidal/non-tidal salinity gradients. *Wetlands* 2009, *29*, 505–519.

129. Rheinhardt, R. A multivariate analysis of vegetation patterns in tidal freshwater swamps of lower Chesapeake Bay, USA. *Bull. Torrey Bot. Club* 1992, *119*, 192–207.

130. Duberstein, J.; Kitchens, W. Community composition of select areas of tidal freshwater forest along the Savannah River. In *Ecology of Tidal Freshwater Forested Wetlands of the Southeastern United States*; Conner, W.H., Doyle, T.W., Krauss, K.W., Eds.; Springer: Dordrecht, 2007; 321–348.

131. Rozas, L.P.; Hackney, C.T. Use of oligohaline marshes by fishes and macrofaunal crustaceans in North Carolina. *Estuaries* 1984, *7*, 213–224.

132. Godfrey, S.T. Herpetofauna occupancy and community composition along a tidal swamp salinity gradient. M.S. thesis, Clemson University, 2018.

133. Cormier, N.; Krauss, K.W.; Conner, W.H. Periodicity in stem growth and litterfall in tidal freshwater forested wetlands: influence of salinity and drought on nitrogen recycling. *Estuar. Coasts* 2013, *36*, 533–546.

134. Effler, R.S.; Shaffer, G.P.; Hoeppner, S.S.; Goyer, R.A. Ecology of the Maurepas swamp: effects of salinity, nutrients, and insect defoliation. In *Ecology of Tidal Freshwater Forested Wetlands of the Southeastern United States*; Conner, W.H., Doyle, T.W., Krauss, K.W., Eds.; Springer: Dordrecht, 2007; 349–384.

135. Day, J.W. Jr.; Butler, T.J.; Conner, W.H. Productivity and nutrient export studies in a cypress swamp and lake system in Louisiana. In *Estuarine Processes*, Volume2; Wiley, M., Ed.; Academic Press: New York, 1977; 255–269.

136. Rodgers, G.C. Jr. *The History of Georgetown County, South Carolina*; The University of South Carolina Press: Columbia, SC, 1970.

137. Gresham, C.A.; Hook, D.D. Rice fields of South Carolina: A resource inventory and management policy evaluation. *Coast. Zone Manage. J.* 1982, *9*, 183–203.

138. Field, D.W.; Reyer, A.J.; Genovese, P.V.; Shearer, B.D. *Coastal Wetlands of The United States: An Accounting of a Valuable National Resource*; Office of Oceanography and Marine Assessment, National Ocean Service, National Oceanic and Atmospheric Administration: Rockville, MD, 1991.

139. Brinson, M.M.; Bradshaw, H.D.; Jones, M.N. Transitions in forested wetlands along gradients of salinity and hydroperiod. *J. Elisha Mitchell Sci. Soc.* 1985, *101*, 76–94.

140. Conner, W.H.; Day, J.W. Jr. Rising water levels in coastal Louisiana: implications for two forested wetland areas in Louisiana. *J. Coast. Res.* 1988, *4*, 589–596.

141. Pezeshki, S.R.; DeLaune, R.D.; Patrick, W.H. Jr. Flooding and saltwater intrusion: potential effects on survival and productivity of wetland forests along the U.S. Gulf Coast. *For. Ecol. Manage.* 1990, *33* (34), 287–301.

142. Shaffer, G.P.; Wood, W.B.; Hoeppner, S.S.; Perkins, T.E.; Zoller, J.; Kandalepas, D. Degradation of bald cypress –water tupelo swamp to marsh and open water in Southeastern Louisiana, USA: An irreversible trajectory? *J. Coast. Res.* 2009, *54*, 152–165.

143. Saenger, P.; Snedaker, S.C. Pantropical trends in mangrove above-ground biomass and annual litterfall. *Oecologia* 1993, *96*, 293–299.

144. Day, J.W.; Day, R.H.; Barreiro, M.T.; Ley-Lou, F.; Madden, C.J. Primary production in the Laguna de Terminos, a tropical estuary in the Southern Gulf of Mexico. *Oceanol. Acta* 1982, SP, 269–276.

145. Steinke, T.D.; Charles, L.M. Litter production by mangroves. I: Mgeni Estuary. *S. Afr. J. Bot.* 1986, *52*, 552–558.

146. Jardel, E.J.; Saldaña A.A.; Barreiro, M.T. Contribución al conocimiento de la ecología de los manglares de la Laguna de Términos, Campeche, México. *Cienc. Mar.* 1987, *13*, 1–22.

147. Hardiwinoto, S., Nakasuga, T.; Igarashi, T. Litter production and decomposition of a mangrove forest at Ohura Bay, Okinawa. *Res. Bull. Coll. Exp. Forests (Hokkaido Univ.)* 1989, *46*, 577–594.

148. Lu, C.; Lin, P. Studies on litter fall and decomposition of *Bruguiera sexangula* (Lour.) Poir. community on Hainan Island. *China. Bull. Mar. Sci.* 1990, *47*, 139–148.

149. Flores-Verdugo, F., González-Farías, F.; Ramírez-Flores, O.; Amezcua-Linares, F.; Yáñez-Arancibia, A.; Alvarez-Rubio, M.; Day, J.W. Mangrove ecology, aquatic primary productivity, and fish community dynamics in the Teacapán-Agua Brava lagoon-estuarine system (Mexican Pacific). *Estuaries* 1990, *13*, 219–230.

150. Schaeffer-Novelli, Y.; Mesquita, H.D.S.L.; Cintrón-Molero, G. The Cananéia lagoon estuarine system, Sao Paulo, Brazil. *Estuaries* 1990, *13*, 193–203.

151. Flores-Verdugo, F., González-Farías, F.; Zamorano, D.S.; Ramirez-Garcia, P. Mangrove ecosystems of the Pacific Coast of Mexico: distribution, structure, litterfall, and detritus dynamics. In *Coastal Plant Communities of Latin America*; Seeliger, U., Ed.; Academic Press: San Diego, CA, 1992; 269–288.

152. Sukardjo, S.; Yamada, I. Biomass and productivity of a *Rhizophora mucronata* Lamarck plantation in Tritih, Central Java, Indonesia. *For. Ecol. Manage.* 1992, *49*, 195–209.

153. Flores-Verdugo, F.; González-Farías, F.; Zaragoza-Araujo, U. Ecological parameters of the mangroves of semi-arid regions of Mexico important for ecosystem management. In *Towards the Rationale Use of High Salinity Tolerant Plants*, Volume 1; Leith, H., Al Masoom, A., Eds.; Kluwer Academic Publishers: the Netherlands, 1993; 123–132.

154. Singh, V.P.; Garge, A.; Mall, L.P. Study of biomass, litter fall and litter decomposition in managed and unmanaged mangrove forests of Andaman Islands. In *Towards the Rationale Use of High Salinity Tolerant Plants*, Volume 1; Leith, H., Al Masoom, A., Eds.; Kluwer Academic Publishers: the Netherlands, 1993; 149–154.

155. Clarke, P.J. Baseline studies of temperate mangrove growth and reproduction; demographic and litterfall measures of leafing and flowering. *Aust. J. Bot.* 1994, *42*, 37–48.
156. Day, J.W., Coronado-Molina, C.; Vera-Herrera, F.R.; Twilley, R.; Rivera-Monroy, V.H.; Alvarez-Guillen, H.; Day, R.; Conner, W. A 7-year record of above-ground net primary production in a southeastern Mexican mangrove forest. *Aquat. Bot.* 1996, *55*, 39–60.
157. Twilley, R.R.; Pozo, M.; Garcia, V.H.; Rivera-Monroy, V.H.; Zambrano, R.; Bodero, A. Litter dynamics in riverine mangrove forests in the Guayas River estuary, Ecuador. *Oecologia* 1997, *111*, 109–122.
158. Lugo, A.E.; Medina, E.; Cuevas, E.; Cintron, G.; Nieves, E.N.L.; Novelli, Y.S. Ecophysiology of a mangrove forest in Jobos Bay, Puerto Rico. *Caribb. J. Sci.* 2007, *43*, 200–219.

III

Wetland
Assessment and
Monitoring

22

Wetlands: Economic Value

Gayatri Acharya
World Bank Institute (WBI)

Introduction

Wetlands have been centers of human civilization throughout history. For example, the Mekong and Red River deltas of Vietnam together comprise 16% of the country but support 42% of the population.[1] Yet, wetlands have been historically transformed or converted and their capacity diminished in some manner by the impacts of land use changes. OECD[2] noted that "The world may have lost 50% of the wetlands that existed since 1900; whilst much of this occurred in the northern countries during the first 50 yr of the century, increasing pressure for conversion to alternative land use has been put on tropical and subtropical wetlands since the 1950s. By 1985, it was estimated that 56–65% of the available wetland had been drained for intensive agriculture in Europe and North America; the figures for tropical and subtropical regions were 27% for Asia, 6% for South America, and 2% for Africa, making a total of 26% worldwide."

The conversion of multiple use ecosystems to single land uses may signify a welfare loss through reduced access to, or availability of, valuable goods and services. Whether or not such conversions are merited can be assessed only with sufficient information on the relative efficiency of converting or conserving the wetland ecosystem vis-à-vis alternative land uses. Economic values are a measure of how much people are willing to pay for, and how much better or worse off they would consider themselves to be as a result of changes in the supply of different goods and services. Valuation provides a means of quantifying the benefits that people receive from wetlands. The net benefits of investing in land uses that are compatible with wetlands conservation, relative to those economic activities, which contribute to wetlands degradation, can then be assessed. Values generated through economic analysis are limited however by a) their inability to capture intrinsic values associated with wetlands and b) the level of understanding, both economists and natural scientists have with regard to the level of use of goods and services, and the variation in their availability with fluxes in the natural system. Furthermore, economic values measure preferences and welfare impacts associated with changes in the natural system or the availability of a service or product; they do not, in themselves, suggest anything about inter- or intra-generational distributional effects of conservation or development investments.

Wetland goods and services are also particularly difficult to value. This is because: a) many goods are not marketed but traded or consumed directly; b) wetland services, such as water quality or groundwater recharge, often occur in areas away from the physical location of the wetland and may not be easily attributable to the wetland; c) many wetlands are transboundary resources and data on the use and consumption of goods and services are difficult to obtain; and d) many wetlands are public property.

For these and other related reasons, the economic benefits generated by wetlands, and the economic costs associated with wetlands degradation or loss, are frequently unknown and omitted in project or policy analysis. As a result, the potential of wetlands to be used as contributors to economic growth, income generating activities, and as sources of goods and services, has been underestimated in many parts of the world, resulting in the loss of valuable species, services, and livelihoods.

Economic Values of Wetlands

Wetlands cover an estimated 6% of the world's land surface and may be of various types such as marshes, flood-plains, freshwater ponds, lakes, swamps, etc. While these ecosystems are associated with a diverse and complex array of direct and indirect uses, the precise functions and benefits of a specific wetland can be determined only after careful study of the hydrological, physical, and biological characteristics of the site, over a period of time.[3] Since the economic value of an ecosystem service (or good) is determined by the contribution it makes to maintaining the present level of human well being, it is calculated as a measure of a change in well being, or *social welfare*. Therefore, valuing a good or service requires us to study the change in a person's welfare due to a change in the availability of the resource. Economic value expressed in currency units is the monetary expression of the tradeoff between identified alternative land uses (see Freeman,[4] among others, for more on valuation theory).

To fully assess these tradeoffs, the ultimate goal of valuation should be to assess the total economic value (TEV) of a system. Total economic value is defined as the sum of use value + existence value where use value is comprised of a) direct use value (consumptive or nonconsumptive); b) indirect use value; and c) option value. Table 22.1 defines each of these values while Figure 22.1 shows how direct and indirect benefits can be identified in a wetland system. While valuation attempts to take account of all the components of the total economic benefit of wetlands, it is generally constrained by data availability and, therefore, partial valuation studies are more common (see Barbier, Adams, et al.[7] for more on the differences in total and partial valuation).

Direct uses include the use of the wetland for water supply and harvesting of wetland products such as fish and plant resources. Commonly recognized values of wetlands include those captured from the use of raw materials and physical products that are generally used in economic activities such as fisheries, water supply, and agriculture. Where wild foods, grasses, bird life, and other naturally occurring species are exploited for significant economic returns, such uses are also easily identifiable and may be valued in existing markets.

Direct uses however represent only a small proportion of the total value of wetlands, which generate economic benefits in excess of just physical products. *Indirect benefits* are derived from environmental functions that wetlands may provide, depending on the type of wetland, soil, and water characteristics, and associated biotic influences.[5] Wetland functions are defined as a process or series of processes that take place within a wetland, including habitat and hydrological and water quality functions. Such functions have an economic value if they contribute to economic activity and enhance human welfare.

TABLE 22.1   Types of Benefits Value by Economic Valuation Techniques

Direct benefits: These include the raw materials and physical products that are used directly for production, consumption, and sale including those providing energy, shelter, foods, water supply, transport, and recreation.

Indirect benefits: These include ecological functions which maintain, protect, and support natural and human systems through services such as maintenance of water quality, flow and storage, flood control and storm protection, nutrient retention, and micro-climate stabilization, and other productive and consumptive activities.

Option benefits: These refer to the premium placed on maintaining a pool of wetlands species and genetic resources for future uses such as for leisure or in commercial, industrial, agricultural and pharmaceutical applications and water-based developments, some of which may not be used or known in the present but may be used in the future.

Existence benefits: The intrinsic value of wetlands species and areas regardless of their current or future use possibilities, such as cultural, aesthetic, heritage and bequest significance.

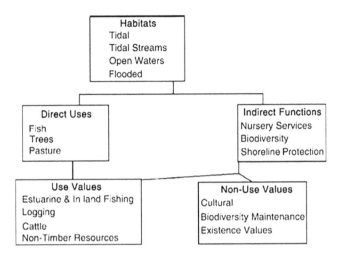

FIGURE 22.1 Identifying direct and indirect benefits in a wetland system.

Valuing Wetlands Goods—Techniques and Examples

A combination of the available techniques used to value environmental goods and services can be used to approximate a partial or total value of any one ecosystem, taking care that we can identify the various components of the system and have adequate economic and ecological data to do so. The following methods have been used in valuing wetland resources:

- Market prices: to estimate economic values for ecosystem products or services that are bought and sold in markets.
- Change in productivity: to estimate economic values for ecosystem products or services that contribute to the production of marketed goods or services.
- Travel cost: to estimate economic values associated with ecosystems/sites used for recreation based on the assumption that the value of a site is reflected in how much people are willing to pay to travel to visit the site.
- Hedonic pricing: to estimate economic values for ecosystem or environmental services that directly affect market prices of some other good, e.g., housing prices may reflect the value of local environmental attributes or amenities.
- Contingent valuation method: to estimate economic values for products and ecosystem services. This can be used to estimate nonuse values by asking people to directly state their willingness to pay for specific environmental services, based on hypothetical scenarios or contingent markets.
- Contingent choice, contingent ranking, contingent behavior: to estimate economic values for products and for ecosystem services by asking people to make tradeoffs between sets of ecosystem or environmental services or characteristics.
- Benefit transfer: to infer economic values by transferring benefit estimates from localized studies to similar areas under similar market conditions.
- Damage cost avoided, replacement cost, and substitute costs: to estimate economic values based on costs of avoided damages resulting from lost or diminished ecosystem services, replacing ecosystem services, or providing substitute services.

Examples of how these techniques have been applied to elicit values of various wetland benefits are given in Tables 22.2 and 22.3 and in Table 22.4. These studies illustrate the range of values associated with different types of goods and services, wetlands provide.

TABLE 22.2 Using Contingent Valuation and Travel Cost Methods to Assess the Recreational Value of Lake Nakuru, Kenya

Lake Nakuru National Park is an important international tourist destination in Kenya. Fees currently charged to enter the park underestimate the total value that tourists place on the wetland and its component species, especially flamingos. A travel cost survey of visitors elicited information about length of stay, travel costs, place of origin, and visitation rates, distinguishing between resident and nonresident tourists. The contingent valuation survey used in this study asked visitors what their personal total costs of travel were; how much they would be willing to increase their expenditures to visit the park; how much they would contribute to a fund to clean-up and control urban pollution affecting the park; and how much they would contribute to a project to conserve flamingos. The survey also asked respondents for the minimum reduction in trip costs that they would be willing to accept should there be no flamingos. Results suggests that the annual recreational value of wildlife viewing in Lake Nakuru National Park was between U.S. $7.5 and $15 million and over a third of this value was accounted for by the willingness to pay for viewing flamingos.

Source: Adapted from Navrud & Mungatana.[6]

TABLE 22.3 Valuing the Hadejia Nguru Wetlands, Nigeria

In the Hadejia–Jama are floodplain region in northern Nigeria, more than one half of the wetlands have already been lost to drought and upstream dams. Ecosystem valuation has been used in this area to weigh the costs and benefits of development projects that would divert some more water away from the floodplain for irrigated agriculture in upstream areas. The net benefits of such a diversion are estimated at U.S. $29 per hectare. In comparison, the floodplain, under the present flooding regime, provides U.S. $167 per hectare in benefit to a wide range of local people engaged in farming, fishing, grazing livestock, or gathering fuelwood and other wild products—benefits that would be greatly diminished by the project. A study of the groundwater recharge function of the wetlands confirms furthermore that the wetlands play an important role by maintaining groundwater recharge in the floodplain. Groundwater recharge supports irrigated agricultural production in the floodplain. Irrigated agriculture using water from the shallow groundwater aquifer has a value of 36,308 Naira (U.S. $413) per hectare for the study area. A value of at least 2863 Naira or U.S. $32.5 per farmer per dry season or U.S. $62 per hectare is attributable to the present rate of groundwater recharge. In terms of maintaining water supply resources, the value of the recharge function is 1,146,588 Naira or U.S. $13,029 per day for the wetlands.

Source: Adapted from Barbier, et al.[7] Acharya & Barbier,[8] Acharya.[9]

TABLE 22.4 Economic Values of Wetlands: Some Examples

Valuation Approach	Good or Service Valued	Reference	Value	Location
Revealed willingness to pay (WTP)	Commercial fishing and trapping	[18]	$486.25 per acre	Louisiana, U.S.
Revealed WTP	Freshwater forested wetlands	[23]	$78,000 per acre	Maryland, U.S.
Mitigation costs	Freshwater emergent wetlands		$49,000 per acre	
Contingent valuation, replacement cost	Nitrogen abatement	[21]	U.S. $59/kgN reduction capacity	Gotland, Sweden
Loss in productivity, market prices	Agriculture, forestry fishing from floodplain	[7]	N109 (U.S. $15)/10^{3}m^3 N381 (U.S. $51)/h	Hadejia–Nguru
Production function, contingent behavior	Groundwater recharge function (drinking water supply)	[9]	U.S. $13,029 per day for the wetlands	Wetlands, Nigeria
Production function	Groundwater recharge function (agriculture)	[8]	$62/h	
Market prices	Artisanal fisheries, aquarium fish, sea cucumbers, shells, tourism	[24]	U.S. $2.14 million/yr	Kisite Marine National Park and Mpunguti Marine National Reserve, Kenya
Preventive expenditure	Coastal zone/shoreline protection	[25]	U.S. $603,248/yr	Mahe, Seychelles

In some parts of the U.S., the benefits of protecting natural watersheds to assure safe and plentiful drinking water supplies are being recognized. The city of New York recently found that it could avoid spending U.S. $6–8 billion in constructing new water treatment plants by protecting the upstate watershed that has traditionally provided these purification services. Based on this assessment, the city invested U.S. $1.5 billion in protecting its watershed by buying land around reservoirs and providing other incentives for land use.[10] This decision ensures that in addition to getting a relatively cheap treatment system for its water supply, the city also maintains the watershed for recreation, wildlife habitat, and other ecological benefits derived from a healthy ecosystem.

Conclusions

Some caveats to valuation also apply. In a number of recent papers,[11–13] researchers have attempted to value the world's ecosystems, some by extrapolating localized studies of specific resource losses to capture the impact of a global loss of such resources. Even if local values are calculated using the best possible data, extrapolating these values to all wetland resources on the planet creates problem not least because there is wide variation in the types of wetlands and their component products. Heimlich et al.[14] presents the results of over 30 studies carried out on wetlands. After accounting for geographical variation and other differences across the studies by calculating the present value over a 50 yr period discounted at 6% for all the studies and then converting these values to an annual basis, they find significant variation among the different studies. For example, general recreational values range from $105 to $9859 per acre; waterfowl hunting from $108 to $3101 per acre; and recreational fishing from $95 to $28,845 per acre. Table 22.4 gives a few examples of various studies and types of goods or services valued. As the figures show, there is a wide variation among the types of wetlands valued and the values attributable to them. Great caution must therefore be exercised when extrapolating localized studies to other geographical areas. Resource values are influenced by local consumptive and productive use patterns, market conditions, and the institutional arrangements affecting the use of the resources. The good news is that there are now hundreds of studies of wetland values (Bardecki[15] for an extensive review) that we can refer to for some guidance on the range of values associated with wetlands.[16–22]

Benefits of some wetlands will always be difficult to quantify and measure primarily because the required scientific, technical, or economic data is difficult to obtain and also that certain intrinsic values are not measurable by existing economic valuation methods. However, as the studies reported in this paper suggest, various wetland goods and services are extremely valuable and in many cases, a measurable economic value can be obtained. The economic value of ecological functions in sustaining the livelihood and cultures of human societies clearly cannot be disregarded in development and conservation policy since these values allow us to make informed decisions about tradeoffs and help in the conservation of natural resources to enhance human welfare. Hence, valuation studies need to be carried out wherever possible and with collaboration between economists and natural scientists to fully capture the economic and ecological linkages that make wetlands valuable.

References

1. Dugan, PJ. *Wetland Conservation: A Review of Current Issues and Required Action*; IUCN: Gland, Switzerland, 1990.
2. OECD/IUCN. *Guidelines for Aid Agencies for Improved Conservation in Sustainable Use of Tropical and Subtropical Wetlands*; OECD: Paris, 1996.
3. Mitsch, W.J.; Gosselink, J.G. *Wetlands,* 2nd Ed.; Van Nostrand Reinhold: New York, 1993.
4. Freeman, A.M. *The Measurement of Environmental and Resource Values: Theory and Methods*; Resources for the Future: Washington, DC, 1993.
5. Maltby, E. *Waterlogged Wealth*; Earthscan: London, 1986.
6. Navrud, S.; Mungatana, E. Environmental valuation in developing countries: the recreation value of wildlife viewing. Ecol. Econ. **1994**, *11,* 135–151.
7. Barbier, E.B.; Adams, W.; Kimmage, K. Economic valuation of wetland benefits. In *The Hadejia-Nguru Wetlands*; Hollis, Ed.; IUCN: Gland, 1993.

8. Acharya, G.; Barbier, E.B. Valuing groundwater recharge through agricultural production in the Hadejia–Nguru wetlands in Northern Nigeria. Agric. Econ. **2000**, *22*, 247–259.

9. Acharya, G. Valuing the hidden hydrological services of wetland ecosystems. Ecol. Econ. **2000**, *35*, 63–74.

10. National Research Council, (NRC). Committee to review the New York City watershed management strategy. In *Watershed Management for Potable Water Supply: Assessing the New York City Strategy*; National Academy Press: Washington, DC, 2000.

11. Costanza, R.; d'Arge, R.; de Groot, R.; Farber, S.; Grasso, M.; Hannon, B.; Limburg, K.; Naeem, S.; O'Neill, R.V.; Paruelo, J.; Raskin, R.G.; Sutton, P.; van den Belt, M. The value of the world's ecosystem services and natural capital. Nature **1997**, *387*, 253–260.

12. Pimentel, D.; Wilson, C.; McCullum, C.; Huang, R.; Dwen, P.; Flack, J.; Tran, Q.; Saltman, T.; Cliff, B. Economic and environmental benefits of biodiversity. BioScience **1997**, *47*(11).

13. Ehrlich, P.R.; Ehrlich, A.H. *Betrayal of Science and Reason: How Anti-Environmental Rhetoric Threatens Our Future*; Putnam: New York, 1996.

14. Heimlich, R.E.; Wiebe, K.D.; Claassen, R.; Gadsby, D.; House, R.M. *Wetlands and Agriculture: Private Interests and Public Benefits;* Agriculture Economic Report No. 765; Resource Economics Division, Economic Research Service, USDA, 1998.

15. Bardecki, M. Wetlands and economics: an annotated review of the literature, 1988–1998 with special reference to the wetlands of the Great Lakes. In *Report Prepared for Environment Canada— Ontario Region;* http://www.on.ec.gc.ca/glimr/data/wetland-valuation/intro.html.; 1998.

16. Hammack, J.; Brown, G.M. *Waterfowl and Wetlands: Toward Bioeconomic Analysis*; Resources for the Future: Washington, DC, 1974.

17. Lynne, G.D.; Conroy, P.; Prochaska, F.J. Economic valuation of marsh areas for marine production processes. J. Environ. Econo. and Manag. **1981**, *8*, 175–186.

18. Farber, S.; Costanza, R. The economic value of wetland systems. J. of Environ. Manag. **1987**, *24*, 41–51.

19. Bell, F. The economic valuation of saltwater marshes supporting marine recreational fishing in south-eastern United States. Ecol. Econ. **1997**, *21*, 243–254.

20. Bergstrom, J.C.; Stoll, J.R.; Titre, J.P.; Wright, V.L. Economic value of wetlands based recreation. Ecol. Econ. **1990**, *2*, 129–147.

21. Gren, I.M.; Folke, C.; Turner, K.; Bateman, I. Primary and secondary values of wetland ecosystems. Environ. Resour. Econ. **1994**, *4*(1), 55–74.

22. Morrison, M.D.; Bennett, J.W.; Blamey, R.K. *Valuing Improved Wetland Quality Using Choice Modelling*; Choice Modelling Research Report No. 6; University College, The University of New South Wales: Canberra, 1998.

23. Bohlen, C.; King, D. *Towards a Sustainable Coastal Watershed: The Chesapeake Experiment*; Proceedings of a conference, CRC Publication No. 149; Nelson, S., Hiss, P., Eds.; Chesapeake Research Consortium: Edgewater, MD, 1995.

24. Emerton, L.; Tessema, Y. *Economic Constraints to the Management of Marine Protected Areas: The Case of Kisite Marine National Park and Mpunguti Marine National Reserve, Kenya*; IUCN—The World Conservation Union: 2001.

25. Emerton, L. *Economic Tools for Valuing Wetlands in Eastern Africa;* IUCN—The World Conservation Union: 1998.

23

Wetlands: Indices of Biotic Integrity

Donald G. Uzarski,
Neil T. Schock,
Ryan L. Wheeler,
Jacob M. Dybiec,
Anna M. Harrison,
and Alexandra
C. Mattingly
CMU Biological Station
Central Michigan
University

Introduction

The assessment of habitat quality has been recognized as an important part of habitat and wildlife management. The goal of this entry is to introduce the framework of indices of biotic integrity (IBIs) and describe how they are constructed, validated, and applied as tools to evaluate habitat quality using biological indicators, particularly as they pertain to wetland habitats. These tools and concepts have proven successful in aquatic habitats, and it is our belief that these same concepts can be applied to numerous biological communities that populate various types of habitats and ecosystems. We hope to clearly outline the value of IBIs as tools for habitat quality assessments and explain the advantages of this process when compared to previously developed approaches to habitat quality assessment.

The Importance of Wetlands

Wetlands maintain biological diversity and are very productive ecosystems.[1,2] Wetlands are unique in that they are transitional zones between the terrestrial upland and true aquatic habitats (Figure 23.1). Wetland types are also diverse and are structured by hydrology, geography, and geology. The high variability in wetland habitat results in many very specialized organisms that are unique to wetlands.[1] Only an estimated 3.5% of the United States is covered by wetlands; yet they support over 50% of the species on the federal endangered species list.[3] High concentrations of nutrients and moisture make wetlands highly diverse and productive systems.[4]

In addition, wetlands provide extensive functions and values that translate to many ecosystem services.[5] Water flowing from terrestrial landscapes is filtered as it passes through wetlands into aquatic environments. Wetland plant communities filter out nutrients by assimilation into biomass, while trapping toxicants and sediments before they enter true aquatic habitats such as oceans or lakes.[6] In this way, wetlands play an important role in carbon and nutrient cycles.[7,8] Other important services include

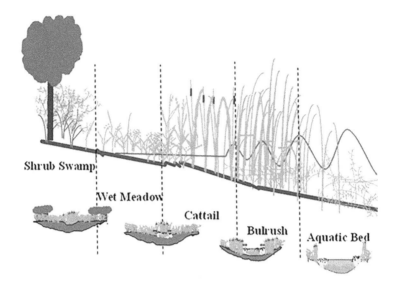

FIGURE 23.1 Example of wetland zones along the transition from terrestrial to aquatic habitats. The zonation of lacustrine wetlands is the spatial equivalent of succession through time in palustrine wetlands.
Source: Taken from Uzarski.[9]

habitat for wildlife, sources of food and medicine, floodwater storage, groundwater recharge, carbon storage, climate stabilization, erosion control, tourism and recreation, building materials, fertilizer, and cultural heritage.[2]

The Need for IBIs

Expanding human populations have given rise to the alteration of millions of acres of land, subsequently resulting in increased agricultural and urbanized land-use, which often corresponds to the degradation of wetlands. In the United States, as much as 50% of wetland habitats have been drained or developed since European settlement.[10] Additionally, it is suspected that all wetlands have been negatively affected by humans in some manner.

Due to the essential services and habitat that wetlands provide, it is important to have methods of evaluating and monitoring the health of existing wetlands to focus restoration and preservation efforts. Catching degradation before it is irreversible is essential. This can prove to be a difficult task due to the highly variable nature of wetlands. Traditional approaches to detect degradation relied heavily on water chemistry. Unfortunately, this approach fails to account for human-induced habitat alteration, introduction of exotic species, and episodic events such as spills and effluent discharge. Biota residing in these wetlands, however, can integrate the overall habitat and water quality of these sites over time and reveal both episodic and cumulative disturbance. Karr introduced the IBI in the 1980s, using fish community characteristics as indicators of environmental health, taking a more holistic approach toward assessing aquatic environmental quality.[11] As such, it is theorized that biological communities can provide information that a single water sample cannot. Episodic disturbances to aquatic ecosystems may quickly become non-detectable as pollutants are diluted or washed away with currents. However, biological communities will maintain a "scar" from the event.[12] IBIs attempt to detect these "scars." A properly developed IBI will use these community "scars" as individual metrics to point toward specific disturbances that may have occurred. Each metric that makes up an IBI is a biological attribute related to specific types of human disturbance. Because of this, IBIs are sometimes referred to as multimetric IBIs. Since the 1980s, IBIs have been expanded to many different habitat types using a variety of biological communities. Currently, wetland IBIs include the use of fish[13–16], macroinvertebrates[17–22], birds[23,24], amphibians[25], zooplankton[26], algae[27,28], microbial communities[29], and vegetation[30–32].

The development of wetland IBIs has been a particular challenge. The majority of the initial work was done in streams and rivers, which relied heavily on detecting community changes as a result of altered temperature and dissolved oxygen regimes.[33] This approach was not applicable to wetlands because substantial temperature and dissolved oxygen fluctuations occur naturally in wetlands regardless of anthropogenic disturbance. Many of these hurdles have been breached, and wetland indices have proven to be cost-effective and accurate tools for assessing wetland condition. Wetland IBIs are currently being used by many government agencies and private organizations to make regulatory and conservation decisions.

IBI Development

Developing indices of biological integrity requires the categorization of wetlands based on habitat type and natural disturbance, to isolate variables directly associated with human disturbance. The delineation of these wetland categories is based upon hydrology and vegetation. Hydrology dictates the amount of natural disturbance shaping communities, and vegetation represents structure for other organisms that also naturally shape communities. The goal in development is to control for this natural variation and isolate community responses only due to human degradation. An example of wetland categories may include riverine, palustrine, and lacustrine wetlands or those associated with rivers, depressions in the landscape, or lakes, respectively. These categories are then further delineated based on hydrology, geography, and geology because this natural variation can create substantial variability in community composition regardless of human disturbance. The process of grouping sites with consistent natural conditions allows researchers to more confidently attribute observed variation among sites to human influences. While building IBIs for specific wetland categories, researchers use known levels of human disturbance as a baseline for building IBI metrics.[34]

IBI Development: Establishment of a Disturbance Gradient

Building IBI metrics involves identifying sites experiencing a range of known human disturbance. Reference sites receiving the least amount of human disturbance represent intact community composition. Those sites experiencing a range of stressors will then be used to determine variation in community composition, and this variation will be used to establish metrics. Ideally, a range of stressor types is also included to establish stressor-response relationships so that individual metrics will indicate specific anthropogenic disturbances to the ecosystem. Metrics can then be combined into an overall multimetric IBI. Known stressors at sites should be linked to chemical-physical characteristics and land use/land cover information. Chemical-physical parameters and land use/land cover data specific to each wetland can be combined using principal components analysis (PCA; Figure 23.2). Sites with known human disturbance are used to identify related variables. Principal components that ordinate sites from the most human impact to the least human impact, or vice versa, can be used to identify chemical-physical and land use/cover variables associated with known human disturbances. These should then be related to variation in community composition. To capture local conditions unique to specific wetland categories, *in situ* chemical-physical variables, as well as adjacent land use and cover, are included in the disturbance gradient.

Disturbance gradients describe the overall cumulative range of known human-induced stressors among all sites in a dataset and allow comparisons of disturbance between those sites, which is vital to creating accurate IBI metrics. One method which has been used in the development and testing of Great Lakes coastal wetland IBIs is referred to as "sumrank". This method combines *in situ* water chemistry (e.g., pH, dissolved oxygen, specific conductance, etc.), *ex situ* water chemistry (such as soluble reactive phosphorus, total nitrate-nitrite, etc.) and land use/land cover to create a robust disturbance gradient.

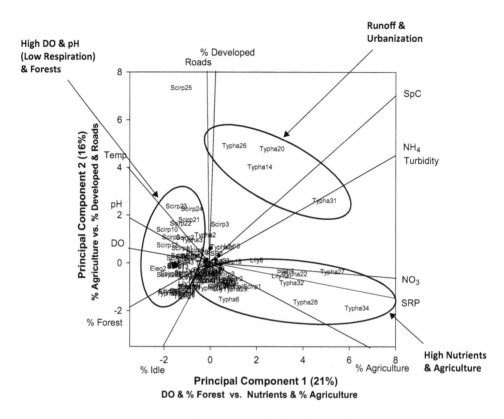

FIGURE 23.2 Principal components analysis showing the distribution of sites based on habitat conditions and land use variables. Abbreviations: % Forest, forested land use; % Idle, inundated land cover; % Agriculture, Agricultural land use; SRP, Soluble reactive phosphorus; NO₃, Nitrate; NH₄, Ammonium; SpC, Specific conductance; % Developed, developed land use; DO, dissolved oxygen.

Values for each variable are "ranked" within the dataset from most disturbed (low rank) to reference condition (high rank), and the ranks are then summed together for each site. These summed ranks are then scaled between 0 and 100 to create easy-to-understand disturbance scores, which can then be used in the process of metric creation and verification. While this is just one example, there are many other potential methods to establish a disturbance gradient, such as water chemistry data only or land use/land cover data only[34].

IBI Development: Metric Creation

Biological community data from the same sites used in PCA can be combined using ordination methods such as nonmetric multidimensional scaling (NMDS) or correspondence analysis (CA). Synthetic community variables or dimensions of these analyses can then be used in Pearson correlation analysis to relate biotic communities to principal components representing chemical-physical and land use/land cover characteristics along a gradient of human disturbance. NMDS and CA dimensions can then be decomposed using species weights or factor loadings to point toward taxa responding to human disturbance. These taxa can then be used to generate metrics. Other commonly used metrics such as diversity, richness, and evenness can also be explored and related to disturbance gradients. Any measures of the biotic communities related to measures of known stress are candidates for metrics to be included in an IBI.

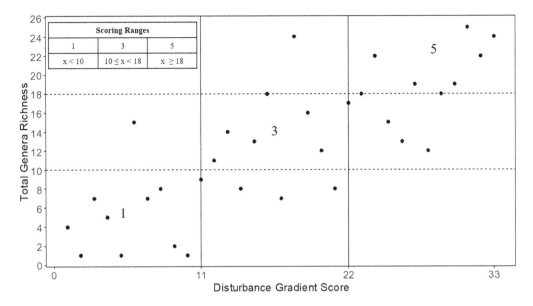

FIGURE 23.3 Example of creating a scoring range for a potential IBI metric across a disturbance gradient. The solid vertical lines represent the bounds of each group along the disturbance gradient axis, while the dashed lines represent the bounds of each scoring range for points within each group. This metric, along with other example metrics, can be found in Table 23.1.

The potential IBI metrics are then used to create "scoring ranges". One such method of scoring is to plot the relationship between a potential metric (y-axis) against the disturbance gradient (x-axis), and then split the disturbance gradient into equal sections (solid lines, Figure 23.3). Points for each site are then binned into groups within each section, such that the majority of points within the group are represented (dashed lines, Figure 23.3). These groups are assigned point values, and the range of values it encapsulates on the y-axis is used to create scoring ranges, as in the table within Figure 23.3. This is done for every potential IBI metric, and then the scores are grouped together into a preliminary multimetric IBI. This IBI can then be applied to any site in question, the score for each metric added together, and a final IBI score assigned.

IBI Development: Validation and Application

Developing an IBI is an iterative process, which involves a great deal of exploratory data analysis. Index scoring systems have been designed to use biological community characteristics such as taxa richness and diversity along with the relative abundance of tolerant and sensitive organisms to evaluate wetland health (Table 23.1).[16,22] Once the metrics are calculated, the performance of the indices is tested on a group of sites that span a known disturbance gradient to validate the process.[22] Index scores from sites with known impacts from human activities are compared to scores from relatively pristine sites.

A confounding factor for IBI use is the interpretation of IBI scores when more than one community index is used to assess the condition of a single wetland. For instance, a fish-based IBI may give a wetland a relatively high score, but a plant-based IBI for the same wetland may result in a relatively low score. The observed differences in assessment scores may be associated with the nature of human influences on the wetland. In this example, the plant-based IBI may be low due to the influence of invasive plant species located too far from water depths where fish could be influenced. An opposite result could be observed if the near-open water portion of the wetland was receiving pollutants from a nearby boat

TABLE 23.1   This Is an Example of the Types of Information Used to Calculate an IBI Score for a Wetland

Example of a Simplified Index of Biotic Integrity Worksheet

Individual Metrics

1	Taxa richness from a specific family (species)			
	0 score = 1	>0 to 3 score = 3	>3 score = 5	
2	Total genera richness			
	<10 score = 1	10 to 18 score = 3	>18 score = 5	
3	Relative abundance of a specific family (%)			
	0 to 1 score = 1	>1 to 25 score = 3	>25 score = 5	
4	Combined taxa richness of two specific families (genera)			
	0 to 2 score = 1	>2 to 4 score = 3	>4 to 6 score = 5	>6 score = 7
5	Shannon diversity index			
	0 to 0.4 score = 1	>0.4 to 0.9 score = 3	>0.9 score = 5	

Category Scores

0% to 15% of possible score
Degraded: In comparison to other wetlands of the same type, this wetland is among the most impacted.
>15% to 50% of possible score
Moderately degraded: The wetland shows obvious signs of anthropogenic disturbance.
>50% to 85% of possible score
Mildly impacted: The wetland is beginning to show signs of anthropogenic disturbance.
>85% to 100% of possible score
Reference conditions: The wetland is among the most pristine of wetlands of a similar type.

These types of categorical calculations are typically performed for specific zones or areas within the wetland depending on the complexity of the habitat. Individual metric scores are added together and category scores are assigned based on the total score's magnitude.

channel or drainage ditch. In this case, the fish-based IBI may indicate a low score, while the plant IBI may be high due to lack of influence from the pollutants. IBI scores should be viewed within the context of on-site observations, which may explain the disturbances and the differences among IBIs.

Taxa-based IBIs may also detect human disturbance at different scales. The distribution, colonization rates, and mobility of each indicator organism determine the scale of detection. For example, the relatively slow life cycle and distribution rates of plant species suggest that they could be indicating larger-scale influences, whereas macroinvertebrates, with their relatively short life cycles and limited range of mobility, might indicate finer scales of influence. As such, all wetlands within a particular region may yield a high quality rating using a plant-based IBI, but a macroinvertebrate-based IBI may yield a wide range of scores for the region and serve as a fine adjustment within the "high quality" category. These possibilities are not absolute, but concepts such as these should be considered when implementing information gathered by IBIs into management decisions.

Conclusion

IBIs are valuable tools that researchers and managers can use to assess aquatic habitat conditions using biological community characteristics. When developed and validated thoroughly, these tools effectively assess the status and trends of wetland habitats and are at the forefront of aquatic habitat assessment techniques. The concepts discussed in this entry can and should be applied toward the development of new IBIs for other ecosystems. The information obtained from these methods will allow for more efficient usage of conservation efforts and funding. Future developments should focus on improving these tools to increase accuracy and ease of use and further decrease the costs of conservation efforts.

Acknowledgment

We would like to dedicate this entry to Dr. Thomas M. Burton, a pioneer in wetland bioassessment. This is Contribution Number XX of the Central Michigan University Institute for Great Lakes Research.

References

1. Leveque, C.; Balian, E.V.; Martens, K. An assessment of animal species diversity in continental waters. Hydrobiologia **2005**, *542*, 39–67.
2. Woodward, R.T.; Wui, Y. The economic value of wetland services: A meta-analysis. Ecol. Econ. **2001**, *37*, 257–270.
3. Mitsch, W.A.; Gosselink, J.G. *Wetlands*, 3rd Ed.; John Wiley and Sons, Inc.: New York, 2000; 920.
4. Wassen, A.J.; Verkroost, A.W.M.; DeRuiter, P.C. Species richness-productivity patterns differ between N-, P-, and K-limited wetlands. Ecology **2003**, *84* (8), 2191–2199.
5. Mitsch, W.J.; Gosselink, J.G. The value of wetlands: Importance of scale and landscape setting. Ecol. Econ. **2000**, *35* (1), 25–33.
6. Chambers, R.M.; Meyerson, L.A.; Saltonstall, K. Expansion of *Phragmites australis* into tidal wetlands of North America. J. Aquat. Bot. **1999**, *64* (3–4), 261–273.
7. Comin, F.A.; Romero, J.A.; Astorga, V.; Garcia, C. Nitrogen removal and cycling in restored wetlands used as filters of nutrients for agricultural run-off. Water Sci. Technol. **1997**, *35* (5), 255–261.
8. Ottovia, V.; Balcarova, J.; Vymazal, J. Microbial characteristics of constructed wetlands. Water Sci. Technol. **1997**, *35* (5), 117–123.
9. Uzarski, D.G. *Wetlands of Large Lakes. Encyclopedia of Inland Waters*; Oxford: Elsevier, 2009; 599–606.
10. Krieger, K.M.; Kalarer, D.M.; Heath, R.T.; Herdendorf, C.E. Coastal wetlands of the Laurentian Great Lakes: Current knowledge and research needs. J. Great Lakes Res. **1992**, *18*, 525–528.
11. Karr, J.R. Assessment of biotic integrity using fish communities. Fisheries **1981**, 6, 21–27.
12. Adams, S.M.; Greeley, M.S. Ecotoxicological indicators of water quality: Using multi-response indicators to assess the health of aquatic ecosystems. Water Air Soil Pollut. **2000**, *123*, 103–115.
13. Bhagat, Y.J.; Ciborowski, J.H.; Johnson, L.B.; Burton, T.M.; Uzarski, D.G.; Timmermans, S.A.; Cooper, M. Testing a fish index of biotic integrity for Great Lakes coastal wetlands: Stratification by plant zones. J. Great Lakes Res. **2007**, *33* (3), 224–235.
14. Cooper, M.J.; Lamberti, G.A.; Moerke, A.H.; et al. An expanded fish-based index of biotic integrity for Great Lakes coastal wetlands. Environmental Monitoring and Assessment **2018**, 190:580.
15. Seilheimer, T.S.; Chow-Fraser P. Development and use of the Wetland Fish Index to assess the quality of coastal wetlands in the Laurentian Great Lakes. Canadian Journal of Fisheries and Aquatic Sciences **2006**, *63*, 354–366.
16. Uzarski, D.G.; Burton, T.M.; Cooper, M.J.; Ingram, J.W.; Timmermans, S. Fish habitat use within and across wetland classes in coastal wetlands of the five Great Lakes: Development of a fish-based index of biotic integrity. J. Great Lakes Res. **2005**, *31* (1), 171–187.
17. Chowdhury, G.W.; Gallardo B.; Aldridge, D.C. Development and testing of a biotic index to assess the ecological quality of lakes in Bangladesh. Hydrobiologia **2016**, *765*, 55–69.
18. Lu, K.L.; Wu, H.T.; Xue, Z.S.; Lu, X.G.; Batzer, D.P. Development of a multi-metric index based on aquatic invertebrates to assess floodplain wetland condition. Hydrobiologia **2019**, *827*, 141–153.
19. Lunde, K.B.; Resh, V.H. Development and validation of a macroinvertebrate index of biotic integrity (IBI) for assessing urban impacts to Northern California freshwater wetlands. Environmental Monitoring and Assessment **2012**, *184*, 3653–3674.
20. Melo, S.; Stenert, C.; Dalzochio, M.S.; Maltchik, L. Development of a multimetric index based on aquatic macroinvertebrate communities to assess water quality of fields in southern Brazil. Hydrobiologia **2015**, *742*, 1–14.

21. Mereta, S.T.; Boets, P.; De Meester, L.; Goethals, P.L.M. Development of a multimetric index based on benthic macroinvertebrates for the assessment of natural wetlands in Southwest Ethiopia. Ecological Indicators **2013**, *29*, 510–521.

22. Uzarski, D.G.; Burton, T.M.; Genet, J.A. Validation and performance of an invertebrate index of biotic integrity for Lakes Huron and Michigan fringing wetlands during a period of lake level decline. Aquat. Health Manag. **2004**, *7*, 269–288. Lougheed, V.L.; Chow-fraser, P. Development and use of a zooplankton index of wetland quality in the Laurentian Great Lakes basin. Ecol. Appl. **2002**, *12* (2), 474–486.

23. Lane, C. R.; Brown, M.T. Diatoms as indicators of isolated herbaceous wetland condition in Florida, USA. Ecological Indicators **2007**, *7*, 521–540.

24. Miller, K.M.; Mitchell, B.R.; McGill, B.J. Constructing multimetric indices and testing ability of landscape metrics to assess condition of freshwater wetlands in the Northeastern US. Ecological Indicators **2016**, *66*, 143–152.

25. Zhu, W.; Liu, Y.; Wang, S.; Yu, M.; Qian, W. Development of microbial community–based index of biotic integrity to evaluate the wetland ecosystem health in Suzhou, China. Environmental Monitoring and Assessment **2019**, 191:377.

26. Gara, B.D.; Stapanian, M.A. A candidate vegetation index of biological integrity based on species dominance and habitat fidelity. Ecological Indicators **2015**, *50*, 225–232.

27. Moges, A.; Beyene A.; Kelbessa E.; Mereta, S.T.; Ambelu, A. Development of a multimetric plant-based index of biotic integrity for assessing the ecological state of forested, urban and agricultural natural wetlands of Jimma Highlands, Ethiopia. Ecological Indicators **2016**, *71*, 208–217.

28. Rothrock, P.E.; Simon, T.P. A Plant Index of Biotic Integrity for Drowned River Mouth Coastal Wetlands of Lake Michigan. In Simon TP and Stewart PM (eds.) *Coastal Wetlands of the Laurentian Great Lakes: Health, Habitat, and Indicators.* **2006**, Authorhouse Press, Bloomington, IN, pp 195–208.

29. Wilhm, J.L.; Dorris, T.C. Biological parameters for water quality criteria. Bioscience **1968**, *18*, 477–481.

30. Burton, T.M.; Uzarski, D.G.; Gathman, J.P.; Genet, J.A.; Keas, B.E.; Stricker, C.A. Development of a preliminary invertebrate index of biotic integrity for Lake Huron coastal wetlands. Wetlands **1999**, *19* (4), 869–882.

31. Houlahan, J.E.; Keddy, P.A.; Makkay, K.; Findlay, C.S. The effects of adjacent land use on wetland species richness and community composition. Wetlands **2006**, *6* (1), 79–96.

32. Uzarski, D.G.; Brady, V.J.; Cooper, M.J.; et al. Standardized Measures of Coastal Wetland Condition: Implementation at a Laurentian Great Lakes Basin-Wide Scale. Wetlands **2017**, *37* (1), 15–32.

33. Kerans, B.L; Karr, J.R. A benthic index of biotic integrity (B-IBI) for rivers of the Tennessee Valley. Ecological Applications **1994**, *4* (4), 768–785.

34. Harrison, A.M; Reisinger, A.J.; Cooper, M.J.; Brady, V.J.; Ciborowski, J.J.H.; O'Reilly, K.E.; Ruetz III, C.R.; Wilcox, D.A.; Uzarski, D.G. A Basin-Wide Survey of Coastal Wetlands of the Laurentian Great Lakes: Development and Comparison of Water Quality Indices. Wetlands. **2019**. https://doi.org/10.1007/s13157-019-01198-z

24

Wetlands: Remote Sensing

Niti B. Mishra
University of Texas
at Austin

Introduction

Wetlands are vital ecosystems for planetary and human health as they provide essential ecological services such as purification of water, cycling of nutrients, and recharge of ground water and also provide wetland products such as fish, rice, and some forest resources.[1,2] Wetlands are often rich in biodiversity and are important sites for biogeochemical processes such as the production of methane and nitrous oxide as well as fixation of carbon and nitrogen.[1,3,4] Historically, wetlands were considered a hindrance to development and were modified following dredge and fill operations, pollutant runoff, eutrophication, and fragmentation by roads.[5] The key role of wetlands in sustaining ecological health has only been recognized recently prompting their monitoring and conservation.[5,6] Over the last four decades, remote sensing has served as an important and powerful tool for mapping, monitoring, and characterization of wetland ecosystems. All types of wetlands including inland freshwater marshes, coastal tidal marshes, mangrove ecosystems, and forested wetlands or swaps have been studied using remotely sensed data available at different spatial, spectral, and temporal resolutions. This entry provides an overview of the data, methods, application areas, and challenges of remote sensing of wetlands.

Types of Remotely Sensed Data Used in Wetland Studies

Aerial Photography and Multispectral Remote Sensing

The earliest remote sensing applications in wetlands utilized color and color-infrared aerial photos, and orthophoto quads acquired over wetlands for applications such as delineating wetland boundaries,[7-9] mapping wetland vegetation,[10-12] classifying wetland types, and assessing vegetation growth and water quality.[13,14] Although aerial photographs provided the high spatial resolution important for characterizing wetland heterogeneity, most of these studies acknowledged the need for improved spectral resolution for enhanced characterization of spatially heterogeneous and spectrally complex wetlands. Furthermore, aerial photography-based studies were site specific and had limited spatial coverage. The availability of multispectral satellite sensors (e.g., Landsat MSS, TM, ETM+, SPOT, IRS) allowed overcoming these limitations by providing larger swath area and repetitive coverage and enabled monitoring at landscape-to-regional scales. The Landsat 1, –2, and –3 had multispectral scanner (MSS) sensors with four spectral bands that were used to study different aspects of wetland ecology in a cost-effective manner.[15–19] Launched in the year 1982, the Landsat Thematic Mapper (TM) represented significant improvement in radiometric and spatial resolution over MSS and was used by several studies for wetland mapping and monitoring.[20–25] The Landsat TM band combination of 2, 4 and 5 is effective for wetland detection; band 5 has the ability to distinguish vegetation and soil moisture levels and band 1 is utilized for determining water quality in wetlands.[23,26,27] Unlike Landsat MSS or TM sensors, Systeme Pour l'Observation de la Terre (SPOT) was the first satellite to have pointable optics that provided increased capabilities for capturing any given area. SPOT High Resolution Visible (HRV) sensor has green, red, and near infrared spectral bands at 20 m spatial resolution while the SPOT-4 sensor launched in 1998 also included a middle infrared band which was intensively utilized in wetland studies.[28–32] Indian Remote Sensing (IRS) satellites equipped with Linear Imaging Self-Scanning Sensor (LISS-II, III) providing multispectral images at medium spatial resolution (i.e., 20–30 m) have also been utilized for wetland studies at landscape scale.[33–36] Multispectral sensors with very high temporal revisit capability (e.g., Moderate Resolution Imaging Spectroradiometer [MODIS], Advanced Very High Resolution Radiometer [AVHRR]) have been utilized for studying temporally dynamic wetland systems (e.g., estuaries, deltas) to characterize the vegetation phenology and rapid vegetation succession.[37,38] However, in spite of high temporal resolution and large swath area, these data sets have not been preferred for wetland studies primarily due to coarse spatial resolution.[26]

High Spatial Resolution Remote Sensing

Most wetlands display high spatial heterogeneity, patchiness, and considerable functional diversity that may require spatial details higher than those offered by medium spatial resolution imagery (Figure 24.1).

The availability of high spatial resolution sensors (e.g., IKONOS, Quickbird, GeoEye with spatial resolution <5m) has enabled more detailed characterization of wetland properties such as identifying multiple classes of wetland vegetation.[39,40] Multitemporal high spatial resolution images have been found useful because many wetland species have overlapping spectral reflectance properties and are spectrally distinguishable at a particular time of year, likely due to differences in biomass and pigments and the rate at which change occurs throughout the growing season.[41] High spatial resolution imagery in wetlands shows high intraclass spectral variability due to which the Object-Based Image Analysis (OBIA) approach has been adopted for producing higher characterization accuracy.[42,43]

Hyperspectral Remote Sensing

Imaging spectroscopy or hyperspectral remote sensing instruments with hundreds of narrow continuous spectral bands between 400 and 2500 nm wavelength offer high spectral sensitivity and have enabled species-level discrimination of wetland vegetation, in particular mapping invasive vegetation species which was not possible with multispectral sensors. Important hyperspectral sensors used in wetland studies include the Airborne Visible/Infrared Imaging Spectrometer (AVIRIS), HyMAP, Compact Airborne Imaging Spectrometer (CASI), and Hyperspectral Digital Imagery Collection Experiment (HYDICE). While all of these are airborne sensors, spaceborne hyperspectral sensors (e.g., EO-1 Hyperion) have a comparatively lower signal-to-noise ratio making them less effective for discriminating vegetation species in wetlands.

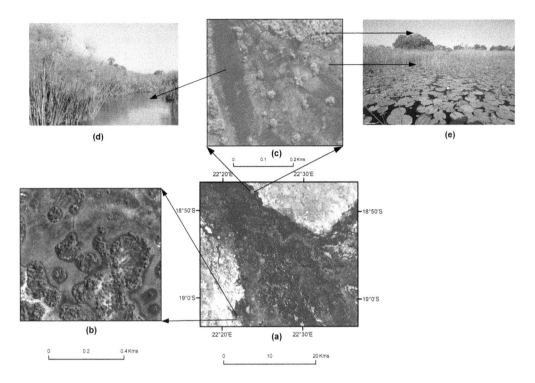

FIGURE 24.1 **(See color insert.)** Spatial heterogeneity and structural diversity of vegetation and land cover in the Okavango Delta, northern Botswana. **(a)** represents part of the delta observed from Landsat TM imagery (RGB:421). **(b)** and **(c)** represent two subset areas as observed from high spatial resolution Quickbird imagery (RGB:421). **(d)** and **(e)** are photographs acquired in the field.

Radar Remote Sensing

Active remote sensors such as radar backscatter provide different information than optical sensors, and radar systems can acquire images at any time of day and almost under any weather conditions. L-band radar data generally provides good distinction between flooded and non-flooded forest and between forest and marsh vegetation. Radar imagery has been used in wetland locations ranging from tropics to boreal regions.[44,45] Radar imaging applications in forested wetlands have found that longer wavelength had greater penetration in forest canopies, and C-band and L-band sensors could clearly differentiate between flooded and non-flooded forests compared to optical Landsat TM sensor, which was unable to detect water under canopy.[46]

Wetland Applications of Remotely Sensed Data

Wetland Inventory/Mapping and Change Detection

Remotely sensed data has been utilized for identifying, describing, and mapping the distribution of wetlands at a range of scales from local to regional to global.[47,48] Remote sensing-based wetland mapping has been cost effective and has minimized the need for extensive fieldwork in often inaccessible or logistically challenging wetland areas or both. At the global level, wetland inventory and mapping based on remotely sensed data have been carried out by notable studies such as the International Geosphere-Biosphere Programme Data and Information System (IGBP-DIS) Wetland Data Initiative,[49] the global wetland inventory promoted through the Ramsar Convention in partnership with international, national, and local organizations,[27] and the incorporation of information from individual projects based on existing and new Earth observation data sources.[50] At the regional and national scales, both government and non-governmental organizations worldwide have been using remotely sensed data for wetland inventory. For the last three decades in the United States, the Fish and Wildlife Service has been using aerial photos and multispectral images at high and medium spatial resolution to support the National Wetland Inventory.[51]

Over the past decades, the dramatic decrease and loss of wetland areas have prompted studies to quantify changes in wetlands using multitemporal remotely sensed data. Both short-term (seasonal) and long-term (>1 year) changes in wetland systems have been studied using remote sensing. Short-term changes in wetland extents are often in response to the seasonal rainfall pattern or flooding cycles, and its characterization is important for understanding wetland productivity, the source–sink distribution for nutrient and greenhouse gases, and associated wildlife habitat and hazard risks. Remote sensing-based studies have examined these seasonal changes in different regions such as wetlands in Australia,[15] Poyang Lake, China,[52,53] Okavango Delta, Botswana,[54,55] and in Brazilian wetlands.[44] Contrary to seasonal dynamics, long-term changes in wetlands are driven by anthropogenic activities as well as climatic variability influencing rainfall patterns or changes in sea level that are mainly impacting coastal wetlands. Satellite images have been utilized in several notable studies for quantifying long-term changes in wetlands at the local-to-regional scales.[26,29,56–59] Additionally, wetland monitoring and change detection programs at the national level have also been initiated by different countries. In the United States, the Coastal Change Analysis Program (C-CAP) uses multitemporal remotely sensed data to provide inventories of coastal intertidal areas, wetlands, and adjacent uplands that is updated every 5 years.

Remote Sensing of Wetland Vegetation

Wetland vegetation is a vital component of the wetland ecosystem that also serves as an indicator for early signs of any physical or chemical degradation. Different types of remotely sensed data have been used for biophysical and biochemical characterization of wetland vegetation. Earlier studies found high spatial resolution of aerial photos suitable for detailed mapping of wetland vegetation but were restricted by both limited swath and spectral dimensionality of these data sets.[60] The availability of consistent multispectral images (e.g., Landsat, SPOT) enabled improved characterization of vegetation types, density, vigor, and moisture content in wetland vegetation, nevertheless with significant challenges and limitations. In wetlands, commonly available multispectral sensors (Landsat TM, ETM+, SPOT) have been successfully utilized for mapping plant functional types where species are grouped based on traits such as life-form, photosynthetic pathway, and leaf longevity.[61–63] However, the spectral dimensionality of multispectral sensors (i.e., <10 spectral bands) has been found insufficient for discriminating individual vegetation species in wetlands. The availability of hundreds of spectral bands with field spectroradiometers and airborne hyperspectral sensors (e.g., AVIRIS, HyMap, CASI) has made it possible to discriminate and classify individual wetland vegetation species (Figure 24.2). Studies based on the analysis of field spectroradiometer-acquired canopy-level spectral reflectance of individual wetland vegetation species found that (i) a single species in different phenological stages showed significant variation in its spectral reflectance and (ii) the difference in the spectral reflectance and absorption features of different wetland vegetation species could be exploited to discriminate them, and these characteristics could be scaled from canopy to airborne remote sensing platforms.[41,64,65] Most studies using either field spectroradiometer or hyperspectral images have found that wetland vegetation has the greatest variation in the near infrared and red-edge regions.[66–68] Using wave bands in these spectral regions, studies have successfully mapped wetland vegetation species.[41,64,68–70]

Estimating Biophysical and Biochemical Parameters of Wetland Vegetation

The most important biophysical and biochemical properties that characterize wetland vegetation are biomass and chlorophyll concentration and leaf water content. Estimating wetland biomass is necessary for understanding productivity, carbon cycle, and nutrient allocation. For biomass estimation,

FIGURE 24.2 Wetland vegetation species map produced from PROBE-1 airborne hyperspectral imagery by combining *in situ* observations with field spectral libraries.
Source: Reproduced from Zomer et al.[106] with permission from Elsevier Publishing.

researchers have examined the correlation between *in situ-derived* biomass estimates with individual reflectance bands as well as various vegetation indices (VIs) derived from multispectral images.[71-74] These studies obtained modest-to-poor results depending upon landscape complexity and species composition and concluded that (i) growing season is the optimal time for biomass estimation in wetlands using remotely sensed data and (ii) near infrared region has the potential for vegetation biomass estimation in wetlands. Although saturation of VIs in densely vegetated areas has posed limitations to VI-based biomass estimation,[75] the development of techniques such as the band depth analysis approach has resolved this saturation issue.[76]

Leaf area index (LAI) is an important biophysical variable for quantifying evapotranspiration, photosynthesis, and primary productivity in terrestrial ecosystems. While research efforts on estimating LAI from remote sensing has intensively focused on forest and agriculture systems, similar studies for wetlands have only been attempted in forest wetlands and mangrove wetlands. Combining Landsat TM- and SPOT XS-derived normalized difference vegetation index (NDVI) with a model based on gap fraction analysis, Green et al.[77] produced a thematic map of LAI for three wetland species in a mangrove forest in West Indies with high accuracy (88%). Another study in a degraded mangrove forest of Mexico found a strong linear relationship between field-derived LAI and high-resolution IKONOS imagery-derived NDVI and sample ratio.[78,79] Hyperspectral remote sensing applications in various terrestrial ecosystems have shown that the narrow band indices can significantly improve biophysical characterization of vegetation. However, very few studies in wetlands have exploited hyperspectral data sets for quantifying biophysical variables such as LAI,[80] and therefore more studies are required.

Plant leaf and canopy water content is an important indicator of the physiological status of plants. There has been rapid growth in studies to access vegetation water content mainly based on laboratory and field-based spectra. However, most of these studies are not conducted in wetlands because remote sensing–based studies in wetlands have mainly focused on discriminating and mapping rather than estimating plant physiognomic properties. Therefore such studies in wetlands are needed that can help quantify the integrated condition of wetland plants and can identify plant water stress across a range of scales.[81]

Remote Sensing of Submerged Aquatic Vegetation

In wetlands, submerged aquatic vegetation (SAV) plays a major role in ecological functioning by providing habitat for many aquatic species, dissipating disturbance, enriching sediments, and improving water quality.[82] The main challenge for remote sensing of SAV is to isolate the weakened plant signal from interference by the water column, the bottom, and the atmospheric effects. Earlier studies showed that in shallow water, SAV density dominates a pixel's reflectance but as the depth increases, dominance of reflectance shifts to water column components.[83] Hence, studies have incorporated bathymetric information to reduce the effect of water column variation.[84,85] Due to the spatial complexity of SAV, high spatial resolution aerial photos have been found useful for monitoring the status and trend of SAV.[86,87] Studies using medium as well as high spatial resolution multispectral imagery have mapped both the various types of SAV and their biomass achieving moderate accuracy.[88–92] The spectral and spatial capabilities offered by airborne hyperspectral images have enabled detailed mapping of SAV types, habitat, and ecological conditions.[69,93] Hyperspectral sensors have been used in combination with LiDAR, field spectroscopy, and ancillary details to identify SAV areas in healthy or degraded conditions for conservation prioritization.

Challenges and Methods Used in Wetland Remote Sensing

The physical complexity, spatial heterogeneity, and diversity of life-forms in wetlands pose a significant challenge for their remote characterization at various spatial and temporal scales. Imagery from traditional multispectral sensors is subject to limitations of spatial and spectral resolution compared to the narrow vegetation units found in wetlands.[42] In wetlands, the biophysical and biochemical estimates derived from VIs obtained from broadband sensors can produce unstable results due to variation in underlying soil type, canopy, and leaf properties.[81] Despite effective performance of hyperspectral sensors in discriminating wetland species, there are still significant challenges because (i) reflectance from different vegetation species could be similar and (ii) within species, spectral variability is often high because of age difference, microclimate, soil and topographic difference. To address these complexities and challenges, researchers are evaluating different data sources and developing improved wetland-specific methods for using newly available and advanced remote sensing data. Recent studies have utilized advanced machine-learning classifiers such as decision trees,[94] artificial neural networks,[95] and fuzzy hybrid classifiers.[96] These techniques have proved useful especially when combined with ancillary information such as topography, soil type, and water level and often provided improved mapping accuracy than more traditionally used parametric classifiers (e.g., maximum likelihood).[97]

 Due to the narrow patch size of wetland vegetation, commonly used multispectral sensors (e.g., Landsat, SPOT) often contain pixels with more than one vegetation or land cover type (mixed pixels). Improving upon the limitations posed by per-pixel analysis approach for such mixed pixels, studies have utilized soft classifiers such as Spectral Mixture Analysis (SMA). SMA models a pixels spectral response as a linear/non-liner proportionally weighted combination of known endmember (pure pixel) spectra and provides subpixel abundance estimates. A modified version of SMA known as the Multiple Endmember SMA (MESMA) allows both number and type of endmembers to vary on a per-pixel basis thus producing more realistic subpixel fractional estimates. Both SMA and MESMA have been successfully utilized for wetland characterization with multispectral[26,98] and hyperspectral images.[64,99]

Contrary to the classical per-pixel analysis approach, more recent studies in wetlands have adopted the OBIA approach. OBIA segments the image into objects representing group of pixels according to desired scale, shape, and compactness criteria followed by classification of objects. The OBIA approach offers several advantages over per-pixel analysis such as absence of the "salt-and-pepper" effect in classification results, ability to utilize shape, contextual and textural features, and treatment of landscapes as consisting of relatively homogeneous objects that increases the signal-to-noise ratio of classification input. Important multispectral satellite sensors with which the OBIA approach has been utilized for wetland studies include Landsat TM, ETM,[61,100,101] and SPOT-5.[102] For tackling the physical complexity and heterogeneity of wetlands and their improved characterization, studies have integrated remotely sensed data from multiple sensors (e.g., optical, radar, lidar) and at multiple spatial scales. Airborne lidar has been integrated with multispectral and hyperspectral images to map SAV.[103,104] Additionally, using only optical sensors, biosensor monitoring of temporally dynamic wetlands by integrating medium spatial resolution imagery (e.g., Landsat) with high temporal resolution imagery (e.g., MODIS) has also been suggested.[105]

Conclusion

The progress in both remote sensing data and analysis methods is offering advanced capabilities and new frontiers for better characterizing the spatial heterogeneity and biophysical complexity of wetlands. Nevertheless, there are several major challenges in remote sensing of wetlands that need to be addressed. Additional research is needed to develop more robust classification techniques considering wetland complexity and utilizing commonly available multispectral images. For the optimum utilization of hyperspectral data, there is an increased need for building comprehensive spectral libraries of wetland vegetation types that could be used to calibrate and validate remotely derived results. Furthermore, the fundamental understanding of the relationship between reflectance and wetland vegetation canopy shape, density, and effects of background needs to be studied in detail. Hyperspectral instruments are considered to hold the future for remote sensing of wetlands as they can identify both species and their quality/conditions in wetlands. Ideally, remote sensing of wetland ecosystems requires high spatial, spectral, and temporal resolutions. However, like all scientific endeavors, in remote sensing, the limited number of photons arriving at a sensor requires trade-offs among bandwidth, pixel size, and noise. More recent studies in wetlands are integrating remotely sensed data at multiple spatial and temporal scales to overcome these challenges. In the future, remote sensing–based characterization of wetlands could lead to the development of an early warning system capable of detecting any subtle changes in wetlands and identifying healthy versus disturbed areas.

References

1. Mitsch, W.J.; Gosselink, J.G. The value of wetlands: Importance of scale and landscape setting. Ecol. Econ. **2000**, *35* (1), 25–33.
2. Houlahan, J.E.; Keddy, P.A.; Makkay, K.; Findlay, C.S. The effects of adjacent land use on wetland species richness and community composition. Wetlands **2006**, *26* (1), 79–96.
3. Liu, D.S.; Xia, F. Assessing object-based classification: advantages and limitations. Remote Sensing Lett. **2010**, *1* (4), 187–194.
4. Hunt, R.J.; Walker, J.F.; Krabbenhoft, D.P. Characterizing hydrology and the importance of ground-water discharge in natural and constructed wetlands. Wetlands **1999**, *19* (2), 458–472.
5. Gibbs, J.P. Wetland loss and biodiversity conservation. Conserv. Biol. **2000**, *14* (1), 314–317.
6. Dudgeon, D.; Arthington, A.H.; Gessner, M.O.; Kawa- bata, Z.-I.; Knowler, D.J.; Lévêque, C.; Naiman, R.J.; Freshwater biodiversity: importance, threats, status and conservation challenges. Biol. Rev. **2006**, *81* (2), 163–182.

7. Stewart, W.R.; Carter, V.; Brooks, P.D. Inland (non-tidal) wetland mapping. Photogram. Eng. Remote Sensing **1980**, *46* (5), 617–628.

8. Carter, V.; Garrett, M.K.; Shima, L.; Gammon, P. The great dismal swamp – Management of a hydrologic resource with aid of remote sensing. Water Resour. Bull. **1977**, *13* (1), 1–12.

9. Mead, R.A.; Gammon, P.T. Mapping wetlands using orthophotoquads and 35-mm aerial photo. Photogram. Eng. Remote Sensing **1981**, *47* (5), 649–652.

10. Shima, L.J.; Anderson, R.R.; Carter, V.P. The use of aerial color infrared photography in mapping the vegetation of a freshwater marsh. Chesapeake Sci. **1976**, *17* (2), 74–85.

11. Lovvorn, J.R.; Kirkpatrick, C.M. Analysis of freshwater wetland vegetation with large-scale color infrared aerial photography. J. Wildl. Manag. **1982**, *46* (1), 61–70.

12. Scarpace, F.L.; Quirk, B.K.; Kiefer, R.W.; Wynn, S.L. Wetland mapping from digitized aerial photography. Photogram. Eng. Remote Sensing **1981**, *47* (6), 829–838.

13. Welch, R.; Remillard, M.M.; Slack, R.B. Remote sensing and geographic information system techniques for aquatic resource evaluation. Photogram. Eng. Remote Sensing **1988**, *54* (2), 177–185.

14. Martyn, R.D.; Noble, R.L.; Bettoli, P.W.; Maggio, R.C. Mapping aquatic weeds with aerial color infrared photography and evaluating their control by grass carp. J. Aquat. Plant Manag. **1986**, *24*, 46–56.

15. Johnston, R.M.; Barson, M.M. Remote sensing of Australian wetlands - An evaluation of Landsat TM data for inventory and classification. Aust. J. Mar. Freshwater Res. **1993**, *44* (2), 235–252.

16. Ackleson, S.G.; Klemas, V.; McKim, H.L.; Merry, C.J. A comparison of SPOT simulator data with Landsat MSS imagery for delineating water masses in Delaware Bay, Broadkill River, and adjacent wetlands. Photogram. Eng. Remote Sensing **1985**, *51* (8), 1123–1129.

17. Jensen, J.R.; Christensen, E.J.; Sharitz, R. Nontidal wetland mapping in South Carolina using airborne multispectral scanner data. Remote Sensing Environ. **1984**, *16* (1), 1–12.

18. Lee, C.T.; Marsh, S.E. The use of archival Landsat MSS and ancillary data in a GIS environment to map historical change in an urban riparian habitat. Photogram. Eng. Remote Sensing **1995**, *61* (8), 999–1008.

19. Wickware, G.M.; Howarth, PJ. Change detection in the Peace—Athabasca delta using digital Landsat data. Remote Sensing Environ. **1981**, *11* (1), 9–25.

20. Ackleson, S.G.; Klemas, V. Remote sensing of submerged aquatic vegetation in lower Chesapeake Bay: A comparison of Landsat MSS to TM imagery. Remote Sensing Environ. **1987**, 22 (2), 235–248.

21. Ramsey, E.W.; Laine, S.C. Comparison of Landsat thematic mapper and high resolution photography to identify change in complex coastal wetlands. J. Coast. Res. **1997**, *13* (2), 281–292.

22. Zainal, A.; Dalby, D.; Robinson, I. Monitoring marine ecological changes on the east coast of Bahrain with Landsat TM. Photogram. Eng. Remote Sensing **1993**, *59* (3), 415–421.

23. Jensen, J.R.; Hodgson, M.E.; Christensen, E.; Mackey, H.E.; Tinney, L.R.; Sharitz, R.R. Remote sensing in inland wetlands - A multispectral approach. Photogram. Eng. Remote Sensing **1986**, *52* (1), 87–100.

24. Dronova, I.; Gong, P.; Clinton, N.E.; Wang, L.; Fu, W.; Qi, S.; Liu, Y. Landscape analysis of wetland plant functional types: The effects of image segmentation scale, vegetation classes and classification methods. Remote Sensing Environ. **2012**, *127*, 357–369.

25. Mishra, N.B.; Crews, K.A.; Neuenschwander, A.L. Sensitivity of EVI-based harmonic regression to temporal resolution in the lower Okavango Delta. Int. J. Remote Sensing **2012**, *33* (24), 7703–7726.

26. Ozesmi, S.L.; Bauer, M.E. Satellite remote sensing of wetlands. Wetlands Ecol. Manag. **2002**, *10* (5), 381–402.

27. Davidson, N.C.; Finlayson, C.M. Earth Observation for wetland inventory, assessment and monitoring. Aquat. Conserv.-Mar. Freshwater Ecosyst. **2007**, *17* (3), 219–228.

28. McCarthy, T.S.; Franey, N.J.; Ellery, W.N.; Ellery, K. The use of SPOT imagery in the study of environmental processes of the Okavango Delta, Botswana. S. Afr. J. Sci. **1993**, *89* (9), 432–436.

29. Houhoulis, P.F.; Michener, W.K. Detecting wetland change: A rule-based approach using NWI and SPOT-XS data. Photogram. Eng. Remote Sensing **2000**, *66* (2), 205–211.
30. Narumalani, S.; Jensen, J.R.; Burkhalter, S.; Mackey, H.E. Aquatic macrophyte modeling using GIS and logistic multiple regression. Photogram. Eng. Remote Sensing **1997**, *63* (1), 41–49.
31. Harvey, K.; Hill, G. Vegetation mapping of a tropical freshwater swamp in the Northern Territory, Australia: a comparison of aerial photography, Landsat TM and SPOT satellite imagery. Int. J. Remote Sensing **2001**, *22* (15), 2911–2925.
32. Lee, J.K.; Park, R.A.; Mausel, P.W. Application of geoprocessing and simulation modeling to estimate impacts of sea level rise on the northeast coast of Florida. Photogram. Eng. Remote Sensing **1992**, *58* (11), 1579–1586.
33. Kushwaha, S.; Dwivedi, R.; Rao, B. Evaluation of various digital image processing techniques for detection of coastal wetlands using ERS-1 SAR data. Int. J. Remote Sensing **2000**, *21* (3), 565–579.
34. Chopra, R.; Verma, V.; Sharma, P. Mapping, monitoring and conservation of Harike wetland ecosystem, Punjab, India, through remote sensing. Int. J. Remote Sensing **2001**, *22* (1), 89–98.
35. Shanmugam, P.; Ahn, Y.-H.; Sanjeevi, S. A comparison of the classification of wetland characteristics by linear spectral mixture modelling and traditional hard classifiers on multispectral remotely sensed imagery in southern India. Ecol. Model. **2006**, *194* (4), 379–394.
36. Prasad, S.N.; Ramachandra, T.V.; Ahalya, N.; Sengupta, T.; Kumar, A.; Tiwari, A.K.; Vijayan, V.S.; Vijayan, L. Conservation of wetlands of India - a review. Trop. Ecol. **2002**, *43* (1), 173–186.
37. Zhao, B.; Yan, Y.; Guo, H.; He, M.; Gu, Y.; Li, B. Monitoring rapid vegetation succession in estuarine wetland using time series MODIS-based indicators: An application in the Yangtze River Delta area. Ecol. Indicators **2009**, *9* (2), 346–356.
38. Yan, Y.-E.; Ouyang, Z.-T.; Guo, H.-Q.; Jin, S.-S.; Zhao, B. Detecting the spatiotemporal changes of tidal flood in the estuarine wetland by using MODIS time series data. J. Hydrol. **2010**, *384* (1), 156–163.
39. Rutchey, K.; Vilchek, L. Air photointerpretation and satellite imagery analysis techniques for mapping cattail coverage in a northern Everglades impoundment. Photogram. Eng. Remote Sensing **1999**, *65* (2), 185–191.
40. Sawaya, K.E.; Olmanson, L.G.; Heinert, N.J.; Brezonik, P.L.; Bauer, M.E. Extending satellite remote sensing to local scales: land and water resource monitoring using high-resolution imagery. Remote Sensing Environ. **2003**, *88* (1–2), 144–156.
41. Schmidt, K.S.; Skidmore, A.K. Spectral discrimination of vegetation types in a coastal wetland. Remote Sensing Environ. **2003**, *85* (1), 92–108.
42. Gilmore, M.S.; Wilson, E.H.; Barrett, N.; Civco, D.L.; Prisloe, S.; Hurd, J.D.; Chadwick, C. Integrating multitemporal spectral and structural information to map wetland vegetation in a lower Connecticut River tidal marsh. Remote Sensing Environ. **2008**, *112* (11), 4048–4060.
43. Laba, M.; Blair, B.; Downs, R.; Monger, B.; Philpot, W.; Smith, S.; Sullivan, P.; Baveye, P.C. Use of textural measurements to map invasive wetland plants in the Hudson River National Estuarine Research Reserve with IKONOS satellite imagery. Remote Sensing Environ. **2010**, *114* (4), 876–886.
44. Hess, L.L.; Melack, J.M.; Novo, E.M.L.M.; Barbosa, C.C.F.; Gastil, M. Dual-season mapping of wetland inundation and vegetation for the central Amazon basin. Remote Sensing Environ **2003**, *87* (4), 404–428.
45. Hess, L.L.; Melack, J.M.; Simonett, D.S. Radar detection of flooding beneath the forest canopy: a review. Int. J. Remote Sensing **1990**, *11* (7), 1313–1325.
46. Townsend, P.A.; Walsh, S.J. Remote sensing of forested wetlands: application of multitemporal and multispectral satellite imagery to determine plant community composition and structure in southeastern USA. Plant Ecol. **2001**, *157* (2), 129A–149A.
47. Lehner, B.; Doll, P. Development and validation of a global database of lakes, reservoirs and wetlands. J.Hydrol. **2004**, *296* (1–4), 1–22.

48. Rebelo, L.M.; Finlayson, C.M.; Nagabhatla, N. Remote sensing and GIS for wetland inventory, mapping and change analysis. J. Environ. Manag. **2009**, *90* (7), 2144–2153.

49. Darras, S. A first step towards identifying a global delineation of wetlands. In *GBP-DIS Wetland Data Initiative;* IGBP-DIS Office: Toulouse, France, 1999.

50. Rosenqvist, A.; Finlayson, C.M.; Lowry, J.; Taylor, D. The potential of long-wavelength satellite-borne radar to support implementation of the Ramsar Wetlands Convention. Aquat. Conserv.-Mar. Freshwater Ecosyst. **2007**, *17* (3), 229–244.

51. Klemas, V. Remote sensing of emergent and submerged wetlands: an overview. Int. J. Remote Sensing **2013**, *34* (18), 6286–6320.

52. Dronova, I.; Gong, P.; Wang, L. Object-based analysis and change detection of major wetland cover types and their classification uncertainty during the low water period at Poyang Lake, China. Remote Sensing Environ. **2011**, *115* (12), 3220–3236.

53. Hui, F.M.; Xu, B.; Huang, H.B.; Yu, Q.; Gong, P. Modelling spatial-temporal change of Poyang Lake using multitemporal Landsat imagery. Int. J. Remote Sensing **2008**, *29* (20), 5767–5784.

54. McCarthy, T.S.; Cooper, G.R.J.; Tyson, P.D.; Ellery, W.N. Seasonal flooding in the Okavango Delta, Botswana - recent history and future prospects. S. Afr. J. Sci. **2000**, *96*, 25–33.

55. Neuenschwander, A.L.; Crawford, M.M.; Ringrose, S. Monitoring of seasonal flooding in the Okavango Delta using EO-1 data. *Igarss: IEEE International Geoscience and Remote Sensing Symposium and 24th Canadian Symposium on Remote Sensing,* Proceedings: Remote Sensing: Integrating Our View of the Planet, New York: IEEE. 2002, Vols. 1–6, 3124–3126.

56. Gong, P.; Niu, Z.; Cheng, X.; Zhao, K.; Zhou, D.; Guo, J.; Liang, L.; Wang, X.; Li, D.; Huang, H.; Wang, Y.; Wang, K.; Li, W.; Wang, X.; Ying, Q.; Yang, Z.; Ye, Y.; Li, Z.; Zhuang, D.; Chi, Y.; Zhou, H.; Yan, J. China's wetland change (1990–2000) determined by remote sensing. Sci. China Earth Sci. **2010**, *53* (7), 1036–1042.

57. Shuman, C.S.; Ambrose, R.F. A comparison of remote sensing and ground-based methods for monitoring wetland restoration success. Restoration Ecol. **2003**, *11* (3), 325–333.

58. Nielsen, E.M.; Prince, S.D.; Koeln, G.T. Wetland change mapping for the U.S. mid-Atlantic region using an outlier detection technique. Remote Sensing Environ. **2008**, *112*, 4061–4074.

59. Munyati, C. Wetland change detection on the Kafue Flats, Zambia, by classification of a multitemporal remote sensing image dataset. Int. J. Remote Sensing **2000**, *21* (9), 1787–1806.

60. Howland, W.G. Multispectral aerial photography for wetland vegetation mapping. Photogram. Eng. Remote Sensing **1980**, *46* (1), 87–99.

61. Dronova, I.; Gong, P.; Clinton, N.E.; Wang, Li.; Fu, W.; Qi, S.; Liu, Y. Landscape analysis of wetland plant functional types: The effects of image segmentation scale, vegetation classes and classification methods. Remote Sensing Environ. **2012**, *127*, 357–369.

62. Khanna, S.; Santos, M.J.; Ustin, S.L.; Haverkamp, PJ. An integrated approach to a biophysiologically based classification of floating aquatic macrophytes. Int. J. Remote Sensing **2011**, *32* (4), 1067–1094.

63. Wang, L.; Dronova, I.; Gong, P.; Yang, W.; Li, Y.; Liu, Q. A new time series vegetation–water index of phenological–hydrological trait across species and functional types for Poyang Lake wetland ecosystem. Remote Sensing Environ. **2012**, *125* (0), 49–63.

64. Rosso, P.H.; Ustin, S.L.; Hastings, A. Mapping marshland vegetation of San Francisco Bay, California, using hyperspectral data. Int. J. Remote Sensing **2005**, *26* (23), 5169–5191.

65. Fyfe, S.K. Spatial and temporal variation in spectral reflectance: Are seagrass species spectrally distinct? Limnol. Oceanogr. **2003**, *48* (1), 464–479.

66. Thenkabail, P.S.; Enclona, E.A.; Ashton, M.S.; Meer, B.V.D. Accuracy assessments of hyperspectral waveband performance for vegetation analysis applications. Remote Sensing Environ. **2004**, *91* (3–4), 354–376.

67. Asner, G.P. Biophysical and biochemical sources of variability in canopy reflectance. Remote Sensing Environ. **1998**, *64* (3), 234–253.

68. Vaiphasa, C.; Ongsomwang, S.; Vaiphasa, T.; Skidmore, A.K. Tropical mangrove species discrimination using hyperspectral data: A laboratory study. Estuarine Coast. Shelf Sci. **2005**, *65* (1–2), 371–379.

69. Belluco, E.; Camuffo, M.; Ferrari, S.; Modenese, L.; Silvestri. S.; Marani, A.; Marani, M. Mapping salt-marsh vegetation by multispectral and hyperspectral remote sensing. Remote Sensing Environ. **2006**, *105* (1), 54–67.

70. Pengra, B.W.; Johnston, C.A.; Loveland, T.R. Mapping an invasive plant, *Phragmites australis,* in coastal wetlands using the EO-1 Hyperion hyperspectral sensor. Remote Sensing Environ. **2007**, *108* (1), 74–81.

71. Tan, Q.; Shao, Y.; Yang, S.; Wei, Q. *Wetland Vegetation Biomass Estimation Using Landsat-7 ETM+ data.* In Geoscience and Remote Sensing Symposium, 2003. IGARSS'03. Proceedings. 2003 IEEE International; 2003.

72. Ren-dong, L.; Ji-yuan, L. Wetland vegetation biomass estimation and mapping from Landsat ETM data: A case study of Poyang Lake. J. Geogr. Sci. **2002**, *12* (1), 35–41.

73. Ramsey, E.W.; Jensen, J.R. Remote sensing of mangrove wetlands: relating canopy spectra to site-specific data. Photogram. Eng. Remote Sensing **1996**, *62* (8), 939–948.

74. Moreau, S.; Bosseno, R.; Gu, X.F.; Baret, F. Assessing the biomass dynamics of Andean bofedal and totora high-protein wetland grasses from NOAA/AVHRR. Remote Sensing Environ. **2003**, *85* (4), 516–529.

75. Thenkabail, P.S.; Smith, R.B.; De Pauw, E. Hyperspectral vegetation indices and their relationships with agricultural crop characteristics. Remote Sensing Environ. **2000**, *71*, 158–182.

76. Mutanga, O.; Skidmore, A.K. Narrow band vegetation indices overcome the saturation problem in biomass estimation. International J. Remote Sensing **2004**, *25* (19), 3999–4014.

77. Green, E.P.; Mumby, P.J.; Edwards, A.J.; Clark, C.D.; Ellis, A.C. Estimating leaf area index of mangroves from satellite data. Aquat. Bot. **1997**, *58* (1), 11–19.

78. Kovacs, J.M.; Flores-Verdugo, F.; Wang, J.F.; Aspden, L.P. Estimating leaf area index of a degraded mangrove forest using high spatial resolution satellite data. Aquat. Bot. **2004**, *80* (1), 13–22.

79. Kovacs, J.M.; Wang, J.F.; Flores-Verdugo, F. Mapping mangrove leaf area index at the species level using IKO- NOS and LAI-2000 sensors for the Agua Brava Lagoon, Mexican Pacific. Estuarine Coast. Shelf Sci. **2005**, *62* (1–2), 377–384.

80. Darvishzadeh, R.; Skidmore, A.; Atzberger, C.; Wieren, S.V. Estimation of vegetation LAI from hyperspectral reflectance data: Effects of soil type and plant architecture. Int. J. Appl. Earth Observation Geoinfo. **2008**, *10* (3), 358–373.

81. Adam, E.; Mutanga, O.; Rugege, D. Multispectral and hyperspectral remote sensing for identification and mapping of wetland vegetation: A review. Wetlands Ecol. Manag. **2010**, *18* (3), 281–296.

82. Hurley, L.M. *Submerged Aquatic Vegetation. Habitat Requirements for Chesapeake Bay Living Resources,* 2nd edn.; Chesapeake Research Consortium, Inc.: Solomons, MD, 1991; 2–1.

83. Silva, T.S.; Costa, M.P.; Melack, J.M.; Novo, E.M.L.M. Remote sensing of aquatic vegetation: theory and applications. Environ. Monitor. Assess. **2008**, *140* (1–3), 131–145.

84. Dierssen, H.M.; Zimmerman, R.C.; Leathers, R.A.; Downes, T.V.; Davis, C.O. Ocean Color Remote Sensing of Seagrass and Bathymetry in the Bahamas Banks by High-Resolution Airborne Imagery; DTIC Document: 2003.

85. Heege, T.; Bogner, A.; Pinnel, N. Mapping of submerged aquatic vegetation with a physically based process chain. In *Remote Sensing*; International Society for Optics and Photonics: 2004.

86. Ferguson, R.L.; Wood, L.; Graham, D. Monitoring spatial change in seagrass habitat with aerial photography. Photogram. Eng. Remote Sensing **1993**, *59* (6), 1033–1038.

87. Finkbeiner, M.; Stevenson, B.; Seaman, R. Guidance for Benthic Habitat Mapping: An Aerial Photographic Approach; 2001.

88. Wabnitz, C.C.; Andrefouet, S.; Torres-Pulliza, D.; MüllerKarger, F.E.; Kramer, P.A. Regional-scale seagrass habitat mapping in the Wider Caribbean region using Landsat sensors: Applications to conservation and ecology. Remote Sensing Environ. **2008**, *112* (8), 3455–3467.

89. Gullström, M.; Lunden, B.; Bodin, M.; Kangwe, J.; Öhman, M.C.; Mtolera, M.S.P.; Björk, M. Assessment of changes in the seagrass-dominated submerged vegetation of tropical Chwaka Bay (Zanzibar) using satellite remote sensing. Estuarine, Coast. Shelf Sci. **2006**, *67* (3), 399–408.

90. Nobi, E.; Thangaradjou, T. Evaluation of the spatial changes in seagrass cover in the lagoons of Lakshadweep islands, India, using IRS LISS III satellite images. Geo-carto Int. **2012**, *27* (8), 647–660.

91. Mishra, D.; Narumalani, S.; Rundquist, D.; Lawson, M. Benthic habitat mapping in tropical marine environments using QuickBird multispectral data. Photogrammetric Eng. Remote Sensing **2006**, *72* (9), 1037.

92. Wolter, P.T.; Johnston, C.A.; Niemi, G.J. Mapping submergent aquatic vegetation in the U.S. Great Lakes using Quickbird satellite data. Int. J. Remote Sensing **2005**, *26* (23), 5255–5274.

93. Williams, D.J.; Rybicki, N.B.; Lombana, A.V.; O'Brien, T.M.; Gomez, R.B. Preliminary investigation of submerged aquatic vegetation mapping using hyperspectral remote sensing. Environ. Monitor. Assess. **2003**, *81* (1–3), 383–392.

94. Xu, M.; Watanachaturaporn, P.; Varshney, P.K.; Arora, M.J. Decision tree regression for soft classification of remote sensing data. Remote Sensing Environ. **2005**, *97*, 322–336.

95. Berberoglu, S.; Lloyd, C.D.; Atkinson, P.M.; Curren, P. The integration of spectral and textural information using neural networks for land cover mapping in the Mediterranean. Comput. Geosci. **2000**, *26* (4), 385–396.

96. Sha, Z.; Bai, Y.; Xie, Y. et al. Using a hybrid fuzzy classifier (HFC) to map typical grassland vegetation in Xilin River Basin, Inner Mongolia, China. Int. J. Remote Sensing **2008**, *29* (8), 2317–2337.

97. Xie, Y.; Sha, Z.; Yu, M. Remote sensing imagery in vegetation mapping: A review. J. Plant Ecol. **2008**, *1* (1), 9–23.

98. Mertes, L.A.; Daniel, D.L.; Melack, J.M.; Nelson, B.; Martinelli, L.A.; Forsberg, B.R. Spatial patterns of hydrology, geomorphology, and vegetation on the floodplain of the Amazon River in Brazil from a remote sensing perspective. Geomorphology **1995**, *13* (1), 215–232.

99. Li, L.; Ustin, S.L.; Lay, M. Application of multiple end- member spectral mixture analysis (MESMA) to AVIRIS imagery for coastal salt marsh mapping: A case study in China Camp, CA, USA. Int. J. Remote Sensing **2005**, *26* (23), 5193–5207.

100. Grenier, M.; Demers, A.-M.; Labrecque, S.; Benoit, M.; Fournire, R.A.; Drolet, B. An object-based method to map wetland using RADARSAT-1 and Landsat ETM images: Test case on two sites in Quebec, Canada. Can. J. Remote Sensing **2007**, *33*, S28–S45.

101. Frohn, R.C.; Autrey, B.C.; Lane, C.R.; Reif, M. Segmentation and object-oriented classification of wetlands in a karst Florida landscape using multi-season Landsat-7 ETM+ imagery. Int. J. Remote Sensing **2011**, *32* (5), 1471–1489.

102. Powers, R.P.; Hay, G.J.; Chen, G. How wetland type and area differ through scale: A GEOBIA case study in Alberta's Boreal Plains. Remote Sensing Environ. **2012**, *117*, 135–145.

103. Yang, J.; Artigas, F.J.; Wang, J. Mapping salt marsh vegetation by integrating hyperspectral and LiDAR remote sensing. Remote Sensing Coast. Environ. CRC: Boca Raton, Florida, **2010**, 173–190.

104. Chust, G.; Galparsoro, I.; Borja, Á.; Javire, F.; Adolfo, U. Coastal and estuarine habitat mapping, using LIDAR height and intensity and multi-spectral imagery. Estuarine, Coast. Shelf Sci. **2008**, *78* (4), 633–643.

105. Michishita, R.; Gong, P.; Xu, B. Spectral mixture analysis for bi-sensor wetland mapping using Landsat TM and Terra MODIS data. Int. J. Remote Sensing **2012**, *33* (11), 3373–3401.

106. Zomer, R.J.; Trabucco, A.; Ustin, S.L. Building spectral libraries for wetlands land cover classification and hyperspectral remote sensing. J. Environ. Manag. **2009**, *90*, 2170–2177.

25

Assessment of Health Status of Lake Wetland by Vegetation-Based Index of Biotic Integrity (V-IBI)

Wenjing Yang
Key Laboratory of Poyang Lake Wetland and Watershed Research (Jiangxi Normal University)

Introduction

Wetlands are important ecosystems that provide numerous beneficial services for people and wildlife [1]. A number of studies indicate that the loss and degradation of wetlands have been accelerated in the past decades [2]. Developing methods to assess the health condition of wetlands has been recognized as a key issue for wetland protection and management [3]. The index of biotic integrity (IBI), originally developed by Karr [4], is a widely used method for evaluating anthropogenic pressures on aquatic and wetland ecosystems. This method integrates selected attributes of biological assemblages into a single index based on the assumption that biological assemblages in human-influenced habitats should be different from those in natural habitats of the region [5,6]. Individual metrics of an IBI reflect biotic condition by measuring aspects of the structure or composition of biological assemblages that respond to anthropogenic pressures in a predictable manner [7,8].

A couple of biological taxa have been proposed for the development of IBIs, including fishes, invertebrates, and planktons [9–11]. Plants are quickly emerging as one of the important indicators of ecological health because they are immobile, relatively easy to sample and identify, and respond to anthropogenic disturbances on an ecological timescale [12,13]. Moreover, as important components of wetland ecosystems, plants intricately shape trophic web dynamics by providing oxygen, food, and shelter to other biological organisms (such as fishes and waterfowls). In addition, a vegetation-based IBI (V-IBI) is a useful tool for monitoring wetland restoration efforts, because seed banks retained in sediment are a viable source for species recolonization that responds to environmental condition [14].

Poyang Lake is the largest freshwater lake in China with one of the most important wetlands in the world [15]. The Poyang Lake wetlands provide critical wintering habitats for Siberian migratory birds where abundant aquatic plants, benthic invertebrates, and fishes serve as critical food resources [16]. This chapter takes Poyang Lake as an example to illustrate the procedure of developing a V-IBI to assess the health status of lake wetlands. Five steps are generally included: (i) field surveys and measurement of environmental variables, (ii) reference site selection, (iii) establishing disturbance gradients, (iv) metric selection, and (v) metric scoring and V-IBI calculation.

Field Surveys and Measurement of Environmental Variables

Figure 25.1 shows the locations of 30 sample sites in Poyang Lake wetlands. The sample sites were selected via a randomized systematic design with a consideration of spatial balance to allow statistically valid inference to reflect the status of the entire wetland [17]. Field surveys for wetland vegetation and water quality were conducted at each site during September 17–October 15, 2015. Wetland vegetation at each site was surveyed along three transects perpendicular to the shoreline to identify physical gradients and corresponding biological gradients. The layout of transects in wetland is shown in Figure 25.2. Relatively discrete zones of wet meadow and emergent/submergent vegetation were identified at all sample sites. Five 1.0 m² quadrats were placed in each zone along a transect and kept an equal distance between neighboring quadrats. All plant species detected in a quadrat were identified and recorded along with approximate coverage values.

Triplicate measures of physicochemical parameters were collected in water at each site before vegetation surveys. Temperature (TEMP), pH, dissolved oxygen (DO), turbidity (TURB), and electrical conductivity (COND) were measured using a portable multi-probe meter at the end of each transect for vegetation surveys. About 2 L water was collected in sterilized plastic bottles and stored in a portable refrigerator at <4°C. Later analyses were conducted in laboratory for total phosphorus (TP), total nitrogen (TN), nitrate-nitrogen (NO_3^--N), ammonium nitrogen (NH_4^+-N), chlorophyll-*a* (Chl *a*), total suspended sediment (TSS), biochemical oxygen demand in manganese (COD_{Mn}), and heavy metals (two major pollutants, Cu and Cd).

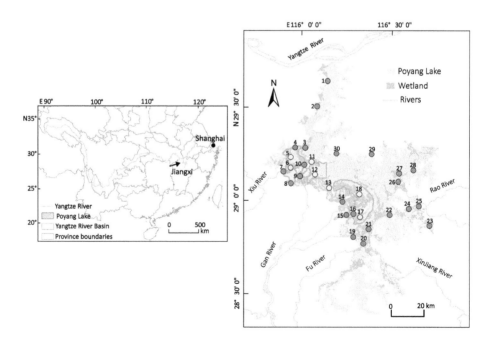

FIGURE 25.1 **(See color insert.)** Locations of 30 sample sites in Poyang Lake wetland. Yellow dots are reference sites, and red dots are impaired sites. Areas enclosed by green colored polygons are national wetland nature reserves. (Modified from Yang, W.; You, Q.; Fang, N.; Xu, L.; Zhou, Y.; Wu, N.; Ni, C.; Liu, Y.; Liu, G.; Yang, T.; Wang, Y. 2018. Assessment of wetland health status of Poyang Lake using vegetation-based indices of biotic integrity. *Ecological Indicators* 90, 79–89.)

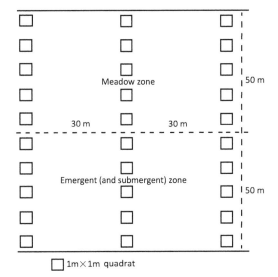

FIGURE 25.2 Design of plant assemblage surveys for Poyang Lake wetland. (Adapted from Yang, W.; You, Q.; Fang, N.; Xu, L.; Zhou, Y.; Wu, N.; Ni, C.; Liu, Y.; Liu, G.; Yang, T.; Wang, Y. 2018. Assessment of wetland health status of Poyang Lake using vegetation-based indices of biotic integrity. *Ecological Indicators* 90, 79–89.)

Land-use data for Poyang Lake region were obtained from an existing land-cover map with a spatial resolution of 30 m derived from the interpretation of Landsat images acquired in late October and early November of 2014 [18]. Other parameters were documented for evaluating habitat quality at each site, including shoreline structure, hydrological connection to lake, habitat structure and complexity, substrate composition, and on-site observable anthropological disturbances such as boat channels and industrial discharge.

Reference Site Selection

The concept of "reference condition" has been used to describe the standard or benchmark against which the current condition is compared [19]. Reference condition originally refers to the naturalness that implies the absence of significant human disturbance or alternation. However, it is impossible to find truly undisturbed wetland in Poyang Lake. For practical purposes, sites in the least disturbed conditions were selected as the reference using the criteria commonly adopted by reported studies [e.g., 20,21]. Specifically, a reference site should have no observable anthropological activities within a 100 m buffer zone. Water quality should be between good and fair according to China's national surface water quality classification standards [22], that is, $TP \leq 0.2$ mg/L, $TN \leq 1$ mg/L, $NO_3^- $-N ≤ 1 mg/L, $DO > 5$ mg/L, $COD_{Mn} \leq 6$ mg/L, $Cu \leq 1.0$ mg/L, and $Cd \leq 0.005$ mg/L.

Establishing Disturbance Gradients

Disturbance gradients were established using four different methods that were commonly used in previous studies. Water quality index (WQI) reduced a wide range of physicochemical variables into a single number and was often used to quantify the intensity of disturbances that an aquatic ecosystem has suffered [23,24]. Principal component analysis (PCA) was applied to convert 11 water variables (except pH, TEMP, and DO) to linearly unrelated principal components (PCs) without the loss of information. Each water physicochemical variable was standardized (mean = 0, standard deviation = 1) to eliminate scale biases before entering into the PCA. PH, TEMP, and DO were excluded from the PCA, because these variables varied in reference ranges and only extreme values (either very high or very low) were considered indicators of disturbance [9]. Table 25.1 indicates that the first three PCs explained

TABLE 25.1 Summary of Pearson's Correlation Coefficients between PC Scores and Environmental Variables and Loadings for Each Parameter in Respective PC Axes

	Variance Explained (%)	Environmental Variable	Loading	r-Value	p-Value
PC1	40.49	TN	0.90	0.75	<0.001
		COND	1.26	0.88	<0.001
		COD_{Mn}	1.25	0.88	<0.001
		Chl *a*	−0.81	−0.60	<0.001
PC2	23.61	Cu	1.16	0.84	<0.001
		Cd	1.17	0.85	<0.001
		TP	−1.08	−0.79	<0.001
PC3	16.39	Chl *a*	0.94	0.68	<0.001
		TURB	−0.82	−0.68	<0.001
Cumulative	80.49	–	–	–	–

Adapted from Yang, W., Q. You, N. Fang, L. Xu, Y. Zhou, N. Wu, C. Ni, Y. Liu, G. Liu, T. Yang, and Y. Wang. 2018. Assessment of Wetland Health Status of Poyang Lake using Vegetation-based Indices of Biotic Integrity. *Ecological Indicators* 90.

TN, total nitrogen; COND, electrical conductivity; COD_{Mn}, biochemical oxygen demand in manganese; Chl *a*, chlorophyll-a; TP, total phosphorus; TURB, turbidity.

80.49% of the variance contained in the original data. PCs were then input into the aggregation function for the calculation of an overall index [25,26]:

$$\text{WQI} = \left(\sum_{i=1}^{4} I_i w_i \right) \times \left(\sum_{j=1}^{3} I_j w_j \right) \times \left(\sum_{k=1}^{2} I_k w_k \right) \tag{25.1}$$

where I_i, I_j, and I_k are the standardized values of variables significantly associated with PCs 1, 2, and 3, respectively. w_i, w_j, and w_k are the loading of each variable in the respective PCs.

The second method quantified human-related disturbances using land-use data [27,28]. Landscape Development Intensity Index (LDI) [29] was used as the indicator for anthropological pressures [30,31]. The basic idea behind LDI is to estimate the amount of supplemental energy needed to maintain the landscape concerned. LDI was calculated using the equation:

$$\text{LDI}_{\text{total}} = \sum_{i=1}^{n} \%\text{LU}_i \times \text{LDI}_i \tag{25.2}$$

where *n* is the number of land-use types within 1 km buffer zone of a sample site, $\%\text{LU}_i$ is the percent of land use *i*, and LDI_i is the landscape development intensity coefficient for land use *i* [32]. LDI_i was derived from a study on the amount of carbon emission by different types of land use in Poyang Lake region [33], and defined as follows: forest, wetland, grassland, water = 1.00; farmland = 4.54; road and highway = 7.81; suburban = 8.66; urban = 9.42.

The third method assessed the intensity of human disturbances through field observations [11,32]. Local Disturbance Index (LOD) is a measure adopted by the Environmental Protection Agency of the United States of America (US-EPA) for wetland quality assessment [34]. This method summarized the amount of evidence observed in or around sample sites for ten types of disturbances that were prevalent in Poyang Lake area, namely, buildings, farmland, roads, pipes, trash and landfill, parks and lawns, pasture, mining, channel revetment, and dikes. LOD was calculated using the following equation:

$$\text{LOD} = \sum_{k=i}^{n} \frac{1}{D_i + 1} \tag{25.3}$$

where n is the number of disturbance types observed in a sample site, and D_i is the distance of disturbance i to the sample site.

Metric Selection

Thirty-eight characteristics of wetland plant assemblages were collected as candidate metrics for V-IBI from published studies [27,28,32,35,36]. These metrics were grouped into five categories based on their ecological properties: richness and composition, abundance, structure, tolerance, and diversity, and are shown in Table 25.2.

Each candidate metric was examined for its scoring range, responsiveness, and redundancy. Scoring range is the distribution of metric values across all of the available data. Range test eliminated metrics that have very small ranges, because a limited range could indicate that a metric might not vary sufficiently across sites to discriminate among sites in different conditions [37]. The coefficient of variation (CV, defined as the ratio of the standard deviation to the mean) was used to measure the relative variability of a metric, and only metrics with CV > 0.15 were retained. Box and whisker plot tests were then conducted on each retained metric by comparing the values of impaired sites with those of reference sites.

TABLE 25.2 Candidate Metrics for Developing a V-IBI for Poyang Lake Wetland

Metrics	
Species Richness and Composition	
Total species richness (↓, 0.31, 0),	Number of submerged species (↓, 1.19, 3)
Number of floating-leaved species (↓, 0.41, 1)	Number of emergent species (↓, 0.39, 1)
Number of weed species (↑, 0.36, 1)	Number of aquatic species (↓, 0.44, 1)
Number of monocot species (↓, 0.32, 1)	Number of Carex spp. (↓, 0.58, 0)
Number of sedge species (↓, 0.40, 1)	Number of grass species (↓, 0.37, 0)
Number of annual species (↑, 0.49, 1)	Number of perennial species (↓, 0.31, 1)
Number of native species (↓, 0.29, 0)	Number of native perennial species (↓, 0.30, 1)
Number of invasive species (↑, 1.81, 3)	
Community Structure	
% Carex spp. (↓, 0.48, 0)	% sedge species (↓, 0.43, 0)
% grass species (↓, 0.42, 0)	% annual species (↑, 0.50, 2)
% perennial species (↓, 0.17, 2)	% native species (↓, 0.04, 0)
% perennial native species (↓, 0.33, 1)	% invasive species (↑, 1.73, 1)
% species with both SSVP (↓, 0.56, 2)	% obligate wetland species (↓, 0.48, 1)
Abundance	
Cover of Carex spp. (↓, 0.58, 0)	Cover of *Phalaris arundinacea L.* (↑, 1.29, 2)
Cover of sedges (↓, 0.57, 0)	Cover of grasses (↓, 0.49, 0)
Cover of annual species (↑, 0.97, 0)	Cover of perennial species (↓, 0.14, 0)
Cover of native species (↓, 0.42, 0)	Cover of invasive species (↑, 2.08, 1)
Cover of monocots (↓, 0.27, 0)	
Tolerance	
Number of tolerant species (↑, 1.28, 3)	Number of sensitive species (↓, 2.71, 3)
Diversity	
Shannon-Wiener diversity index (↓, 0.28, 0)	Evenness (↓, 0.18, 0)

Adapted from Yang, W.; You, Q.; Fang, N.; Xu, L.; Zhou, Y.; Wu, N.; Ni, C.; Liu, Y.; Liu, G.; Yang, T.; Wang, Y. 2018. Assessment of wetland health status of Poyang Lake using vegetation-based indices of biotic integrity. *Ecological Indicators* 90.
Percentage (%) metrics are calculated based on the total number of plant species recorded in a sample site. ↓, decrease with disturbance intensity; ↑, increase with disturbance intensity. The first number in parentheses indicates the CV, and the second number is the interquartile range score.

The discrimination power of each metric was judged according to the degree of interquartile overlap in the box plots. A discrimination power of 3 was assigned when boxes did not overlap between the two site groups. Value of 2 was assigned when interquartile ranges overlapped but did not reach medians. Value of 1 was assigned when only one median was within the interquartile range of the other box. Value of 0 was assigned when both medians were within the range of the other box [11]. Metrics with a high discrimination power (interquartile range, IQ ≥ 2) were further screened [38]. The redundancy of the retained metrics was evaluated using Spearman's correlations. Metrics were considered redundant if the Spearman correlation coefficient was ≥ 0.75 [$p < 0.05$; 39]. Maximally two metrics per category remained for the purpose of including approximately equal number of metrics from each category in the final V-IBI [37].

In addition, candidate metrics were correlated with each of the disturbance gradients using Spearman's correlations. Metrics were selected when they showed significant correlations ($p < 0.05$) with any of the disturbance gradients. If two or more metrics were highly correlated ($r ≥ 7.50$) with each other, we only retained the metric that had higher or the highest correlation with disturbance gradients.

Metric Scoring and V-IBI Calculation

For metrics that decreased with disturbance, the 95th and 5th percentiles of all values were set as the "upper anchor" (expected value) and the "lower anchor" (threshold value), respectively [21]. The metric was rescaled as:

$$\text{Metric value} = \frac{\text{Site value} - \text{Lower anchor}}{\text{Upper anchor} - \text{Lower anchor}}$$

For metrics increasing with disturbance, the 5th and 95th percentiles of all values were set as the "upper anchor" and "lower anchor," respectively. These increasing metrics were scaled as:

$$\text{Metric value} = \frac{\text{Lower anchor} - \text{Site value}}{\text{Lower anchor} - \text{Upper anchor}}$$

Any values < 0 were given a value of 0, and any values > 1 were given a value of 1. Final V-IBIs were the sums of the scores for each individual metric. Natural breaks in V-IBI scores were then used as cutoffs for five ordinal rating categories, namely, excellent, good, fair, poor, and very poor, to assess wetland health condition [9]. The natural breaks method determined the best arrangement of V-IBI scores into different classes by seeking to reduce the variance within classes and maximize the variance between classes. Comparing with other methods for rating category classification [40], this method did not require to set arbitrary thresholds for the category division.

Site Classification and Environmental Characteristics

The site classification screening procedure identified 7 reference and 23 impaired sites, as shown in Figure 25.1. Six sites (5, 6, 11, 12, 17, and 18) classified as references were located in national natural reserves where human activities have been strictly controlled. One reference site (13) was close to a reserve and surrounded by open water and natural wetland. All the seven reference sites had fair (level III) water quality, while the impaired sites had fair ($n = 7$), poor ($n = 10$), and very poor ($n = 6$) water quality and were characterized by high levels of agricultural land uses.

Selected Metrics

Two of the initial 38 candidate metrics with CV < 0.15, that is, percentage of (%) native species and cover of perennial species, were rejected by the range test (Table 25.1). Eight of the remaining metrics passed the test for responsiveness by demonstrating a strong ability to discriminate reference sites from impaired (IQ ≥ 2), as indicated in Figure 25.3. The following redundancy test indicated that numbers

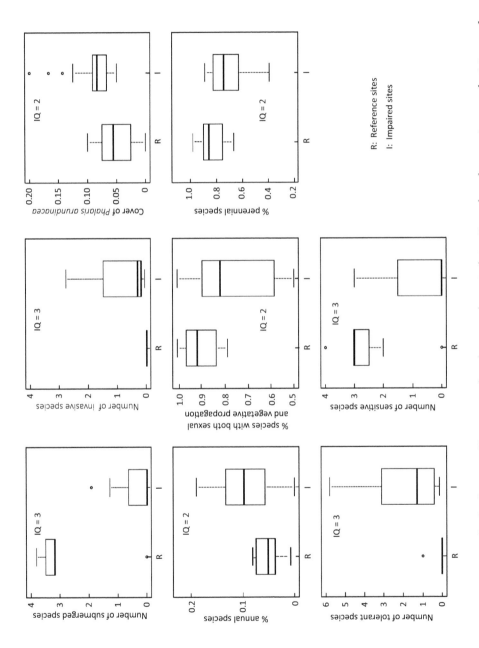

FIGURE 25.3 Discriminatory power of eight plant assemblage metrics for reference and impaired sites. Range bars show maximum and minimum of non-outliers; boxes are interquartile ranges (25 percentile to 75 percentile); bold bars are medians; circles are outliers. IQ, interquartile range score. (Adapted from Yang, W.; You, Q.; Fang, N.; Xu, L.; Zhou, Y.; Wu, N.; Ni, C.; Liu, Y.; Liu, G.; Yang, T.; Wang, Y. 2018. Assessment of wetland health status of Poyang Lake using vegetation-based Indices of biotic integrity. *Ecological Indicators* 90, 79–89.)

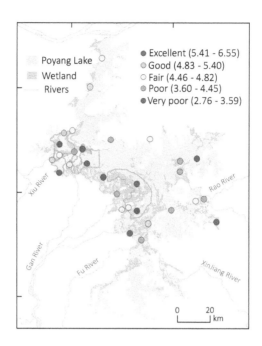

FIGURE 25.4 (See color insert.) Health status classifications of 30 sample sites in Poyang Lake wetland based on the V-IBI. (Modified from Yang, W.; You, Q.; Fang, N.; Xu, L.; Zhou, Y.; Wu, N.; Ni, C.; Liu, Y.; Liu, G.; Yang, T.; Wang, Y. 2018. Assessment of wetland health status of Poyang Lake using vegetation-based indices of biotic integrity. *Ecological Indicators* 90, 79–89.)

of submerged and sensitive species were highly correlated (Spearman's $r = 0.90$, $p < 0.001$); % annual species, % perennial species, and % species with both sexual and vegetative propagation (% SSVP) were mutually strongly correlated ($|r|$ ranging between 0.75 and 0.78, $p < 0.001$). The number of submerged species and % SSVP were retained, because they showed slightly stronger discriminatory power than the other three metrics. Two other metrics (i.e., % perennial species and % obligate wetland species) were significantly correlated with LOD and LDI, respectively. The core metrics therefore included numbers of invasive species, submerged species, tolerant species, and % SSVP, cover of *Phalaris arundinacea* L., % perennial species, and % obligate wetland species.

V-IBI Scores and the Assessment of Wetland Health Status

V-IBI scores ranged between 2.80 and 6.44, with a mean of 4.82, as shown in Figure 25.4. Based on PMS3 V-IBI scores, six sites were classified as "excellent" (5.35–6.44), five as "good" (4.78–5.34), eight as "fair" (4.34–4.77), seven as "poor" (3.65–4.33), and four as "very poor" (2.80–3.64). The overall health condition of Poyang Lake wetland is rated as good, with the wetland in national nature reserves in a better condition while estuarine wetland in the other respect, suggesting that controlling the pollutant input from tributary rivers would be extremely important for the wetland conservation.

Conclusions

The development of a V-IBI generally includes six steps: field surveys and measurement of environmental variables, reference site selection, establishing disturbance gradients, metric selection, metric scoring, and V-IBI calculation. Seven metrics are selected to construct the V-IBI for Poyang Lake wetland: numbers of invasive, sensitive, tolerant, and perennial species; percentages of obligate wetland species;

species with both SSVP; and cover of *Phalaris arundinacea* L. The assessment indicated that the overall health condition of Poyang Lake wetland was in a "good" category, while the estuarine wetland was in either "poor" or "very poor" categories.

References

1. Zedler, J.B.; Kercher, S. Wetland resources: status, trends, ecosystem services, and restorability. *Annual Review of Environment and Resources* 2005, 30, 39–74.
2. Davidson, N.C. How much wetland has the world lost? Long-term and recent trends in global wetland area. *Marine and Freshwater Research* 2014, 65, 934–941.
3. Fennessy, S.; Jacobs, A.; Kentula, M.E. An evaluation of rapid methods for assessing the ecological condition of wetlands. *Wetlands* 2009, 27, 543–560.
4. Karr, J.R. Assessment of biotic integrity using fish communities. *Fisheries* 1981, 6, 21–27.
5. Karr, J.R.; Dudley, D.R. Ecological perspective on water quality goals. *Environmental Management* 1981, 5, 55–68.
6. Ruaro, R.; Gubiani, É.A. A scientometric assessment of 30 years of the Index of Biotic Integrity in aquatic ecosystems: applications and main flaws. *Ecological Indicators* 2013, 29, 105–110.
7. Vondracek, B.; Koch, J.D.; Beck, M.W. A comparison of survey methods to evaluate macrophyte index of biotic integrity performance in Minnesota lakes. *Ecological Indicators* 2014, 36, 178–185.
8. Yang, W.; You, Q.; Fang, N.; Xu, L.; Zhou, Y.; Wu, N.; Ni, C.; Liu, Y.; Liu, G.; Yang, T.; Wang, Y. Assessment of wetland health status of Poyang Lake using vegetation-based indices of biotic integrity. *Ecological Indicators* 2018, 90, 79–89.
9. Uzarski, D.G.; Burton, T.M.; Cooper, M.J.; Ingram, J.W.; Timmermans, S.T. Fish habitat use within and across wetland classes in coastal wetlands of the five Great Lakes: development of a fish-based index of biotic integrity. *Journal of Great Lakes Research* 2005, 31, 171–187.
10. Baek, S.H.; Son, M.; Kim, D.; Choi, H.-W.; Kim, Y.-O. Assessing the ecosystem health status of Korea Gwangyang and Jinhae bays based on a planktonic index of biotic integrity (P-IBI). *Ocean Science Journal* 2014, 49, 291–311.
11. Barbour, M.; Gerritsen, J.; Griffith, G.; Frydenborg, R.; McCarron, E.; White, J.; Bastian, M. A framework for biological criteria for Florida streams using benthic macroinvertebrates. *Journal of the North American Benthological Society* 1996, 15, 185–211.
12. Beck, M.W.; Hatch, L.K.; Vondracek, B.; Valley, R.D. Development of a macrophyte-based index of biotic integrity for Minnesota lakes. *Ecological Indicators* 2010, 10, 968–979.
13. Brieda, J.T.; Jogb, S.K.; Matthewsc, J.W. Floristic quality assessment signals human disturbance over natural variability in a wetland system. *Ecological Indicators* 2013, 34, 260–267.
14. Jurik, T.W.; Wang, S.; Der Valk, A.G.V. Effects of sediment load on seedling emergence from wetland seed banks. *Wetlands* 1994, 14, 159–165.
15. Han, X.; Chen, X.; Feng, L. Four decades of winter wetland changes in Poyang Lake based on Landsat observations between 1973 and 2013. *Remote Sensing of Environment* 2015, 156, 426–437.
16. Zhang, B. *Research of Poyang Lake*. Shanghai scientific & Technical Publishers, Shanghai, 1988.
17. Herlihy, A.T.; Larsen, D.P.; Paulsen, S.G.; Urquhart, N.S.; Rosenbaum, B. Designing a spatially balanced, randomized site selection process for regional stream surveys: the Emap mid-Atlantic pilot study. *Environmental Monitoring and Assessment* 2000, 63, 95–113.
18. Tang, X.; Li, H.; Xu, X.; Yang, G.; Liu, G.; Li, X.; Chen, D. Changing land use and its impact on the habitat suitability for wintering Anseriformes in China's Poyang Lake region. *Science of the Total Environment* 2016, 557–558, 296–306.
19. Stoddard, J.L.; Larsen, D.P.; Hawkins, C.P.; Johnson, R.K.; Norris, R.H. Setting expectations for the ecological condition of streams: the concept of reference condition. *Ecological Applications* 2006, 16, 1267–1276.

20. Jun, Y.; Won, D.; Lee, S.; Kong, D.; Hwang, S. A multimetric benthic macroinvertebrate index for the assessment of stream biotic integrity in Korea. *International Journal of Environmental Research and Public Health* 2012, 9, 3599–3628.

21. Wang, B.; Yang, L.; Hu, B.; Shan, L. A preliminary study on the assessment of stream ecosystem health in south of anhui province using benthic-index of biotic integrity. *Acta Ecologica Sinica* 2005, 25, 1481–1490.

22. Ministry of Environmental Protection of the People's Republic of China. 2002. Environmental quality standard for surface water (GB3838-2002). http://kjs.mep.gov.cn/hjbhbz/bzwb/shjbh/shjzlbz/200206/t20020601_66497.htm.

23. Parinet, B.; Lhote, A.; Legube, B. Principal component analysis: an appropriate tool for water quality evaluation and management—application to a tropical lake system. *Ecological Modelling* 2004, 178, 295–311.

24. Gharibi, H.; Mahvi, A.H.; Nabizadeh, R.; Arabalibeik, H.; Yunesian, M.; Sowlat, M.H. A novel approach in water quality assessment based on fuzzy logic. *Journal of Environmental Management* 2012, 112, 87–95.

25. Liou, S.C.; Lo, S.; Wang, S. A generalized water quality index for Taiwan. *Environmental Monitoring and Assessment* 2004, 96, 35–52.

26. Grabas, G.P.; Blukacz-Richards, E.A.; Pernanen, S. Development of a submerged aquatic vegetation community index of biotic integrity for use in Lake Ontario coastal wetlands. *Journal of Great Lakes Research* 2012, 38, 243–250.

27. Miller, S.J.; Wardrop, D.H.; Mahaney, W.M.; Brooks, R.P. A plant-based index of biological integrity (IBI) for headwater wetlands in central Pennsylvania. *Ecological Indicators* 2006, 6, 290–312.

28. DeKeyser, E.S.; Kirby, D.R.; Ell, M.J. An index of plant community integrity: development of the methodology for assessing prairie wetland plant communities. *Ecological Indicators* 2003, 3, 119–133.

29. Brown, M.T.; Vivas, M.B. Landscape development intensity index. *Environmental Monitoring and Assessment* 2005, 101, 289–309.

30. Mack, J.J. Developing a wetland IBI with statewide application after multiple testing iterations. *Ecological Indicators* 2007, 7, 864–881.

31. Micacchion, M.; Stapanian, M.A.; Adams, J.V. Site-scale disturbance and habitat development best predict an index of amphibian biotic integrity in Ohio shrub and forested Wetlands. *Wetlands* 2015, 35, 509–519.

32. Mack, J.J.; Avdis, N.H.; Braig IV, E.C.; Johnson, D.L. Application of a vegetation-based index of biotic integrity for Lake Erie coastal marshes in Ohio. *Aquatic Ecosystem Health & Management* 2008, 11, 91–104.

33. Luo, Z.; Shi, X.; Han, L.; Nie, L. A study on carbon emission effects of changes in land use in poyang lake region. *Acta Agriculturae Universitatis Jiangxiensis* 2013, 35, 1074–1081.

34. Kaufmann, P.R.; Levine, P.; Robison, E.G.; Seeliger, C.; Peck, D.V. *Quantifying physical habitat in wadeable streams EPA/620/R-99/003.* US Environmental Protection Agency, Washington DC, 1999.

35. Chen, Z.; Lin, B.; Shang, H.; Li, Y. An index of biological integrity: developing the methodology for assessing the health of the Baiyangdian wetland. *Acta Ecologica Sinica* 2012, 32, 6619–6627.

36. Simon, T.P.; Stewart, P.M.; Rothrock, P.E. Development of multimetric indices of biotic integrity for riverine and palustrine wetland plant communities along southern lake Michigan. *Aquatic Ecosystem Health & Management* 2001, 4, 293–309.

37. Stoddard, J.L.; Herlihy, A.T.; Peck, D.V.; Hughes, R.M.; Whittier, T.R.; Tarquinio, E. A process for creating multimetric indices for large-scale aquatic surveys. *Journal of the North American Benthological Society* 2008, 27, 878–891.

38. Wu, N.; Cai, Q.; Fohrer, N. Development and evaluation of a diatom-based index of biotic integrity (D-IBI) for rivers impacted by run-of-river dams. *Ecological Indicators* 2012, 18, 108–117.

39. Whittier, T.R.; Stoddard, J.L.; Larsen, D.P.; Herlihy, A.T. Selecting reference sites for stream biological assessments: best professional judgment or objective criteria. *Journal of the North American Benthological Society* 2007, 26, 349–360.
40. Hering, D.; Feld, C.K.; Moog, O.; Ofenbock, T. Cook book for the development of a Multimetric Index for biological condition of aquatic ecosystems: experiences from the European AQEM and STAR projects and related initiatives. *Hydrobiologia* 2006, 566, 311–324.

26

Remote Sensing for Wetland Indices

Jason Yang
Ball State University

Introduction

A wetland is land where the water table is at, near, or above the surface or which is saturated for a long enough period to promote such features as wet-altered soils and water-tolerant vegetation (Warner et al. 1986). Other names for wetlands include marshes, estuaries, mangroves, mudflats, mires, fens, swamps, billabongs, lagoons, and bogs. Wetlands are transition zones where the flow of water and the cycling of nutrients and the energy of the sun meet to produce a unique ecosystem—making these areas very important features of a watershed. There are many different kinds of wetlands and many ways to categorize them. NOAA classifies wetlands into five general types: marine (ocean), estuarine (estuary), riverine (river), lacustrine (lake), and palustrine (marsh) (NOAA 2019), while U.S. EPA classifies them into two general categories: tidal (coastal) wetlands and nontidal (inland) wetlands (EPA 2018).

Tidal wetlands can be found along protected coastlines in middle and high latitudes worldwide. They are most prevalent in the United States on the eastern coast from Maine to Florida and continuing on to Louisiana and Texas along the Gulf of Mexico. Some are freshwater marshes, others are brackish (somewhat salty), and still others are saline (salty), but they are all influenced by the motion of ocean tides. Tidal marshes are normally categorized into two distinct zones, the low (intertidal) marsh and the high (upper) marsh.

Nontidal wetlands are the most prevalent and widely distributed wetlands in North America. They are mostly freshwater marshes and frequently occur along streams in poorly drained depressions and in the shallow water along the boundaries of lakes, ponds, and rivers. Based on the increasing level of water and decreasing level of vegetation, a riverine system can be classified as the following five general wetland habitats: forested wetland (with vegetation on land), scrub-shrub wetland (with vegetation above water), emergent wetland (with vegetation near water surface), aquatic bed (with vegetation under water), and open water or unconsolidated bottom (with pure water and no vegetation) (Committee 2013).

Remote Sensing of Wetlands

The real-time monitoring of wetlands is essential work to control surface water pollution and protect ecological environment. Due to the advantages of fast information updates, and the ability to produce real-time observations for large geographic areas, remote sensing has been widely applied on investigating and monitoring Earth's water resources in the past several decades (Guo et al. 2017). In addition, if a remote sensor is passively recording the electromagnetic signals, the unobtrusive images obtained can greatly reduce the biases and improve the scientific analysis (Bi et al. 2012).

After the launch of the first Earth Resources Observation Satellite (Landsat-1) in 1972, Landsat images have become an important data resource in environmental remote sensing due to its moderate spatial resolution, multiple spectral resolution, and continuous observation (Kuenzer et al. 2011). One of the many fields where Landsat images are used is in the detection of land-use and land-cover changes over time (Laba et al. 1997; Wang et al. 2004; Yuan et al. 2005). Wetlands, as one of the important land covers on the Earth's surface, have also been studied by many scholars using remote sensing during the past five decades (Schmidt and Skidmore 2003; Amler et al. 2015; Dronova 2015), and many methods and algorithms have been developed to address issues of wetland identification, classification, change detection, and biomass (Rundquist et al. 2001). For example, Son et al. used Landsat data from 1979 to 2013 to monitor the mangrove forest changes of Ca Mau Peninsula in South Vietnam (Son et al. 2015). Ruizluna used Landsat data to detect landscape changes of six land types (mangrove, lagoon, salt marsh, dry forest, secondary succession, and agriculture) in the Majahual coastal system of Mexico (Berlangarobles and Ruiz-Luna 2002). Jia et al. used Landsat data to map the distribution of mangrove forests in China to advance the management and protection of wetlands (Jia et al. 2014). Rapinel et al. used Landsat 8 OLI (Operational Land Imager) images to map plant communities in coastal marshlands at regional scale along the French Atlantic Coast (Rapinel et al. 2015).

Remote Sensing for Wetland Indices

Since the 1960s, scientists have studied biophysical variables of the Earth's surface using remotely sensed data. Much of this effort has involved the use of wetland indices—dimensionless and radiometric measures that indicate the relative abundance and activity of green vegetation, soil, and water due to being simple and inexpensive implementation (McFeeters 1996; Xu 2006; Zhong 2007; Jiang et al. 2014). Based on the spectral reflectance profiles of five wetland habitats in a riverine system on the six multispectral bands in Landsat TM (Figure 26.1), theoretically, it is possible to map wetland types with different levels of water and vegetation (Committee 2013).

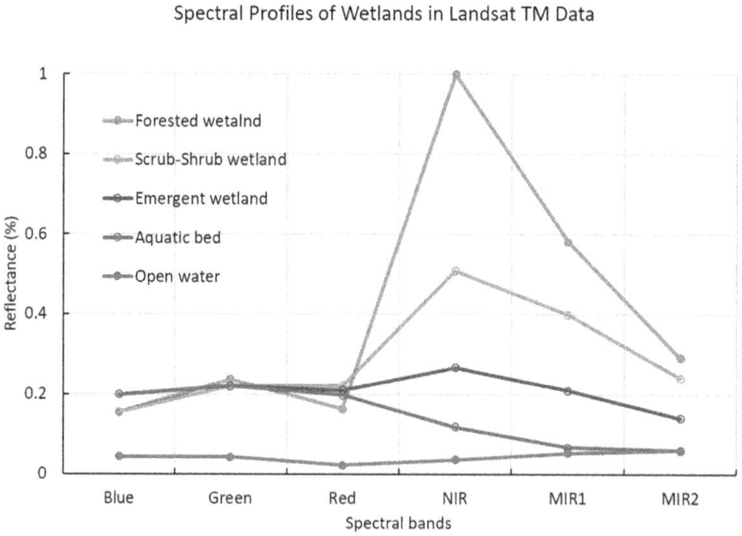

FIGURE 26.1   Spectral profiles of some typical wetlands habitats in riverine system in Landsat TM data.

TABLE 26.1 Selected Wetlands Indices in Remote Sensing

Wetlands Index	Equation	Selected References	Landsat TM Bands
NDVI	$\mathrm{NDVI} = \dfrac{\rho_{\mathrm{NIR}} - \rho_{\mathrm{Red}}}{\rho_{\mathrm{NIR}} + \rho_{\mathrm{Red}}}$	Rouse 1974	ρ_{Blue} : TM1
			ρ_{Green} : TM2
Wetness (W)	$W = 0.1446\rho_{\mathrm{Blue}} + 0.1761\rho_{\mathrm{Green}} +$ $0.3322\rho_{\mathrm{Red}} + 0.3396\rho_{\mathrm{NIR}} -$ $0.6210\rho_{\mathrm{MIR1}} - 0.4186\rho_{\mathrm{MIR2}}$	Kauth and Thoams 1976 Crist and Kauth 1986	ρ_{Red} : TM3 ρ_{NIR} : TM4
NDWI	$\mathrm{NDWI} = \dfrac{\rho_{\mathrm{Green}} - \rho_{\mathrm{NIR}}}{\rho_{\mathrm{Green}} + \rho_{\mathrm{NIR}}}$	McFeeters 1996	ρ_{MIR1} : TM5 ρ_{MIR2} : TM7
NDMI	$\mathrm{NDMI} = \dfrac{\rho_{\mathrm{NIR}} - \rho_{\mathrm{MIR}}}{\rho_{\mathrm{NIR}} + \rho_{\mathrm{MIR}}}$	Hardisky et al. 1983 Gao 1996	
GWI	$\mathrm{GWI} = \left(\rho_{\mathrm{Green}} + \rho_{\mathrm{Red}}\right) - \left(\rho_{\mathrm{NIR}} + \rho_{\mathrm{MIR}}\right)$	Yang and Xu 1998	
MNDWI	$\mathrm{MNDWI} = \dfrac{\rho_{\mathrm{Green}} - \rho_{\mathrm{MIR}}}{\rho_{\mathrm{Green}} + \rho_{\mathrm{MIR}}}$	Xu 2006	

Forested wetland is dominated by trees, where reflectance profile from the tree crown is almost the same as it from typical green vegetation; scrub-shrub wetland includes areas dominated by woody vegetation less than 6 m tall, which has lower reflectance on the near-infrared band. Emergent wetland is usually dominated by perennial plants, which has a mixed reflectance from plants and water. Aquatic bed is dominated by plants that grow principally on or below the surface of the water for much of the growing season in most years, with a reflectance profile closer to pure water. Open water absorbs most of the incident radiance and has the lowest reflectance on all six spectral bands. Table 26.1 provides a summary of some most widely adopted wetland indices in remote sensing.

Rouse et al. were some of the first to use what has become known as the normalized difference vegetation index (NDVI) using the near-infrared band $\left(\rho_{\mathrm{NIR}}\right)$ and red band $\left(\rho_{\mathrm{Red}}\right)$ in a remotely sensed image to study vegetation (Rouse 1974):

$$\mathrm{NDVI} = \frac{\rho_{\mathrm{NIR}} - \rho_{\mathrm{Red}}}{\rho_{\mathrm{NIR}} + \rho_{\mathrm{Red}}}$$

The NDVI has become an important vegetation index and applied on many satellite images such as Landsat and MODIS to study the vegetation distribution at regional and global levels (Huete et al. 2002).

Wetness (W) image is one of the indices concerning the moisture status of the wetland environment through the tasseled cap transformation (Kauth and Thoams 1976). This transformation converts six Landsat multispectral bands into three variables: soil brightness (B), vegetation greenness (G), and wetness (W). The wetness image can be calculated using the following coefficients (Crist and Kauth, 1986):

$$W = 0.1446\rho_{\mathrm{Blue}} + 0.1761\rho_{\mathrm{Green}} + 0.3322\rho_{\mathrm{Red}} + 0.3396\rho_{\mathrm{NIR}} - 0.6210\rho_{\mathrm{MIR1}} - 0.4186\rho_{\mathrm{MIR2}}$$

The tasseled cap transformation is a global vegetation index, which can be used anywhere in the world to disaggregate the amount of soil, brightness, vegetation, and moisture content in individual pixels in Landsat MSS, TM, ETM+, Landsat 8, or other types of multispectral data (Jensen 2015). However, it is better to compute the coefficients based on local conditions if possible in practice.

The normalized difference moisture index (NDMI), also expressed as Infrared Index (II), was developed and used to detect vegetation moisture based on near-infrared band (ρ_{NIR}) and mid-infrared band (ρ_{MIR}) in multispectral remote sensing data (Hardisky et al. 1983):

$$NDMI = II = \frac{\rho_{NIR} - \rho_{MIR}}{\rho_{NIR} + \rho_{MIR}}$$

This index was found to be highly correlated with canopy water content and more closely tracked changes in plant biomass and water stress than did the NDVI (Gao 1996).

The NDMI can also be applied on hyperspectral remotely sensed data to determine vegetation liquid water content. Two infrared bands are used in computation of the NDMI: One is centered at approximately 860 nm and the other at 1,240 nm (Gray and Song 2011).

The normalized difference water index (NDWI) was first proposed and used by McFeeters (McFeeters 1996) to extract water bodies:

$$NDWI = \frac{\rho_{Green} - \rho_{NIR}}{\rho_{Green} + \rho_{NIR}}$$

This index uses the green band (ρ_{Green}) and near-infrared band (ρ_{NIR}) to maximize the limits of vegetation, so as to outstand the water information. However, the water bodies extracted using NDWI from Landsat TM data are often mixed with many non-water features such as shadows, and the result is even worse when it is applied on mountain areas.

Another water index called General Water Index (GWI) was proposed by Yang and Xu in 1998 to extract the water bodies from Landsat data in the following equation (Yang and Xu 1998):

$$GWI = (\rho_{Green} + \rho_{Red}) - (\rho_{NIR} + \rho_{MIR})$$

Instead of using one visible band and one infrared band in calculating NDWI, GWI is calculated using two visible bands and two infrared bands. By comparing the results calculated from NDWI and GWI, no significant difference has been found between these two indices in detection of water bodies.

Xu proposed a modified NDWI (MNDWI) in 2006 to enhance the extraction of water bodies using mid-infrared band (ρ_{MIR}) instead of near-infrared band (ρ_{NIR}) for normalization (Xu 2006):

$$MNDWI = \frac{\rho_{Green} - \rho_{MIR}}{\rho_{Green} + \rho_{MIR}}$$

Since the MNDWI uses mid-infrared band to calculate the normalized ratio, it can eliminate the effect of topographical differences, specifically resolving the shadow issues in the extraction of water bodies. Therefore, the MNDWI is efficient in extracting water bodies from urban and mountain areas with many shadows.

Figure 26.2 presents some wetland index images calculated from a Landsat TM image obtained on August 28, 2010, for New Jersey Meadowland and New York City area. The NDVI image presents the most vegetated areas, including forested wetland and scrub-shrub wetland, with no water information provided in the image. The wetness image presents moist surfaces, including emergent wetland and open water, with some urban areas mixed. The NDWI image presents more information for open water, but less information for wetlands. The MNDWI is the best index to separate water bodies from land, with little information for other wetlands. The NDMI image presents information for all five wetlands types from forested wetland (pure vegetation) to open water (pure water).

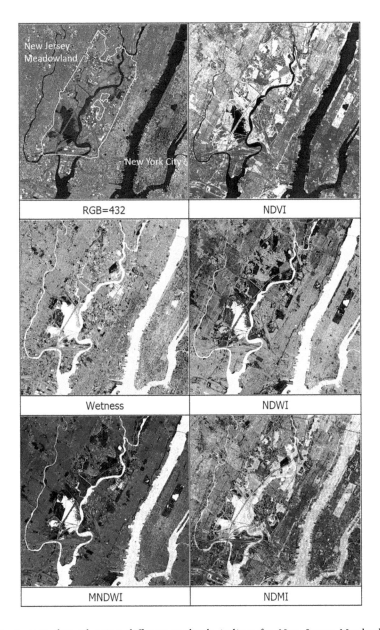

FIGURE 26.2 Images derived using different wetlands indices for New Jersey Meadowlands and the surrounding area.

Conclusions

Wetland indices are easy and quick ways to get water body and wetland information using multispectral remote sensing data; however, most wetland indices are only good to detect wetland areas and extents. In order to study wetland types or even species of wetland vegetation, more classification algorithms and hyperspectral images are needed. For extracting the areas and extents of wetland habitats from multispectral remote sensing data, the NDVI is good for forested wetland and scrub-shrub wetland; wetness, NDWI, and GWI are good for aquatic bed and open water; the MNDWI is the best for open water; and the NDMI is the best for all five wetland habitats.

References

Amler, E., et al. (2015). "Definitions and mapping of east African wetlands: A review." *Remote Sensing* 7 SRC - BaiduScholar: 5256–5282.

Berlangarobles, C. A., Ruiz-Luna, A. (2002). "Land use mapping and change detection in the coastal zone of Northwest Mexico using remote sensing techniques." *Journal of Coastal Research* 18: 514–522.

Bi, H., et al. (2012). "Comparison and analysis of several common water extraction methods based on TM Image." *Remote Sensing Information* 27(5): 77–82.

Committee, F. G. D. (2013). *Classification of wetlands and deepwater habitats of the United States*. F. G. Wetlands Subcommittee and D. C. a. U. S. F. a. W. Service: Washington, DC.

Crist, E. P., Kauth, R.J. (1986). "The tasseled cap de-mystified." *Photogrammetric Engineering and Remote Sensing* 52: 81–86.

Dronova, I. (2015). "Object-based image analysis in wetland research: A review." *Remote Sens.* 7 SRC - BaiduScholar: 6380–6413.

EPA, U. S. (2018, 07/05/2018). "Classification and Types of Wetlands." Retrieved 06/18, 2019, from www.epa.gov/wetlands/classification-and-types-wetlands#marshes.

Gao, B.-C. (1996). "NDWI—A normalized difference water index for remote sensing of vegetation liquid water from space." *Remote Sensing of Environment* 58(3): 257–266.

Gray, J., Song, C. (2011). "Mapping leaf area index using spatial, spectral, and temporal information from multiple sensors." *Remote Sensing of Environment* 119: 173–183.

Guo, M., et al. (2017). "A review of wetland remote sensing." *Sensors* 17(4): 777.

Hardisky, M., et al. (1983). "The influence of soil salinity, growth form, and leaf moisture on the spectral radiance of Spartina Alterniflora canopies." *Photogrammetric Engineering and Remote Sensing* 48: 77–84.

Huete, A., et al. (2002). "Overview of the radiometric and biophysical performance of the MODIS vegetation indices." *Remote Sensing of Environment* 83(1): 195–213.

Jensen, J. R. (2015). *Introductory digital image processing: A remote sensing perspective*. Prentice Hall Press and Pearson: Upper Saddle River, NJ, USA.

Jia, M., et al. (2014). "Mapping China's mangroves based on an object-oriented classification of Landsat imagery." *Wetlands* 34(2): 277–283.

Jiang, H., et al. (2014). "An automated method for extracting rivers and lakes from landsat imagery." *Remote Sensing* 6: 5067–5089.

Kauth, R. J., Thomas, G. S. (1976). "The Tasselled Cap -- A Graphic Description of the Spectral-Temporal Development of Agricultural Crops as Seen by LANDSAT." *LARS Symposia*, 159.

Kuenzer, C., et al. (2011). "Remote sensing of mangrove ecosystems: A review." *Remote Sensing* 3 SRC - BaiduScholar: 878–928.

Laba, M., et al. (1997). "Landsat-based land cover mapping in the lower Yuna River watershed in the Dominican Republic." *International Journal of Remote Sensing* 18(14): 3011–3025.

McFeeters, S. (1996). "The use of the normalized difference water index (NDWI) in the delineation of open water feature." *International Journal of Remote Sensing* 17(7): 1425–1432.

NOAA (2019, 08/08/19). "What Is a Wetland?". Retrieved 08/10/19, 2019, from https://oceanservice.noaa.gov/facts/wetland.html.

Rapinel, S., et al. (2015). "Use of bi-seasonal landsat-8 imagery for mapping marshland plant community combinations at the regional scale." *Wetlands* 35(6): 1043–1054.

Rouse, J. W., Eds. (1974). Monitoring Vegetation Systems in the Great Plains with ERTS. *3rd Earth Resources Technology Satellite-1 Symposium*. Washington, D.C., NASA.

Rundquist, D. C., et al. (2001). "A review of wetlands remote sensing and defining new considerations." *Remote Sensing Reviews* 20(3): 207–226.

Schmidt, K., Skidmore, A. (2003). "Spectral discrimination of vegetation types in a coastal wetland." *Remote Sensing of Environment* 85 SRC - BaiduScholar: 92–108.

Son, N., et al. (2015). "Mangrove mapping and change detection in Ca Mau peninsula, Vietnam, using landsat data and object-based image analysis." *IEEE Journal of Selected Topics in Applied Earth Observations and Remote Sensing* **8**(2): 503–510.

Wang, L. S., et al. (2004). "Integration of object-based and pixel-based classification for mapping mangroves with IKONOS imagery." *International Journal of Remote Sensing* **25**: 5655–5668.

Warner, B. G., et al. (1986). "National Wetlands Working Group." In: Warner, B.G. and C.D.A. Rubec (eds.) *The Canadian Wetland Classification System* 2nd Edition **27**(5 SRC - BaiduScholar): 77–82.

Xu, H. (2006). "Modification of normalized difference water index (NDWI) to enhance open water features in remotely sensed imagery." *International Journal of Remote Sensing* **27**: 3025–3033.

Yang, C., Xu, M. (1998). "Discussion on water extraction based on remote sensing information mechanism." *Geographical Research* **7**: 86–89.

Yuan, F., et al. (2005). "Land cover classification and change analysis of the Twin Cities (Minnesota) Metropolitan area by multitemporal landsat remote sensing." *Remote Sensing of Environment* **98** SRC - BaiduScholar: 317–328.

Zhong, C. (2007). "Study of water extraction in wetland automatically from TM images." *Water Resources Research* **28**(4): 589–595.

27

Wetland Mapping: Poyang Lake, Remote Sensing

Jian Xu and
Dan Gao
Jiangxi Normal University

Introduction

Wetlands refer to areas of marshland, peatland, or water, whether natural or artificial, permanent or temporary, with water that is static or flowing, fresh, brackish, or salt, including areas of marine water less than 6 m in depth at low tide [1,2]. Wetlands are essential natural resources for human beings in terms of providing multiple ecosystem services [3,4]. However, wetlands are threatened by many problems caused by various anthropogenic and natural factors, such as environmental pollution, spread of invasive species, and climate change. Wetland degradation has aroused widespread concerns. Therefore, monitoring of wetlands is critically important for the protection of related environments and ecosystems [5]. Remote sensing science and technologies, with the ability to cover large spatial areas at frequent temporal intervals, have been broadly applied in monitoring of wetland [6]. In particular, remote sensing is the most effective in the monitoring of wetland environment with significant dynamic fluctuation and inundation hydrological patterns.

Landsat and SPOT images are among major data sources that have been used in the monitoring of water fluctuations in wetland areas. Multitemporal data are very effective in the extraction of wetland information when combined with elevation and topography data [7]. Early research employed Landsat and SPOT multispectral data to evaluate aquatic macrophyte changes within the Florida Everglades [8]. Changed areas in China's wetland between 1990 and 2000 had been identified based on Landsat data [9].

Poyang Lake is situated at the lower Yangtze River Basin, and it is the largest freshwater lake in China. Poyang Lake is fed by tributaries of five rivers of Gan, Fu, Xin, Rao, and Xiu, and it is connected and exchanges water with Yangtze River through lake mouth in the north. As controlled by water from the five tributary rivers as well as the Yangtze River, the lake's highly dynamic and seasonal variations in water level present a unique landscape of freshwater lake-wetland ecosystem. The variation of the size of the lake is illustrated as an ocean when flooded during the wet season and as a line of river when withered during the dry season. The lake plays an irreplaceable role for flood control, river shipping, city water supply, and conservation of biological diversity of middle and lower reaches of Yangtze River [10]. There are two national wetland natural reserves (Poyang Lake National Reserve and Nanji Wetland National Reserve) established in Poyang Lake. The Poyang Lake wetland is a key habitat site for wintering migratory birds with global importance. The Poyang Lake wetland was first selected as the protected area under the international Ramsar Convention in China because of its biological productivity, species richness, and being a critical wintering habitat for rare and endangered migratory bird species, such as the Siberian crane (*Grus leucogeranus*). In this chapter, the fluctuations of Poyang Lake wetland areas in different seasons were extracted based on multitemporal Landsat-8 Operational Land Imager (OLI) images.

Remote Sensing Data Processing

Four scenes of Landsat-8 OLI images (path 121/Row 40) with little or no cloud cover were downloaded from the United States Geological Survey (USGS) website (https://earthexplorer.usgs.gov/) for mapping wetland landscape in different seasons (Table 27.1).

Preprocessing of Landsat data includes mainly radiance calibration, atmospheric correction, mosaic, and subset. Radiance calibration was performed to convert the digital numbers value to radiance. FLAASH algorithm was applied for atmospheric correction to obtain the surface spectral reflectance. Subset data was obtained from a selected Landsat-5 Thematic Mapper (TM) image that recorded the area of the largest water coverage of Poyang Lake in a flooding season. The boundary of the water area was then used to extract the Landsat-8 OLI data as the study area for wetland mapping (Figure 27.1).

Supervised image classification by support vector machine (SVM) algorithm was applied. The radial basis function (RBF) was selected as the kernel function for the SVM classifiers. Based on the knowledge of land-cover distribution characteristics, four land-cover categories, namely, water, grassland, sand, and bare soil, were identified as the final class types. With the dynamic nature of Poyang Lake, the associated wetlands were included in the water category. Samples were randomly selected from known sites using the region of interest (ROI) tools provided by ENVI 5.3 software with the assistance of Google Earth tool and ground knowledge to recognize the land-cover categories. These ROI samples were easily identified using visual interpretation on the Landsat-8 OLI image. The distribution of the samples was considered to cover the entire study area. The characteristics of the ROI samples of Landsat-8 OLI data acquired on March 14, 2014, for training and validating the classifiers are summarized in Table 27.2. The total ROI samples used for training SVM classifiers were 64 (16 samples in each of the four categories). Another 100 samples were used for classification accuracy assessment. The overall classification accuracy, producer's accuracy, user's accuracy, and kappa statistics were estimated for quantitative classification performance analysis.

Wetland Dynamics in Poyang Lake

The land-cover classification results of four Landsat-8 OLI scenes are illustrated in Figure 27.2. The classification accuracy and kappa statistics were estimated based on the validation samples for Landsat-8 OLI data of March 14, 2014. The confusion matrix of the classification results is shown in Table 27.3. Accuracy assessment results indicated 98.8% overall accuracy and 0.96 kappa coefficient. The grassland class had the lowest user accuracy and the maximum confusion with the bare-soil category. The bare-soil class had the lowest producer accuracy and the maximum confusion with the grassland category. Other classes all had a better separation from each other and higher accuracies.

Visually, each class type could be identified using the SVM classifiers based on an expert's knowledge. The landscape of Poyang Lake wetland, especially two national wetland reserves, presented obvious seasonal variation (Figure 27.2). The results indicated that the landscape composition and configuration of Poyang Lake wetland was mainly affected by water level. Overall, the area of water was increased with increasing water level, while the areas of grassland, sand, and bare soil were decreased with increasing water level (Table 27.4).

TABLE 27.1 Landsat Images and Water Level in Xingzi Gauging Station (Wusong Base Level)

Date of Imagery Acquisition (DD-MM-YYYY)	Satellite/Sensor	Water Level (m)	Purpose of Usage
14-03-2014	Landsat8/OLI	10.7	Wetland landscape in spring
23-06-2016	Landsat8/OLI	18.3	Wetland landscape in summer
24-10-2014	Landsat8/OLI	11.5	Wetland landscape in autumn
13-02-2015	Landsat8/OLI	7.7	Wetland landscape in winter

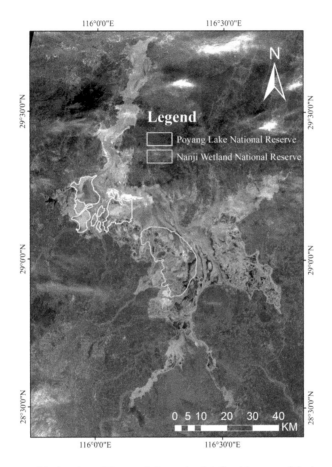

FIGURE 27.1 The geographical region of Poyang Lake wetland defined by area of the largest water coverage in flooding season. The Poyang Lake area is displayed in standard pseudo color of Landsat-8 OLI image acquired on March 14, 2014. The background display is Landsat-8 panchromatic image acquired the same day.

TABLE 27.2 Number of ROIs and Pixels in Each Class Type Used for Training and Validating the SVM Classifiers (Landsat-8 OLI Data of March 14, 2014)

	Water	Grassland	Sand	Bare Soil
Number of ROIs for training	16	16	16	16
Number of pixels for training	5,995	1,484	438	406
Number of ROIs for validation	43	31	9	18
Number of pixels for validation	3,079	395	198	183

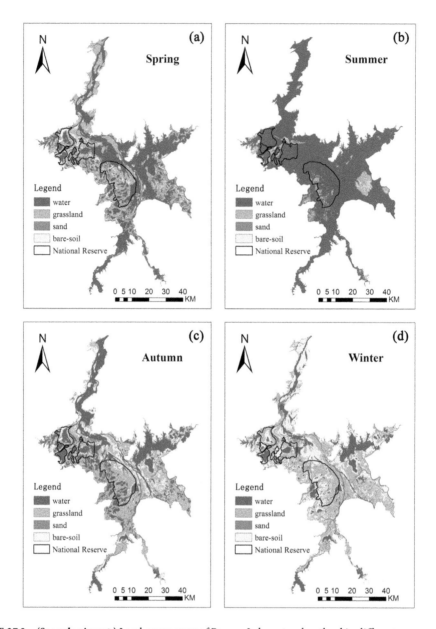

FIGURE 27.2 **(See color insert.)** Land-cover maps of Poyang Lake natural wetland in different seasons.

TABLE 27.3 Error Matrix and Classification Accuracy Assessment for Landsat-8 OLI Data of March 14, 2014

| Mapped Class | Ground Truth (Pixels) | | | | Row Total | User Accuracy (%) |
	Water	Grassland	Sand	Bare Soil		
Water	3,079	5	0	2	3,086	99.81
Grassland	0	390	0	39	429	90.91
Sand	0	0	198	0	198	100
Bare soil	0	0	0	142	142	100
Column total	3,079	359	198	183	3,855	
Producer accuracy	100%	98.73%	100%	77.6%		

TABLE 27.4 Area of Land-Cover Classes with Different Seasons in Poyang Lake Wetland (km²)

		Water	Grassland	Sand	Bare Soil
All Poyang Lake area	**Spring**	1,723.75	1,205.67	17.08	348.40
	Summer	2,923.08	304.85	7.70	59.27
	Autumn	1,289.77	1,401.12	14.88	589.13
	Winter	732.90	1,202.00	42.03	1,317.97
Poyang Lake National Reserve	**Spring**	80.94	103.26	0	37.65
	Summer	219.11	2.74	0	0
	Autumn	113.81	86.96	0	21.08
	Winter	57.63	60.4	0	103.82
Nanji Wetland National Reserve	**Spring**	127.48	156.23	0.30	37.74
	Summer	283.91	37.01	0.79	0.04
	Autumn	86.93	198.58	0.01	36.23
	Winter	32.49	119.89	0.02	169.35

Conclusions

As the largest freshwater lake in China shows the greatest variation in water level and inundation extent, multitemporal remote sensing plays a key role in monitoring of wetlands as critical habitats of a global significance in biodiversity conservation. The mapping results reveal that the landscape of Poyang Lake wetland presents dramatic seasonal variation. Finer spatial resolution multitemporal Landsat-8 OLI data are much appreciated for monitoring of the wetlands that are routinely affected by the dynamics of water levels of the Poyang Lake. The data process and analysis approaches are applicable to most of the situations for monitoring of the changing environment, in particular, for the subjects of inland water and wetland monitoring.

References

1. Matthews, G.V.T., *The ramsar convention on wetlands: Its history and development*. Ramsar Convention Bureau: Gland, 1993.
2. van der Valk, A.G. Wetlands: Classification. In *Encyclopedia of natural resources: Land*, Wang, Y., Ed. Taylor and Francis: New York, Published online: 21 Oct 2014; pp. 538–545.
3. Costanza, R.; d'Arge, R.; De Groot, R.; Farber, S.; Grasso, M.; Hannon, B.; Limburg, K.; Naeem, S.; O'neill, R.V.; Paruelo, J., The value of the world's ecosystem services and natural capital. *Nature* 1997, 387, 253–260.
4. Fu, B.; Li, Y.; Wang, Y.; Zhang, B.; Yin, S.; Zhu, H.; Xing, Z., Evaluation of ecosystem service value of riparian zone using land use data from 1986 to 2012. *Ecological Indicators* 2016, 69, 873–881.
5. Wang, Y., *Remote sensing of protected lands*. CRC Press: Boca Raton, 2011.
6. Wang, Y., *Remote sensing of coastal environments*. CRC Press: Boca Raton, 2009.
7. Ozesmi, S.L.; Bauer, M.E., Satellite remote sensing of wetlands. *Wetlands Ecology and Management* 2002, 10, 381–402.
8. Jensen, J.R.; Rutchey, K.; Koch, M.S.; Narumalani, S., Inland wetland change detection in the everglades water conservation area 2a using a time series of normalized remotely sensed data. *Photogrammetric Engineering and Remote Sensing* 1995, 61, 199–209.
9. Gong, P.; Niu, Z.; Cheng, X.; Zhao, K.; Zhou, D.; Guo, J.; Liang, L.; Wang, X.; Li, D.; Huang, H., China's wetland change (1990–2000) determined by remote sensing. *Science China Earth Sciences* 2010, 53, 1036–1042.
10. Gao, J.H.; Jia, J.; Kettner, A.J.; Xing, F.; Wang, Y.P.; Xu, X.N.; Yang, Y.; Zou, X.Q.; Gao, S.; Qi, S., Changes in water and sediment exchange between the changjiang river and poyang lake under natural and anthropogenic conditions, china. *Science of the Total Environment* 2014, 481, 542–553.

28

Wetland Conservation Efforts and Policy Impacts in China

Zongming Wang,
Dehua Mao, and
Mingming Jia
*Northeast Institute
of Geography and
Agroecology, Chinese
Academy of Sciences*

Introduction

Wetlands cover approximately 6% of the terrestrial surface and provide important and diverse benefits to people around the world [1]. Their loss is of particular concern. Wetlands have been among the fastest of any ecosystem type to show loss rates worldwide [2]. The Organization of Economic Cooperation and Development (OECD) (1996) estimated that the world might have lost 50% of its wetlands since 1900 [3]. Since the Ramsar Convention on Wetlands in 1971, wetland conservation (maintenance and sustainable use) and restoration (recovery of degraded natural wetlands) have been considered high priorities for many countries.

China has the world's fourth largest wetland area. However, massive wetland loss has occurred in China due to climate change and human activities. Over the past 60 years, China has lost 23.0% of marshes and swamps, 16.1% of lakes, 15.3% of rivers, and 51.2% of coastal wetlands [4]. While the Chinese government has increasingly recognized the importance of wetland protection, particularly after joining the Ramsar Convention in 1992, natural wetlands in China have suffered great loss and degradation [5]. To address this problem, China has implemented the National Wetland Conservation Program (NWCP), one of the largest of its kind in the world with ambitious goals, massive investments, and potentially enormous impacts [6]. Furthermore, NWCP has global implications because it aims to rehabilitate habitats for water birds of international importance, enhance carbon sequestration, conserve soil and water, and protect important headwaters of international rivers and lakes.

General Situation of China's Wetlands

The Second National Wetland Inventory (2009–2013) revealed that China has extensive wetlands that account for approximately 5.58% of the nation's territorial area. Of them, natural and human-made wetlands occupy 87.4% and 12.6% of China's wetlands, respectively. Of the natural wetlands' total area of 466,747 km², 217,329 km² are marshes and swamps, 85,938 km² lakes, 105,521 km² rivers, and 57,959 km² marine and coastal wetlands. Qinghai, Tibet, Inner Mongolia, and Heilongjiang are the four major provinces in which wetlands are distributed. China's wetlands provide a significant amount of ecosystem services, including freshwater supply, flood control, water purification, wildlife habitat support, and aquatic life preserves. More than 0.3 billion people's livelihood depends on China's wetlands, and the total value of these wetlands could account for 54.9% of the annual ecosystem services in China. For example, 82% of freshwater resources in China are contained in wetlands. China's wetlands are home to approximately 5,000 plant species, 3,200 animal species, and 700 fish species, providing habitats for 60% of the species of cranes and 26% of the species of geese and ducks in the world [7]. In addition, wetlands contribute to the gene bank for wildlife. For example, a number of high-yield hybrid rice varieties were developed from wild rice growing in natural wetlands [6].

China's wetlands are facing many types of threats. According to the First National Wetland Inventory (1995–2003), of the 323 investigated wetland sites, pollution, agricultural encroachment, and illegal poaching are the main threats. In contrast, based on the Second National Wetland Inventory (2009–2013), the major threats have been pollution, agricultural encroachment, builtup land expansion, overfishing and harvesting, and exotic species invasion. It was estimated that 60% of wetland decline in marsh and swamp during 1990–2010 could be attributed to the expansion of agricultural lands (Figure 28.1). It dominantly occurred in Northeast China and the middle and lower reaches of Yangtze River [8]. Moreover, in the 20 years, China lost 2,883 km² of wetlands to urban expansion (Figure 28.2), of which about 2,394 km² took place in the eastern regions, including Northeast, North, Southeast, and South China. Of all wetland categories, reservoirs/ponds and marshes suffered the most severe losses. Four hot spots of urbanization-induced wetland loss in China were identified: the

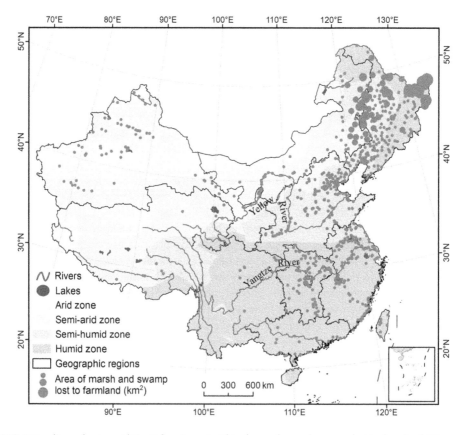

FIGURE 28.1   (See color insert.) Spatial variance in China's marshes and swamps lost to farmland (bigger points denote larger areas of marshes and swamps lost to farmland within a county) [8].

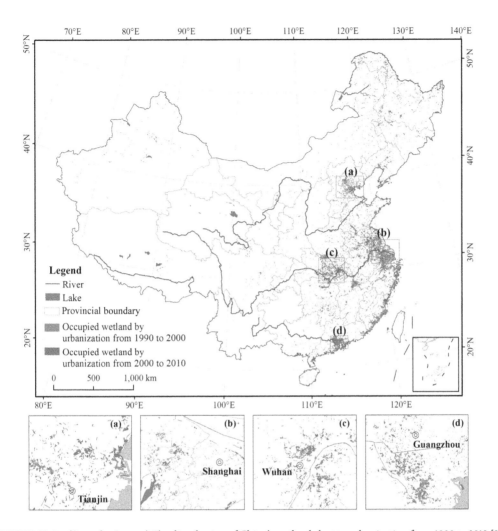

FIGURE 28.2 **(See color insert.)** The distribution of China's wetlands lost to urbanization from 1990 to 2010 [9].

Beijing-Tianjin metropolitan region, the Yangtze River Delta, the Jianghan Plain, and the Pearl River Delta [9]. Exotic species invasion is also one of the top five factors threatening wetlands. Northeast Institute of Geography and Agroecology, Chinese Academy of Sciences, established a new dataset on the *Spartina alterniflora* (*S. alterniflora*) invasion, which was called CAS *S. alterniflora*, and reported that the area of the *S. alterniflora* invasion was estimated to 546 km² [10]. Considered as an exotic species in China's coastal wetlands, *S. alterniflora* expanded by more than 500 km² from 1990 to 2015 and was mostly observed in Jiangsu, Zhejiang, Shanghai, and Fujian provinces (Figure 28.3). Three particular hot spots for *S. alterniflora* invasion were Yancheng of Jiangsu, Chongming of Shanghai, and Ningbo of Zhejiang, and each had a net area increase larger than 5,000 ha. Moreover, an obvious shrinkage of *S. alterniflora* was identified in three coastal cities, namely, Cangzhou of Hebei, Dongguan and Jiangmen of Guangdong. *S. alterniflora* invaded mostly into mudflats and shrank primarily due to aquaculture [11].

The shrinkage and degradation of wetlands have led to loss of biodiversity, damage to ecological services, as well as increased environmental disasters, such as severe droughts in 1997, 2009, and 2010; catastrophic floods in 1998; and the Taihu Lake algal bloom and drinking water crisis in 2007. These events directly affect public health and social order in China [6].

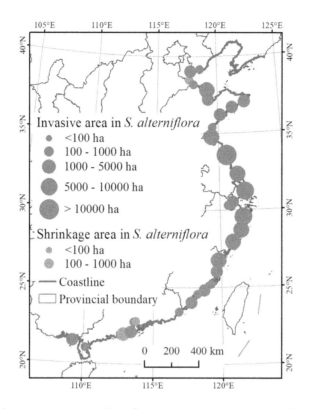

FIGURE 28.3 Spatial variance in *Spartina alterniflora* invasion over the coastal mainland China [11].

Policies for Wetlands Conservation in China

From the late 1970s, the ecological services of natural wetlands were universally recognized (Figure 28.4). China established its first national wetland nature reserve in 1978 and joined the Ramsar Convention on the Protection of Wetlands in 1992. Subsequently, some ecological projects and wetland inventories were developed to protect and restore natural wetlands in China. Two National Wetland Inventories were launched to quantify wetland areas and distributions. Natural wetland conservation was also promoted by legislation in different provinces. The Chinese government developed the China National Wetland Conservation Action Plan in 2000 and implemented the 2002–2030 NWCP in 2005.

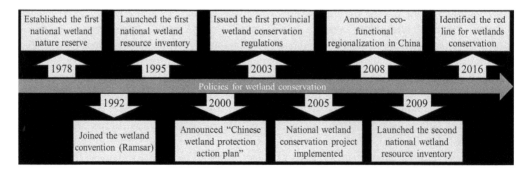

FIGURE 28.4 China's wetland conservation policies.

The NWCP had a set of ambitious goals, including establishing 713 wetland reserves and with more than 90% of natural wetlands effectively protected by 2030, restoring 1.4 million ha of natural wetlands, and building 53 national pilot zones for wetland protection and prudent use. To accomplish the NWCP, four subprograms were prioritized: the Wetland Reserve Construction Program (WRCP), the Degraded Wetland Rehabilitation Program (DWRP), the Sustainable Use Pilot Program (SUPP), and the Management Capacity Building Program (MCBP).

During the recent decade, China has developed an eco-functional regionalization plan, which delineates the country's land into regions with different dominant designated land uses for the purpose of optimizing the geospatial pattern of ecological security. Eleven of such regions are covered by extensive natural wetlands. The Sanjiang Plain, Liao River Delta, Huanghe River Delta, North Jiangsu Coastal Zone, Jianghan Plain, Dongting Lake, Poyang Lake, Wanjiang River Basin, Middle Reaches of Huaihe River Basin, and the Hongze Lake were planned as national key eco-functional regions. To protect specific wetland type, in May 2017, the State Forestry Administration and National Development and Reform Commission released a 10-year plan as the Planning of Establishing National Coastal Shelter Forests (2016–2025) [12]. In this plan, mangrove afforestation was listed as a key project with an aim to plant 48,650 ha mangrove forests in 62 counties. In addition, from 2017, the national park system has been established in China. As the important component of natural ecosystem, wetland is one of the main protection objects.

Conservation Efforts and Achievements

Great progresses have been made toward achieving the goals of NWCP. The Chinese State Council included wetland conservation as one of the key areas in the national Eleventh (2006–2010), Twelfth (2011–2015), and Thirteenth (2016–2020) Five-Year Plans. By the end of 2017, China has established 602 wetland protected areas and 1,699 wetland parks. Of these, 898 parks and more than 100 protected areas were designated at national level (www.shidi.org/). By May 2019, China has 57 sites designated as Wetlands of International Importance (i.e., Ramsar Sites), with a surface area of 6.9 million ha (The Ramsar Convention on Wetlands 2019: www.ramsar.org).

The Chinese government has spent on average 1.6 billion RMB (about $235 million US dollars) per year to implement the policy of Returning Farmland to Wetland (www.mof.gov.cn). The restoration of marshes and swamps from farmland was primarily implemented in Northeast China (Figure 28.5). Between 1990 and 2010, a total of 1,369 km² of marshes and swamps were recovered from farmland, of which 66.3% occurred in the second decade. Compared to the 461 km² marshes and swamps of restoration from farmland in the period 1990–2000, China made additional efforts to restore the two types of natural wetlands from farmland during 2000–2010 (908 km²). This suggests that policies and management measures be continually protecting and restoring natural wetlands in China [8].

Since the early 1990s, a number of regulations to protect coastal wetlands have been formulated, including the Action Plan for China Biodiversity Protection, the Forestry Action Plan for China's Agenda of the 21st Century (1995; 1996), the Plan for China Ecological Environment Conservation (1998), and the Action Plan for China Wetland Protection (2000) [13]. Ultimately, two very beneficial consequences were directly caused by these laws and regulations: One is the boom of establishing coastal wetland reserves, and the other is large-scale coastal wetland restoration. Mangrove forests are valuable coastal wetlands that are located at tropical and subtropical areas. According to a dataset of China's mangrove forest changes from 1973 to 2015, mangrove forests declined from 48,801 to 18,702 ha between 1973 and 2000, then partially recovered to 22,419 ha in 2015 [14]. Spatial distribution of China's mangroves in 2015 and areal changes at the provincial level are shown in Figure 28.6. China's mangrove forests have been continuously increasing since 2000, which indicates that China's mangrove forests conservation is effective.

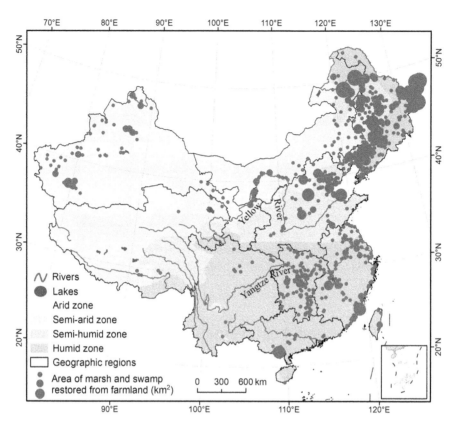

FIGURE 28.5   (See color insert.) Spatial pattern of marshes and swamps restored from farmland from 1990 to 2010 (bigger points denote larger areas of marshes and swamps restored from farmland within a county) [8].

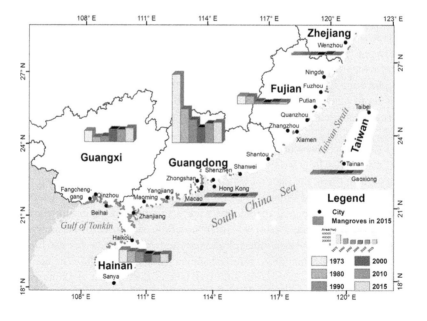

FIGURE 28.6   (See color insert.) Spatial pattern of mangrove forests in 2015 and areal extent changes between 1973 and 2015 [14].

Challenges and Recommendations

NWCP, however, has not been successful in all areas. Only 34% and 52% of the planned investments for the periods 2005–2010 and 2010–2015, respectively, were in place (People Daily Agency 2011, National Forestry Administration 2017) [15,16]. In addition, the achievements of NWCP in natural wetland conservation and restoration have been compromised by rapid urbanization, excessive land reclamation for polders, preference of constructed wetland parks to preservation of natural wetlands, and insufficient legislation to protect wetlands. China will be one of the most impacted regions in the world by future climate changes (IPCC 2007). Severe droughts like those in 2011 have already seriously threatened some wetland sites of international importance. More active mitigation efforts must be made to cope with climate change. Despite increasing threats of climate change to wetlands, specific mitigation and adaption policies are still lacking.

The greatest challenge is to find effective approaches to regulating behavior of and relationships between the various stakeholders (e.g., different levels of government, the industrial sector, the public, and even other countries), which often have conflicting objectives and expectations. To effectively conserve wetlands in China, one problem that must be addressed is the scattering of authority across different agencies. While the National Forestry and Grassland Administration was authorized to organize, coordinate, and supervise wetland conservation and utilization, it is not authorized for integrated wetland managements. At present, wetlands are jointly administrated by the Ministry of Natural Resources, the National Forestry and Grassland Administration, the Ministry of Ecology and Environment, the Ministry of Water Resources, and the Ministry of Agriculture and Rural Affairs. It will be formidable to efficiently implement wetland conservation plans without breaking down these bureaucratic barriers.

To execute integrated management and fulfill long-term goals for conservation, a national wetland protection law is most crucial and urgent at this time. China has established a number of environmental laws and policies covering wetlands, but each focuses only on a specific aspect of wetlands, often ignoring the interactive effects among different components of these ecosystems. For example, there are separate national laws and policies regarding the protection of forest, water, grassland, and oceanic ecosystems. Although the National Forestry and Grassland Administration and 26 provinces have issued their own regulations for wetland protection, it will be difficult to enforce them because of the lack of national-level laws.

The success of wetland conservation programs depends on the timing, duration, and efficacy of their implementation, as well as the complex interactions among existing and emerging forces (e.g., socioeconomic, political, demographic, and technological) [6]. Enhanced and diversified funding from various project beneficiaries is needed. Enforcement of existing laws and regulations and effective management needs to be greatly strengthened. Moreover, systematic planning and monitoring, integrative wetland protection at the watershed level as well as conservation education can help ensure the success of future endeavors. For NWCP to succeed, interdisciplinary research, technical innovation, policy and institutional design, and wetland conservation legislation and enforcement are all crucial. Finally, it is also important to change the way local people use wetlands and have them involved directly in the governance of these resources.

These are grand challenges, but there is hope. Based on the investigated results from the Second National Wetland Inventory, the National Forestry and Grassland Administration announced in 2016 that China must have no less than 53.3 million ha of wetlands by 2020. With the ideology of "Green hills and clear waters are gold & silver mountains," China is continually enhancing the wetlands conservation.

References

1. MA (Assessment, Millennium Ecosystem), 2005. *Ecosystems and Human Well-Being: Wetlands and Water.* World resources institute, Washington, DC, 5.
2. Balmford, A., Bruner, A., Copper, P., Costanza, R., Farber, S., Green, R. E., Jenkins, M., Jefferiss, P., Jessamy, V., Madden, J., Munro, K., Myers, N., Naeem, S., Paavola, J., Rayment, M., Rosendo, S., Roughgarden, J., Trumper, K., Kerry-Tuner, R., 2002. Economic reasons for conserving wild nature. *Science* 297: 950–953.

3. OECD/IUCN, 1996. *Guidelines for Aid Agencies for Improved Conservation and Sustainable Use of Tropical and Sub-Tropical Wetlands.* OECD, Paris, France.

4. Gong, P., Niu, Z., Cheng, X., Zhao, K., Zhou, D., Guo, J., Liang, L., Wang, X., Li, D., Huang, H., Wang, Y., Wang, K., Li, W., Wang, X., Ying, Q., Yang, Z., Ye, Y., Li, Z., Zhuang, D., Chi, Y., Zhou, H., Yan, J., 2010. China's wetland change (1990–2000) determined by remote sensing. *Science China Earth Sciences* 53(1): 1–7.

5. An, S., Li, H., Guan, B., Zhou, C., Wang, Z., Deng, Z., Zhi, Y., Liu, Y., Xu, C., Fang, S., Jiang, J., Li, H., 2007. China's natural wetlands: Past problems, current status, and future challenges. *AMBIO* 36: 335–342.

6. Wang, Z., Wu, J., Madden, M., Mao, D., 2012. China's wetlands: Conservation plans and policy impacts. *AMBIO* 41: 782–786.

7. State Forestry Administration (SFA). 2006. *Implementation Plan for National Wetland Conservation Program.* State Forestry Administration (in Chinese): Beijing, China.

8. Mao, D., Luo, L., Wang, Z., Wilson, M.C., Zeng, Y., Wu, B., Wu, J., 2018a. Conversions between natural wetlands and farmland in China: A multiscale geospatial analysis. *Science of the Total Environment* 634: 550–560.

9. Mao, D., Wang, Z., Wu, J., Wu, B., Zeng, Y., Song, K., Yi, K., Luo, L., 2018b. China's wetlands loss to urban expansion. *Land Degradation & Development* 29: 2644–2657.

10. Liu, M., Mao, D., Wang, Z., Li, L., Man, W., Jia, M., Ren, C., Zhang, Y., 2018. Rapid invasion of *Spartina alterniflora* in the coastal zone of mainland China: New observations from Landsat OLI images. *Remote Sensing* 10: 1933. doi:10.3390/rs10121933.

11. Mao, D., Liu, M., Wang, Z., Li, L., Man, W., Jia, M., Zhang, Y., 2019. Rapid invasion of *Spartina alterniflora* in the coastal zone of mainland China: Spatiotemporal patterns and human prevention. *Sensors* 19: 2308. doi:10.3390/s19102308.

12. State Forestry Administration (SFA). 2014. Results of the Second National Wetland Resource Inventory (2009–2013), www.forestry.gov.cn/, (in Chinese).

13. Jia, M., Wang, Z., Zhang, Y., Ren, C., Song, K., 2015. Landsat-based estimation of mangrove forest loss and restoration in Guangxi province, China, influenced by human and natural factors. *IEEE Journal of Selected Topics in Applied Earth Observations and Remote Sensin* 8(1): 311–323

14. Jia, M., Wang, Z., Zhang, Y., Mao, D., Wang, C., 2018. Monitoring loss and recovery of mangrove forests during 42 years: The achievements of mangrove conservation in China. *International Journal of Applied Earth Observations and Geoinformation* 73: 535–545.

15. National Forestry Administration. 2017. National Wetland Protection Implementation Plan for the 13th Five-Year Plan. www.forestry.gov.cn/uploadfile/main/2017-4/file/2017-4-19-e3e8e6c738d6 4e10a36a5cd57b054d31.pdf (in Chinese).

16. People Daily Agency. 2011. Loss of Wetlands. http://env.people.com.cn/ GB/211746/213647/13850391.html (in Chinese).

29

Water Dynamics of Floodpath Lakes and Wetlands: Remote Sensing

Yeqiao Wang
University of Rhode Island

Introduction

Water exists in various states on the Earth: freshwater, saltwater, water vapor, rain, snow, and ice. It is common to obtain *in situ* measurements of various hydrologic (water) parameters such as precipitation, water depth, temperature, salinity, velocity, and volume, at very specific locations on the Earth. It is always challenging to obtain spatial variation using *in situ* observation for a number of the most important hydrologic variables, such as water surface area (streams, rivers, ponds, lakes, reservoirs, and wetlands), water constituents (organic and inorganic), water surface temperature, snow surface area and snow water equivalent, ice surface area and ice water equivalent, cloud cover, precipitation, and water vapor. Remote sensing has the advantages of obtaining such types of information.

For monitoring the surface extent of water bodies, the best wavelength region for discriminating land from water is the near- and mid-infrared (IR) (0.74–2.50 μm). In the near- and mid-IR regions, water bodies absorb almost all of the incident radiant fluxes, especially when water is deep and contains little suspended sediment or organic matter. Care must be exercised when there are organic and inorganic constituents in the water column, especially those near the surface, because those materials can cause near-IR surface reflection.

Minerals such as silicon, aluminum, and iron oxides are found in suspension in most natural water bodies. Sediments come from a variety of sources, including agriculture erosion, weathering of mountainous terrain, shore erosion caused by waves or boat traffic, and volcanic eruptions. Most suspended mineral sediment is concentrated in inland and near-shore water bodies.

Monitoring of terrestrial aquatic ecosystem using space and airborne remote sensing has long been in attentions of the world [1–7]. Timely monitoring of seasonal water-level dynamics and the effects on lake and wetland environments is necessary for understanding flooding hazards, sediment transport, nutrient exchange, and habitat conditions [8–11]. Affected by the seasonal variations of water level and surface area, lakes and wetlands are extremely sensitive to natural and anthropogenic impacts, such as climate change, human-induced intervention on hydrological regimes, and land-use and land-cover change [12–16].

Water Extent Monitoring

Remote sensing from optical sensors (e.g., Landsat, SPOT, MODIS, MERIS, AVHRR, and Sentinels) and active sensors (e.g., ERS, ENVISAT, J-ERS, PALSAR RADARSAT, COSMO SKYMED, and the TerraSAR-X) has been proven effective in monitoring the change of flooding and inundation conditions from global to local environments [17–20]. Technology improvements provide multiple capacities: Satellite altimetry, synthetic aperture radar (SAR), and interferometric SAR (InSAR), hyperspectral, and high spatial resolution remote sensing have been applied in monitoring water extents, heights, and flows [21–22]. Rapid developments of remote sensing capacities have expanded their applications into ecological, hydrological, geomorphological, and societal interests in inundated situations. Sentinel satellites have demonstrated enhanced capacities in the monitoring of changing water and wetland environments, in recent years [23–27].

Water Quality Monitoring

The spectral reflectance of suspended sediment in surface waters is a function of both the quantity and characteristics, for example, particle size and absorption of the materials in water. Phytoplankton in water bodies contains the photosynthetically active pigment chlorophyll *a* (chl-*a*), which changes water's spectral reflectance characteristics. The scattering/absorption features of chlorophyll *a* are evident in the algae-laden water. The spectral patterns include strong chl-*a* absorptions of blue light between 0.40 and 0.50 μm and red light at approximately 0.675 μm. The spectral patterns also include the reflectance maximum around 0.55 μm (green peak) caused by relatively lower absorption of green light by algae, and prominent reflectance peak around 0.69–0.70 μm caused by an interaction of algae-cell scattering and minimum combined effect of pigment and water absorption. The height of this peak above the baseline can be used to accurately measure chlorophyll amount.

Remote Sensing Applications

Remote sensing applications have been widely reported in lake and wetland mappings, in particular, for representative large lakes and associated wetlands. For example, the Poyang and Dongting lakes in the middle and lower reaches of the Yangtze River are among the representative floodpath lakes with dramatic spatial and temporal variation patterns in water surface areas and the associated wetlands (Figure 29.1). The lakes play a crucial role in the accommodation of floodwater from its tributaries and the Yangtze, as well as for the regulation of sedimentation in the lower reach of the Yangtze [16]. Poyang Lake wetland is recognized to be among the most important wetlands of the world for its extraordinary biodiversity and conservation value. Different types of remote sensing data have been applied to reveal the spatial and temporal patterns of water extents and levels and the responses of vegetation and habitats, as well as the effects of sand dragging, sedimentation, and the contamination of this unique floodpath lake and wetland combination [28–30].

Remote sensing has been effectively applied in the monitoring of waters and wetlands around the world, for example, in the Amazon Basin [31], in the African Great Lakes [32], in the greater Everglades [33], in coastal Louisiana [34], in Alpine lakes on the Tibetan Plateau [35], in tropical lakes [36], in river deltas [37,38], in Lake Baikal [39,40], and in large lakes in Europe and China [41–43]. Remote sensing of floodpath lakes and wetlands has become routine in scientific research and in management practices, in flood and flood-prone areas [44]. Challenges remain to be addressed for the advancement of science, technologies, and management practices.

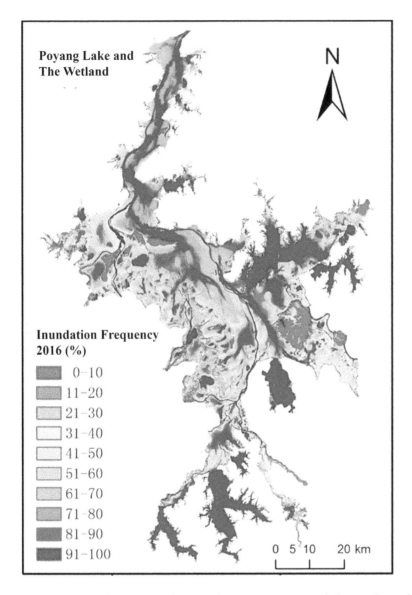

Poyang Lake and The Wetland

Inundation Frequency 2016 (%)

- 0–10
- 11–20
- 21–30
- 31–40
- 41–50
- 51–60
- 61–70
- 71–80
- 81–90
- 91–100

0 5 10 20 km

FIGURE 29.1 (**See color insert.**) Variation of 2016 surface water coverage and the inundation frequency of associated wetland in Poyang Lake, China, as derived from GF-1 satellite wide-field view (WFV) sensor.

Challenges in Remote Sensing of Water Dynamics

A growing scarcity of freshwater resources and their security, under the changing environment, has long been recognized as a primary challenge [45,46]. For remote sensing, the challenges mostly exist in obtaining time-sensitive data and generating precise and accurate information [47,48]. Improvements of capacities in data acquisition and methodologies, in information extraction, are deemed necessary. Also, there still exists the challenges and uncertainties, in data continuity and availability, which might be introduced by potential governmental action [49].

Water quality monitoring is challenging because of complexities of water environment, in connection to their contributing tributaries within the watersheds. Remote sensing of inland waters has faced challenges in the retrieval of physical and biogeochemical properties [50–52]. In particular, the optical complexity of inland waters is typically characterized by high concentrations of phytoplankton biomass, chl-*a*, and turbidities with suspended materials that typically do not covary over space and time. The development and validation of atmospheric and in-water models for optically complex waters can only be properly advanced through rigorous testing and refinement of candidate algorithms, across the full spectrum of optical water types. Comprehensive validations in remote sensing of water quality are much needed through collaborative team studies. For example, approaches in integration of remote sensing and field-based monitoring have been constantly explored for improving the understanding of spatial and temporal patterns of water constituents [53,54], such as chl-*a* concentration, colored dissolved organic matter (CDOM), dissolved organic carbon (DOC), total suspended materials (TSM), intensity of the sedimentation, and effects of the human-induced hydrologic engineering projects on the water qualities for both inland and coastal waters.

Floodpath wetlands provide unique ecosystem services that are invaluable to the well-being of life on the planet Earth. Degradation and loss of wetlands are caused by land conversion, water eutrophication and pollution, and the introduction of invasive alien species, and are indirectly fostered by economic development and population growth, as well. Yangtze reaches and the connected lakes, such as Poyang and Dongting lakes, are the most important wintering zones of waterfowl in East Asia, hosting significant proportions of the populations of cranes, geese, and swans. The Yangtze floodplain holds the highest diversities of Anatidae in the world. Challenges for monitoring of floodpath wetlands are associated with dynamics of water levels, extents, and quality. In particular, monitoring and explaining the spatiotemporal changes in wetland biodiversity caused by biotic and abiotic factors remain to be precisely mapped and understood. The timing of exposure of recessional vegetation leads to changes in the landscape composition and configuration patterns, which in turn affect the biodiversity, abundance, and habitats of the migratory wildfowl. Species that depend on seasonal submerged aquatic vegetation for their migratory foraging requirements have experienced challenges due to the impacts on variations of water level and extents. These impacts, in particular, include alterations of hydrological patterns from human activities, such as sand dredging and engineering measures of water control. Improved remote sensing capabilities in data acquisition and processing facilitate precision and accuracy of change detection, for example, in monitoring of specific vegetation dynamics and their influence on sensitive wetland areas in floodpath environments [55].

Real-time remote monitoring of water environments demonstrated the advancement of sensor networks for automated *in situ* data acquisition, wireless data transmitting, and information extraction [56,57]. Remote sensing practices need support from *in situ* data observed from field-based monitoring. Integration of automated *in situ* data acquisition, with space-borne and airborne remote sensing data, can improve the validation requirements in dynamic water environments from GPS-guided field survey, GPS-based wildlife telemetry, and time-series field-based observations.

Unmanned aerial vehicles (UAVs) extend the capacities and potentials for *in situ* water quality measurement and integration with remote sensing, for improved spatial resolution, flexibility, and frequency in data acquisition [58,59]. Seamless integration of UAV data, with remote sensing and field-based observations, presents technical challenges to streamline the process.

References

1. Ozesmi, S.L.; Bauer, M.E. Satellite remote sensing of wetlands. *Wetl. Ecol. Manag.* **2002**, *10*, 381–402.
2. Alsdorf, D.E.; Melack, J.M.; Dunne, T.; Mertes, L.A.; Hess, L.L.; Smith, L.C. Interferometric radar measurements of water level changes on the Amazon flood plain. *Nature* **2000**, *404*, 174.
3. Sippel, S.; Hamilton, S.; Melack, J. Inundation area and morphometry of lakes on the amazon river floodplain, brazil. *Arch. Fur. Hydrobiol. Stuttg.* **1992**, *123*, 385–400.
4. Birkett, C. Synergistic remote sensing of lake chad: Variability of basin inundation. *Remote Sens. Environ.* **2000**, *72*, 218–236.

5. Fluet-Chouinard, E.; Lehner, B.; Rebelo, L.-M.; Papa, F.; Hamilton, S.K. Development of a global inundation map at high spatial resolution from topographic downscaling of coarse-scale remote sensing data. *Remote Sens. Environ.* **2015**, *158*, 348–361.

6. Berry, P.; Garlick, J.; Freeman, J.; Mathers, E. Global inland water monitoring from multi-mission altimetry. *Geophys. Res. Lett.* **2005**, *32*. doi:10.1029/2005GL022814.

7. Jarihani, A.A.; Callow, J.N.; Johansen, K.; Gouweleeuw, B. Evaluation of multiple satellite altimetry data for studying inland water bodies and river floods. *J. Hydrol.* **2013**, *505*, 78–90.

8. Smith, L.C. Satellite remote sensing of river inundation area, stage, and discharge: A review. *Hydrol. Process.* **1997**, *11*, 1427–1439.

9. Dörnhöfer, K.; Oppelt, N. Remote sensing for lake research and monitoring–recent advances. *Ecol. Indic.* **2016**, *64*, 105–122.

10. Da Silva, J.S.; Seyler, F.; Calmant, S.; Rotunno Filho, O.C.; Roux, E.; Araújo, A.A.M.; Guyot, J.L. Water level dynamics of amazon wetlands at the watershed scale by satellite altimetry. *Int. J. Remote Sens.* **2012**, *33*, 3323–3353.

11. Phan, V.H.; Lindenbergh, R.; Menenti, M. Icesat derived elevation changes of Tibetan lakes between 2003 and 2009. *Int. J. Appl. Earth Obs. Geoinf.* **2012**, *17*, 12–22.

12. Feng, L.; Han, X.; Hu, C.; Chen, X. Four decades of wetland changes of the largest freshwater lake in China: Possible linkage to the three gorges dam? *Remote Sens. Environ.* **2016**, *176*, 43–55.

13. Wang, J.; Sheng, Y.; Tong, T.S.D. Monitoring decadal lake dynamics across the Yangtze basin downstream of three gorges dam. *Remote Sens. Environ.* **2014**, *152*, 251–269.

14. Campbell, A.; Wang, Y.; Christiano, M.; Stevens, S. Salt marsh monitoring in Jamaica bay, New York from 2003 to 2013: A decade of change from restoration to hurricane sandy. *Remote Sens.* **2017**, *9*, 131.

15. Song, C.; Huang, B.; Ke, L.; Richards, K.S. Remote sensing of alpine lake water environment changes on the Tibetan plateau and surroundings: A review. *ISPRS J. Photogramm. Remote Sens.* **2014**, *92*, 26–37.

16. Hervé, Y.; Claire, H.; Xijun, L.; Stephane, A.; Jiren, L.; Sylviane, D.; Muriel, B.N.; Xiaoling, C.; Shifeng, H.; Burnham, J.; Jean-François, C. Nine years of water resources monitoring over the middle reaches of the Yangtze river, with Envisat, Modis, Beijing-1 time series, altimetric data and field measurements. *Lakes Reserv. Res. Manag.* **2011**, *16*, 231–247.

17. Kim, J.-W.; Lu, Z.; Lee, H.; Shum, C.; Swarzenski, C.M.; Doyle, T.W.; Baek, S.-H. Integrated analysis of Palsar/Radarsat-1 InSAR and Envisat altimeter data for mapping of absolute water level changes in Louisiana wetlands. *Remote Sens. Environ.* **2009**, *113*, 2356–2365.

18. Crowley, J.W.; Mitrovica, J.X.; Bailey, R.C.; Tamisiea, M.E.; Davis, J.L. Land water storage within the Congo basin inferred from GRACE satellite gravity data. *Geophys. Res. Lett.* **2006**, *33*. doi:10.1029/2006GL027070.

19. Kasischke, E.S.; Smith, K.B.; Bourgeau-Chavez, L.L.; Romanowicz, E.A.; Brunzell, S.; Richardson, C.J. Effects of seasonal hydrologic patterns in South Florida wetlands on radar backscatter measured from ERS-2 SAR imagery. *Remote Sens. Environ.* **2003**, *88*, 423–441.

20. Shen, G.; Liao, J.; Guo, H.; Liu, J. Poyang Lake wetland vegetation biomass inversion using Polarimetric Radarsat-2 synthetic aperture radar data. *J. Appl. Remote Sens.* **2015**, *9*, 096077.

21. Lu, Z.; Kwoun, O.; Rykhus, R. Interferometric synthetic aperture radar (InSAR): Its past, present and future. *Photogramm. Eng. Remote Sens.* **2007**, *73*, 217.

22. Turpie, K.R.; Klemas, V.V.; Byrd, K.; Kelly, M.; Jo, Y.-H. Prospective Hyspiri global observations of tidal wetlands. *Remote Sens. Environ.* **2015**, *167*, 206–217.

23. Zeng, L.; Schmitt, M.; Li, L.; Zhu, X.X. Analysing changes of the Poyang Lake water area using Sentinel-1 synthetic aperture radar imagery. *Int. J. Remote Sens.* **2017**, *38*, 7041–7069.

24. Tian, H.; Wu, M.; Wang, L.; Niu, Z. Mapping early, middle and late rice extent using Sentinel-1A and landsat-8 data in the Poyang Lake plain, China. *Sensors* **2018**, *18*, 185.

25. Drusch, M.; Del Bello, U.; Carlier, S.; Colin, O.; Fernandez, V.; Gascon, F.; Hoersch, B.; Isola, C.; Laberinti, P.; Martimort, P. Sentinel-2: ESA's optical high-resolution mission for gmes operational services. *Remote Sens. Environ.* **2012**, *120*, 25–36.

26. Toming, K.; Kutser, T.; Uiboupin, R.; Arikas, A.; Vahter, K.; Paavel, B. Mapping water quality parameters with sentinel-3 ocean and land colour instrument imagery in the Baltic sea. *Remote Sens.* **2017**, *9*, 1070.

27. Shen, M.; Duan, H.; Cao, Z.; Xue, K.; Loiselle, S.; Yesou, H. Determination of the downwelling diffuse attenuation coefficient of lake water with the sentinel-3A OLCI. *Remote Sens.* **2017**, *9*, 1246.

28. Han, X.; Chen, X.; Feng, L. Four decades of winter wetland changes in Poyang Lake based on landsat observations between 1973 and 2013. *Remote Sens. Environ.* **2015**, *156*, 426–437.

29. Mei, X.; Dai, Z.; Fagherazzi, S.; Chen, J. Dramatic variations in emergent wetland area in China's largest freshwater lake, Poyang lake. *Adv. Water Resour.* **2016**, *96*, 1–10.

30. Ding, X.; Li, X. Monitoring of the water-area variations of lake Dongting in China with envisat asar images. *Int. J. Appl. Earth Obs. Geoinf.* **2011**, *13*, 894–901.

31. Hess, L.L.; Melack, J.M.; Novo, E.M.; Barbosa, C.C.; Gastil, M. Dual-season mapping of wetland inundation and vegetation for the central Amazon Basin. *Remote Sens. Environ.* **2003**, *87*, 404–428.

32. Ramillien, G.; Frappart, F.; Seoane, L. Application of the regional water mass variations from grace satellite gravimetry to large-scale water management in Africa. *Remote Sens.* **2014**, *6*, 7379–7405.

33. Hong, S.-H.; Wdowinski, S.; Kim, S.-W.; Won, J.-S. Multi-temporal monitoring of wetland water levels in the Florida Everglades using interferometric synthetic aperture radar (InSAR). *Remote Sens. Environ.* **2010**, *114*, 2436–2447.

34. Lu, Z.; Kwoun, O.-I. Radarsat-1 and ers insar analysis over southeastern coastal Louisiana: Implications for mapping water-level changes beneath swamp forests. *IEEE Trans. Geosci. Remote Sens.* **2008**, *46*, 2167–2184.

35. Jiang, L.; Nielsen, K.; Andersen, O.B.; Bauer-Gottwein, P. Monitoring recent lake level variations on the Tibetan plateau using Cryosat-2 sarin mode data. *J. Hydrol.* **2017**, *544*, 109–124.

36. Ricko, M.; Carton, J.A.; Birkett, C. Climatic effects on lake basins. Part I: Modeling tropical lake levels. *J. Clim.* **2011**, *24*, 2983–2999.

37. Kuenzer, C.; Klein, I.; Ullmann, T.; Georgiou, E.F.; Baumhauer, R.; Dech, S. Remote sensing of river delta inundation: Exploiting the potential of coarse spatial resolution, temporally-dense MODIS time series. *Remote Sens.* **2015**, *7*, 8516–8542.

38. Kuenzer, C.; Guo, H.; Huth, J.; Leinenkugel, P.; Li, X.; Dech, S. Flood mapping and flood dynamics of the Mekong delta: Envisat-ASAR-WSM based time series analyses. *Remote Sens.* **2013**, *5*, 687–715.

39. Bolgrien, D.W.; Granin, N.G.; Levin, L. Surface temperature dynamics of lake baikal observed from AVHRR images. *Photogramm. Eng. Remote Sens.* **1995**, *61*, 211–216.

40. Kouraev, A.V.; Semovski, S.V.; Shimaraev, M.N.; Mognard, N.M.; Légresy, B.; Remy, F. Observations of lake Baikal ice from satellite altimetry and radiometry. *Remote Sens. Environ.* **2007**, *108*, 240–253.

41. Mleczko, M.; Mróz, M. Wetland mapping using SAR data from the Sentinel-1A and tandem-x missions: A comparative study in the biebrza floodplain (Poland). *Remote Sens.* **2018**, *10*, 78.

42. Wang, X.; Gong, P.; Zhao, Y.; Xu, Y.; Cheng, X.; Niu, Z.; Luo, Z.; Huang, H.; Sun, F.; Li, X. Water-level changes in China's large lakes determined from ICESAT/GLAS data. *Remote Sens. Environ.* **2013**, *132*, 131–144.

43. Luo, J.; Li, X.; Ma, R.; Li, F.; Duan, H.; Hu, W.; Qin, B.; Huang, W. Applying remote sensing techniques to monitoring seasonal and interannual changes of aquatic vegetation in Taihu lake, China. *Ecol. Indic.* **2016**, *60*, 503–513.

44. Klemas, V. Remote sensing of floods and flood-prone areas: An overview. *J. Coast. Res.* **2014**, *31*, 1005–1013.

45. Postel, S.L. Entering an era of water scarcity: The challenges ahead. *Ecol. Appl.* **2000**, *10*, 941–948.

46. Williamson, C.E.; Saros, J.E.; Vincent, W.F.; Smol, J.P. Lakes and reservoirs as sentinels, integrators, and regulators of climate change. *Limnol. Oceanogr.* **2009**, *54*, 2273–2282.

47. Palmer, S.C.; Kutser, T.; Hunter, P.D. Remote sensing of inland waters: Challenges, progress and future directions. *Remote Sens. Environ.* **2015**, *157*, 1–8.

48. Mouw, C.B.; Greb, S.; Aurin, D.; DiGiacomo, P.M.; Lee, Z.; Twardowski, M.; Binding, C.; Hu, C.; Ma, R.; Moore, T. Aquatic color radiometry remote sensing of coastal and inland waters: Challenges and recommendations for future satellite missions. *Remote Sens. Environ.* **2015**, *160*, 15–30.

49. Popkin, G. Us government considers charging for popular earth-observing data. *Nature* **2018**, *556*, 417–418.

50. Kallio, K.; Kutser, T.; Hannonen, T.; Koponen, S.; Pulliainen, J.; Vepsäläinen, J.; Pyhälahti, T. Retrieval of water quality from airborne imaging spectrometry of various lake types in different seasons. *Sci. Total Environ.* **2001**, *268*, 59–77.

51. Kutser, T.; Pierson, D.C.; Kallio, K.Y.; Reinart, A.; Sobek, S. Mapping lake CDOM by satellite remote sensing. *Remote Sens. Environ.* **2005**, *94*, 535–540.

52. Brezonik, P.L.; Olmanson, L.G.; Finlay, J.C.; Bauer, M.E. Factors affecting the measurement of CDOM by remote sensing of optically complex inland waters. *Remote Sens. Environ.* **2015**, *157*, 199–215.

53. Bhatti, A.M.; Rundquist, D.; Schalles, J.; Ramirez, L.; Nasu, S. A comparison between above-water surface and subsurface spectral reflectances collected over inland waters. *Geocarto Int.* **2009**, *24*, 133–141.

54. Busch, J.A.; Hedley, J.D.; Zielinski, O. Correction of hyperspectral reflectance measurements for surface objects and direct sun reflection on surface waters. *Int. J. Remote Sens.* **2013**, *34*, 6651–6667.

55. Wang, Y.; Hervé, Y. Remote sensing of floodpath lakes and wetlands: A challenging frontier in the monitoring of changing environments. *Remote Sens.* **2018**, *10*, 1955.

56. Glasgow, H.B.; Burkholder, J.M.; Reed, R.E.; Lewitus, A.J.; Kleinman, J.E. Real-time remote monitoring of water quality: A review of current applications, and advancements in sensor, telemetry, and computing technologies. *J. Exp. Mar. Boil. Ecol.* **2004**, *300*, 409–448.

57. Li, X.; Cheng, X.; Gong, P.; Yan, K. Design and implementation of a wireless sensor network-based remote water-level monitoring system. *Sensors* **2011**, *11*, 1706–1720

58. Pajares, G. Overview and current status of remote sensing applications based on unmanned aerial vehicles (UAVs). *Photogramm. Eng. Remote Sens.* **2015**, *81*, 281–330.

59. Koparan, C.; Koc, A.B.; Privette, C.V.; Sawyer, C.B. In situ water quality measurements using an unmanned aerial vehicle (UAV) system. *Water* **2018**, *10*, 264.

30

Water-Level Variation of Poyang Lake: Remote Sensing

Shuhua Qi
Key Laboratory of Poyang Lake Wetland and Watershed Research (Jiangxi Normal University)

Yeqiao Wang
University of Rhode Island

Introduction

Remote sensing is extremely effective in monitoring dynamics of water surface areas [1]. Remote sensing has been applied in the Amazon Basin [2–4], in the African Great Lakes [5–12], in the greater Everglades [13–15], in coastal Louisiana [16–20], in Alpine lakes on the Tibetan Plateau [21–26], in river deltas [27,28], and in large lakes in Europe and China [29–34]. Landsat images have been used to monitor water environments [35,36]. Moderate Resolution Imaging Spectroradiometer (MODIS) data have been used to study short- and long-term characteristics of Poyang Lake inundation [37,38]. Active sensors, such as synthetic aperture radar (SAR) and interferometry SAR (InSAR), have been applied in water-level and wetland mapping [39–43]. Monitoring results provide critical information for water management, in particular for decision support in preparation for extreme harmful hydrological disasters.

Remote Sensing of Poyang Lake

Poyang Lake is situated at the lower Yangtze River Basin, and it is the largest freshwater lake in China. Poyang Lake is fed by tributaries and is connected and exchanges water with the Yangtze River through the lake mouth in the north (Figure 30.1). As controlled by water from five major tributary rivers as well as the Yangtze River, seasonal variations in water level present a unique landscape of freshwater lake-wetland ecosystem. Poyang Lake wetland is a key habitat site for wintering migratory birds with global importance. The lake plays an irreplaceable role in flood control, river shipping, city water supply, and conservation of biodiversity in the middle and lower reaches of the Yangtze River [44].

Poyang Lake is affected by subtropical monsoon climate with a mean annual precipitation of 1,632 mm. About 60% of the annual rainfall happened in flood season from April to August. It was estimated that approximately 1.43×10^7 tons of sediments with nutritive materials were carried from

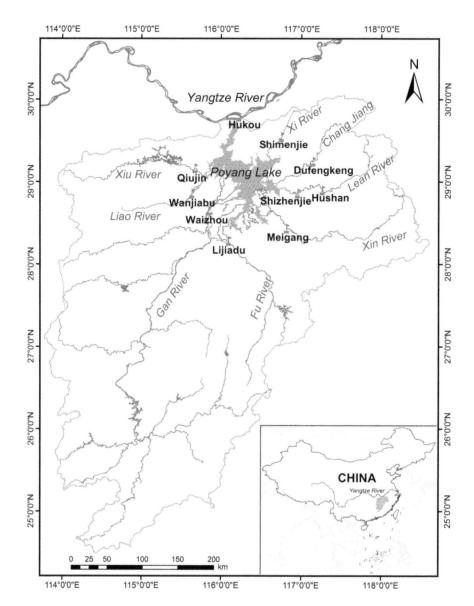

FIGURE 30.1 (See color insert.) Location of Poyang Lake, the tributaries, and hydrological stations as marked (🔺).

tributary rivers and deposit in the floodplain each year, which formed fertile deltas. There are about 102 vegetation species of aquatic vascular plants and freshwater organisms in the floodplain. Poyang Lake wetland was first selected as the protected area under the international Ramsar Convention in China because of its biological productivity and species richness. Poyang Lake is one of the most important wintering zones of waterfowl in East Asia, hosting significant proportions of the populations of cranes, geese, and swans, including rare and endangered species such as the Siberian crane (*Grus leucogeranus*) (Figure 30.2).

The area and shape of Poyang Lake were affected by natural deposition and erosion in the past decades. Increased human population and economic growth induced activities such as sand mining,

FIGURE 30.2 **(See color insert.)** Poyang Lake is one of the most important wintering zones of waterfowl in East Asia, hosting significant proportions of the populations of cranes, geese, and swans, including rare and endangered species such as the Siberian crane (*Grus leucogeranus*). (Photos: Yeqiao Wang.)

reclamation for agriculture, fishery, aquaculture, and settlements [45,46], which also affected areas and surrounding landscape. It is estimated that the inundated area of Poyang Lake in flood season was reduced from 5,160 km² in 1954 to 3,860 km² in 1998 [47]. Reclaiming farmland was the most significant activity changing the morphology of Poyang Lake before 1998. Sand dredging in the Poyang Lake water system was intensified since 2001 because of the demand of raw building materials with rapid urbanization in lower Yangtze River Valley, and sand dredging was banned in the Yangtze River in 2000. Because of high profits, sand dredging business developed quickly with hundreds of large sand vessels assembled and operated in the Poyang Lake water system. Poyang Lake has attracted wide attention of the international and scientific communities [48–51].

Mapping of Inundated Area

In this study, time series MODIS, Sentinel-1 C band SAR, and Landsat-8 multispectral Operational Land Imager (OLI) imagery data were employed to map inundation areas. The 8-day MODIS surface reflectance data (MOD09Q1) collected between 2001 and 2018 were obtained from an open source (https://ladsweb.nascom.nasa.gov/data.html). The MOD09Q1 includes 46 scenes of data product each year; that is, the images from every 8 days were used to composite one scene. A total of 828 scenes of MOD09Q1 images were acquired between 2001 and 2018. MOD09Q1 contains three data layers: surface reflectance for band 1 (620–670 nm), surface reflectance for band 2 (841–876 nm), and surface reflectance quality control flags, all with 250 m spatial resolution. In addition, five cloud-free multispectral images acquired by Landsat-8 OLI (www.usgs.gov/land-resources/nli/landsat) and 31 Sentinel-1 SAR images (https://sentinel.esa.int/web/sentinel/user-guides/sentinel-1-sar/) were downloaded for analyzing the inundation dynamics in Poyang Lake in 2018. All collected MOD09Q1, Landsat-8/OLI, and Sentinel-1 SAR images were resampled using the nearest neighbor method and geometrically rectified to WGS84 datum with Universal Transverse Mercator (UTM) coordinate system. All the images were clipped by the boundary of the historical largest water area of the Poyang Lake using mask calculation. The water surface areas in different years were extracted from other features using the normalized difference vegetation index (NDVI) threshold with MODIS and Landsat images:

$$NDVI = \frac{(NIR - VIS)}{(NIR + VIS)} \tag{30.1}$$

where VIS and NIR stand for the spectral reflectance measurements acquired in the visible (red) and near-infrared regions, respectively. Normally, the value of NDVI for water is less than 0. However, the existence of large amount of aquatic vegetation in Poyang Lake affects the absorption, reflection, and transmission of visible and near-infrared spectrum on water surface.

A modified NDVI threshold of less than 0.1 was applied to extract water surface areas. For those MOD09Q1 images that were affected by cloud covers, images acquired in similar dates were applied instead. At last, the extracted water surface images were added together for each year to obtain the inundation variations in Poyang Lake from 2001 to 2018 (Figures 30.3 and 30.4). For Sentitnel-1 SAR images with interferometric wide (IW) swath imaging mode and VV polarization, the digital number-17 was used as the segment threshold to extract the water surface areas.

According to the extents of inundation areas, the lake water area at the maximum flooding time was about 3,400 km², while the minimum inundate area was only about 320 km². The largest annual variability ratio between the maximum and minimum water surface areas was 7.85, which occurred in 2017. Together with water-level records measured at a gauging station on the lake, a strong correlation existed between inundation areas and water levels. A decreased trend is evident between maximum and minimum water levels since 2001 (Table 30.1).

Figures 30.5 and 30.6 reveal monthly inundation area and spatial pattern of water variation in 2018 by Sentinel-1 SAR imageries.

Water-level variation of Poyang Lake governs the dynamics of associated wetlands. NDVI provides information and measurements about vegetation dynamics and biomass accumulation corresponding

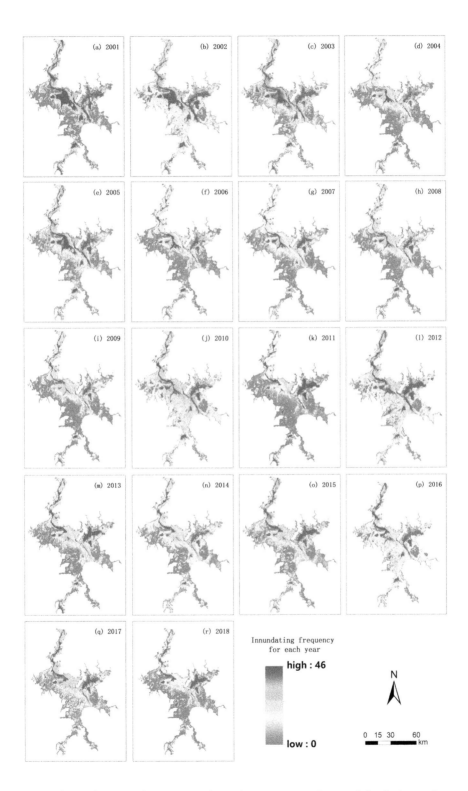

FIGURE 30.3 **(See color insert.)** Variation of inundation areas in Poyang Lake during each year from 2001 to 2018.

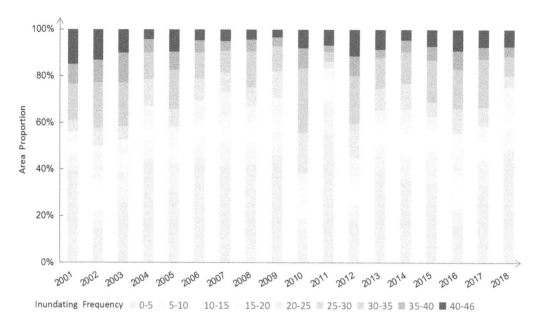

Inundating Frequency   0-5   5-10   10-15   15-20   20-25   25-30   30-35   35-40   40-46

FIGURE 30.4   (See color insert.) Inundation area of Poyang Lake at different lake water levels from 2001 to 2018.

TABLE 30.1   Annual Maximum and Minimum Inundation Area of Poyang Lake from 2001 to 2018

Year	Annual Max. Inundation Area (km²)	Annual Min. Inundation Area (km²)	Annual Max. Water Level (m)	Annual Min. Water Level (m)
2001	2,650.50	1,183.38	15.14	7.67
2002	3,396.88	941.94	18.36	6.51
2003	3,241.31	799.38	17.49	6.07
2004	2,526.44	614.75	15.53	5.23
2005	3,104.50	1,045.38	17.15	6.37
2006	2,915.75	617.69	14.82	5.91
2007	3,045.63	623.81	16.59	5.40
2008	2,399.44	560.94	15.78	5.48
2009	2,137.13	472.25	15.27	5.60
2010	2,881.31	526.25	18.38	5.85
2011	2,457.69	512.19	15.50	6.22
2012	2,905.13	556.69	17.75	5.90
2013	2,280.31	585	15.06	5.54
2014	3,040.56	616.81	16.70	5.41
2015	2,492.5	320.5	15.53	5.62
2016	2,729.88	427.94	17.10	7.75
2017	3,034.11	386.69	18.88	6.46
2018	2,498.25	472.74	13.93	8.94

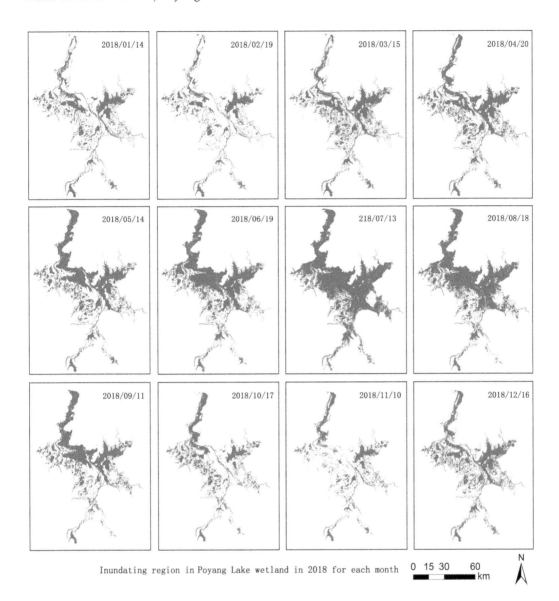

Inundating region in Poyang Lake wetland in 2018 for each month

FIGURE 30.5 Monthly inundation area and spatial distribution of water variation in 2018 derived from Sentinel-1 SAR data.

to water-level variation in the wetland and grassland areas in preflooding and post-flooding seasons (Figures 30.7 and 30.8). A strong correlation between NDVI and annual average inundation frequency reflects that water-level change is the controlling factor of variation of associated wetlands and habitat conditions in Poyang Lake (Figure 30.9).

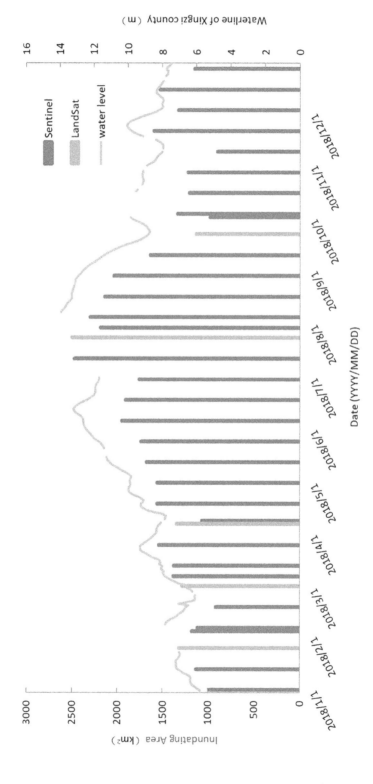

FIGURE 30.6 Monthly inundation variation of lake water levels and areas in 2018 derived from Sentinel-1 SAR data.

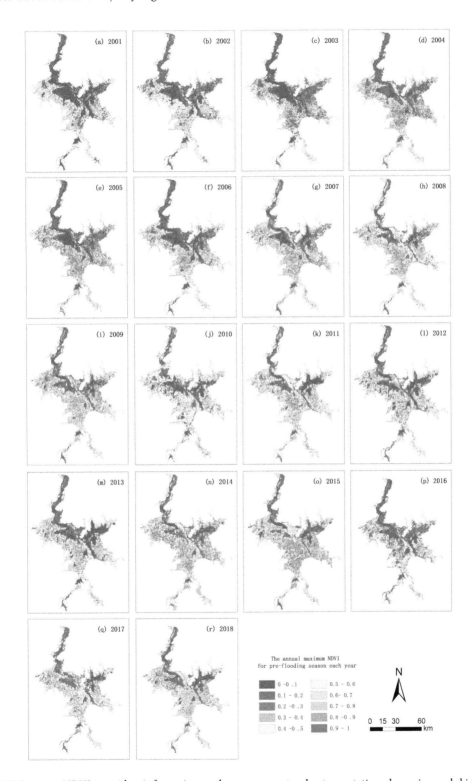

FIGURE 30.7 NDVI provides information and measurements about vegetation dynamics and biomass accumulation corresponding to water-level variation in the wetland and grassland areas in preflooding season between October and March of the following year from 2001 to 2018.

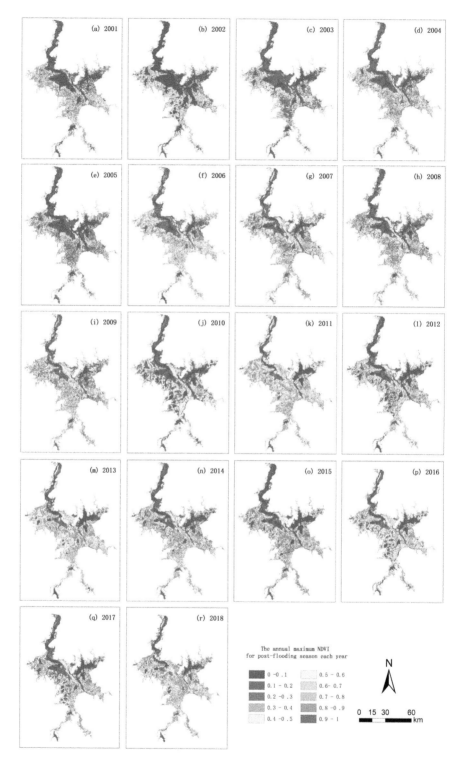

FIGURE 30.8 NDVI provides information and measurements about vegetation dynamics and biomass accumulation corresponding to water-level variation in the wetland and grassland areas in post-flooding season between April and September from 2001 to 2018.

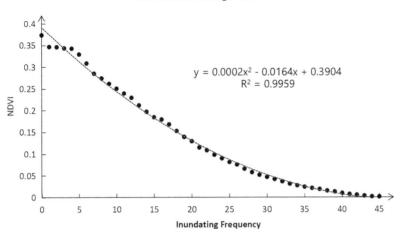

FIGURE 30.9   Relationship between NDVI and annual average inundation frequency in Poyang Lake.

Conclusion

Remote sensing plays a key role in monitoring of water dynamics. Time series MOD09Q1 data products and high spatial resolution Sentinel-1 SAR data were very effective in capturing the change of water surface areas. MOD09Q1 data revealed that the largest annual variability ratio from 2001 to 2018 and between maximum and minimum water surface areas was 7.85, which occurred in 2017. Together with water-level records measured at gauging stations around the lake, a strong correlation existed between inundation areas and water levels. A decreased trend is evident between the maximum and minimum water levels since 2001. The data process and analysis approaches are applicable to most of the situations for mapping the changing water environment, in particular for inland water and wetland monitoring.

References

1. Wang, Y.; Yésou, H. Remote sensing of floodpath lakes and wetlands: A challenging frontier in the monitoring of changing environments. *Remote Sens.* 2018, 10, 1955.
2. Hess, L.L.; Melack, J.M.; Filoso, S.; Wang, Y. Delineation of inundated area and vegetation along the amazon floodplain with the sir-c synthetic aperture radar. *IEEE Trans. Geosci. Remote Sens.* 1995, 33, 896–904.
3. Alsdorf, D.E.; Smith, L.C.; Melack, J.M. Amazon floodplain water level changes measured with interferometric sir-c radar. *IEEE Trans. Geosci. Remote Sens.* 2001, 39, 423–431.
4. Kasischke, E.S.; Smith, K.B.; Bourgeau-Chavez, L.L.; Romanowicz, E.A.; Brunzell, S.; Richardson, C.J. Effects of seasonal hydrologic patterns in south florida wetlands on radar backscatter measured from ers-2 sar imagery. *Remote Sens. Environ.* 2003, 88, 423–441.
5. Birkett, C. Synergistic remote sensing of lake chad: Variability of basin inundation. *Remote Sens. Environ.* 2000, 72, 218–236.
6. Crowley, J.W.; Mitrovica, J.X.; Bailey, R.C.; Tamisiea, M.E.; Davis, J.L. Land water storage within the congo basin inferred from grace satellite gravity data. *Geophys. Res. Lett.* 2006, 33. doi:10.1029/2006GL027070.
7. Rebelo, L.-M.; Finlayson, C.M.; Nagabhatla, N. Remote sensing and GIS for wetland inventory, mapping and change analysis. *J. Environ. Manag.* 2009, 90, 2144–2153.

8. Lemoalle, J.; Bader, J.-C.; Leblanc, M.; Sedick, A. Recent changes in lake chad: Observations, simulations and management options (1973–2011). *Glob. Planet. Chang.* 2012, 80, 247–254.
9. Ramillien, G.; Frappart, F.; Seoane, L. Application of the regional water mass variations from grace satellite gravimetry to large-scale water management in Africa. *Remote Sens.* 2014, 6, 7379–7405.
10. Musopole, A. Analyzing periodicity in remote sensing images for Lake Malawi. *J. Clim. Weather Forecast.* 2016, 4, 2.
11. Onamuti, O.Y.; Okogbue, E.C.; Orimoloye, I.R. Remote sensing appraisal of lake chad shrinkage connotes severe impacts on green economics and socio-economics of the Catchment area. *R. Soc. Open Sci.* 2017, 4, 171120.
12. Policelli, F.; Hubbard, A.; Jung, H.C.; Zaitchik, B.; Ichoku, C. Lake chad total surface water area as derived from land surface temperature and radar remote sensing data. *Remote Sens.* 2018, 10, 252.
13. Wdowinski, S.; Amelung, F.; Miralles-Wilhelm, F.; Dixon, T.H.; Carande, R. Space-based measurements of sheet-flow characteristics in the everglades wetland, Florida. *Geophys. Res. Lett.* 2004, 31. doi:10.1029/2004GL020383.
14. Bourgeau-Chavez, L.L.; Smith, K.B.; Brunzell, S.M.; Kasischke, E.S.; Romanowicz, E.A.; Richardson, C.J. Remote monitoring of regional inundation patterns and hydroperiod in the greater everglades using synthetic aperture radar. *Wetlands* 2005, 25, 176–191.
15. Frappart, F.; Calmant, S.; Cauhopé, M.; Seyler, F.; Cazenave, A. Preliminary results of envisat ra-2-derived water levels validation over the Amazon basin. *Remote Sens. Environ.* 2006, 100, 252–264.
16. Lu, Z.; Kwoun, O.-I. Radarsat-1 and ers insar analysis over southeastern coastal louisiana: Implications for mapping water-level changes beneath swamp forests. *IEEE Trans. Geosci. Remote Sens.* 2008, 46, 2167–2184.
17. Kim, J.-W.; Lu, Z.; Lee, H.; Shum, C.; Swarzenski, C.M.; Doyle, T.W.; Baek, S.-H. Integrated analysis of palsar/radarsat-1 insar and envisat altimeter data for mapping of absolute water level changes in louisiana wetlands. *Remote Sens. Environ.* 2009, 113, 2356–2365.
18. Kwoun, O.-I.; Lu, Z. Multi-temporal radarsat-1 and ers backscattering signatures of coastal wetlands in southeastern louisiana. *Photogramm. Eng. Remote Sens.* 2009, 75, 607–617.
19. Hong, S.-H.; Wdowinski, S.; Kim, S.-W.; Won, J.-S. Multi-temporal monitoring of wetland water levels in the Florida everglades using interferometric synthetic aperture radar (insar). *Remote Sens. Environ.* 2010, 114, 2436–2447.
20. Oliver-Cabrera, T.; Wdowinski, S. Insar-based mapping of tidal inundation extent and amplitude in Louisiana coastal wetlands. *Remote Sens.* 2016, 8, 393.
21. Crétaux, J.-F.; Abarca-del-Río, R.; Berge-Nguyen, M.; Arsen, A.; Drolon, V.; Clos, G.; Maisongrande, P. Lake volume monitoring from space. *Surv. Geophys.* 2016, 37, 269–305.
22. Phan, V.H.; Lindenbergh, R.; Menenti, M. Icesat derived elevation changes of Tibetan lakes between 2003 and 2009. *Int. J. Appl. Earth Obs. Geoinf.* 2012, 17, 12–22.
23. Song, C.; Huang, B.; Ke, L.; Richards, K.S. Remote sensing of alpine lake water environment changes on the Tibetan plateau and surroundings: A review. *ISPRS J. Photogramm. Remote Sens.* 2014, 92, 26–37.
24. Jiang, L.; Nielsen, K.; Andersen, O.B.; Bauer-Gottwein, P. Monitoring recent lake level variations on the tibetan plateau using cryosat-2 sarin mode data. *J. Hydrol.* 2017, 544, 109–124.
25. Ashraf, A.; Naz, R.; Iqbal, M.B. Altitudinal dynamics of glacial lakes under changing climate in the Hindu Kush, Karakoram, and Himalaya ranges. *Geomorphology* 2017, 283, 72–79.
26. Cai, Y.; Ke, C.-Q.; Duan, Z. Monitoring ice variations in Qinghai lake from 1979 to 2016 using passive microwave remote sensing data. *Sci. Total Environ.* 2017, 607, 120–131.
27. Kuenzer, C.; Guo, H.; Huth, J.; Leinenkugel, P.; Li, X.; Dech, S. Flood mapping and flood dynamics of the mekong delta: Envisat-asar-wsm based time series analyses. *Remote Sens.* 2013, 5, 687–715.
28. Kuenzer, C.; Klein, I.; Ullmann, T.; Georgiou, E.F.; Baumhauer, R.; Dech, S. Remote sensing of river delta inundation: Exploiting the potential of coarse spatial resolution, temporally-dense modis time series. *Remote Sens.* 2015, 7, 8516–8542.

29. Tyler, A.; Svab, E.; Preston, T.; Présing, M.; Kovács, W. Remote sensing of the water quality of shallow lakes: A mixture modelling approach to quantifying phytoplankton in water characterized by high-suspended sediment. *Int. J. Remote Sens.* 2006, 27, 1521–1537.

30. Wang, J.; Sheng, Y.; Tong, T.S.D. Monitoring decadal lake dynamics across the yangtze basin downstream of three gorges dam. *Remote Sens. Environ.* 2014, 152, 251–269.

31. Wang, X.; Gong, P.; Zhao, Y.; Xu, Y.; Cheng, X.; Niu, Z.; Luo, Z.; Huang, H.; Sun, F.; Li, X. Water-level changes in China's large lakes determined from icesat/glas data. *Remote Sens. Environ.* 2013, 132, 131–144.

32. Giardino, C.; Bresciani, M.; Stroppiana, D.; Oggioni, A.; Morabito, G. Optical remote sensing of lakes: An overview on lake Maggiore. *J. Limnol.* 2013, 73. doi:10.4081/jlimnol.2014.817.

33. Doña, C.; Chang, N.-B.; Caselles, V.; Sánchez, J.M.; Pérez-Planells, L.; Bisquert, M.D.M.; García-Santos, V.; Imen, S.; Camacho, A. Monitoring hydrological patterns of temporary lakes using remote sensing and machine learning models: Case study of la mancha húmeda biosphere reserve in central spain. *Remote Sens.* 2016, 8, 618.

34. Luo, J.; Li, X.; Ma, R.; Li, F.; Duan, H.; Hu, W.; Qin, B.; Huang, W. Applying remote sensing techniques to monitoring seasonal and interannual changes of aquatic vegetation in taihu lake, China. *Ecol. Indic.* 2016, 60, 503–513.

35. Gong, P.; Niu, Z.; Cheng, X.; Zhao, K.; Zhou, D.; Guo, J.; Liang, L.; Wang, X.; Li, D.; Huang, H.; Wang, Y. China's wetland change (1990–2000) determined by remote sensing. *Sci. China Earth Sci.* 2010, 53 (7), 1036–1042.

36. Ma, R.; Duan, H.; Hu, C.; Feng, X.; Li, A.; Ju, W.; Jiang, J.; Yang, G. A half-century of changes in China's lakes: Global warming or human influence? *Geophys. Res. Lett.* 2010, 37 (24). doi:10.1029/2010GL045514.

37. Feng, L.; Hu, C.; Chen, X.; Tian, L.; Chen, L. Human induced turbidity changes in Poyang lake between 2000 and 2010: Observations from MODIS. *J. Geophys. Res.* 2012a, 117. doi:10.1029/2011JC007864.

38. Feng, L.; Hu, C.; Chen, X.; Cai, X.; Tian, L.; Gan, W. Assessment of inundation changes of Poyang lake using MODIS observations between 2000 and 2010. *Remote Sens. Environ.* 2012b, 121, 80–92.

39. Pottier, E.; Marechal, C.; Allain-Bailhache, S.; Meric, S.; Hubert-Moy, L.; Corgne, S. On the use of fully polarimetric radarsat-2 time-series datasets for delineating and monitoring the seasonal dynamics of wetland ecosystem. *Proceedings of the 2012 IEEE International Geoscience and Remote Sensing Symposium (IGARSS), Munich, Germany*, 22–27 July 2012; pp. 107–110.

40. Brisco, B.; Schmitt, A.; Murnaghan, K.; Kaya, S.; Roth, A. Sar polarimetric change detection for flooded vegetation. *Int. J. Digit. Earth* 2013, 6, 103–114.

41. Dabrowska-Zielinska, K.; Budzynska, M.; Tomaszewska, M.; Bartold, M.; Gatkowska, M.; Malek, I.; Turlej, K.; Napiorkowska, M. Monitoring wetlands ecosystems using alos palsar (l-band, hv) supplemented by optical data: A case study of biebrza wetlands in Northeast Poland. *Remote Sens.* 2014, 6, 1605–1633.

42. Shen, G.; Liao, J.; Guo, H.; Liu, J. Poyang lake wetland vegetation biomass inversion using polarimetric radarsat-2 synthetic aperture radar data. *J. Appl. Remote Sens.* 2015, 9, 096077.

43. Xie, C.; Shao, Y.; Xu, J.; Wan, Z.; Fang, L. Analysis of alos palsar insar data for mapping water level changes in Yellow river delta wetlands. *Int. J. Remote Sens.* 2013, 34, 2047–2056.

44. Gao, J.H.; Jia, J.; Kettner, A.J.; Xing, F.; Wang, Y.P.; Xu, X.N.; Yang, Y.; Zou, X.Q.; Gao, S.; Qi, S.; Liao, F. Changes in water and sediment exchange between the Changjiang river and Poyang lake under natural and anthropogenic conditions, China. *Sci. Total Environ.* 2014, 481, 542–553.

45. Qi, S.; Brown, D.G.; Tian, Q.; Jiang, L.; Zhao, T.; Bergen, K.M. Inundation extent and flood fre-quency mapping using LANDSAT imagery and digital elevation models. *GISci. Remote Sens.* 2009, 46 (1), 101–127.

46. Min, Q. Evaluation of the effects of Poyang lake reclamation on floods. *Yangtze River* 1999, 30 (7), 30–32.

47. Shankman, D.; Liang, Q. Landscape changes and increasing flood frequency in China's Poyang lake region. *Prof. Geogr.* 2003, 55 (4), 434–445

48. Yésou, H.; Huber, C.; Lai, X. et al (2011) Nine years of water resources monitoring over the middle reaches of the Yangtze River, with ENVISAT, MODIS, Beijing-1 time se-ries, Altimetric data and field measurements. *Lakes Reservoirs Res. Manage.* 16(3), 231–247

49. Dronova, I.; Gong, P.; Wang, L. Object-based analysis and change detection of major wet-land cover types and their classification uncertainty during the low water period at Poyang lake, China. *Remote Sens. Environ.* 2011, 115 (12), 3220–3236.

50. Zhang, Q.; Ye, X.; Werner, A.D.; Li, Y.; Yao, J.; Li, X.; Xu, C. An investigation of enhanced re-cessions in Poyang lake: Comparison of Yangtze River and local catchment impacts. *J. Hydrol.* 2014, 517, 425–434.

51. Jiao, L. Scientists line up against dam that would alter protected wetlands. *Science* 2009, 326 (5952), 508–509

Index

Note: *Italic* page numbers refer to figures and tables.